The Nature of the Chemical Concept
Re-constructing Chemical Knowledge in Teaching and Learning

Advances in Chemistry Education Series

Editor-in-chief:
Keith S. Taber, *University of Cambridge, UK*

Series editors:
Avi Hofstein, *Weizmann Institute of Science, Israel*
Vicente Talanquer, *University of Arizona, USA*
David Treagust, *Curtin University, Australia*

Editorial Advisory Board:
George Bodner, Purdue University, USA, **Mei-Hung Chiu**, National Taiwan Normal University, Taiwan, **Richard Coll**, The University of Fiji, Fiji Islands, **Rosaria da Silva Justi**, Universidade Federal de Minas Gerais, Brazil, **Onno De Jong**, Utrecht University, Netherlands, **Ingo Eilks**, University of Bremen, Germany, **John Gilbert**, University of Reading, UK, **Murat Kahveci**, Çanakkale Onsekiz Mart University, Turkey, **Vanessa Kind**, Durham University, UK, **Stacey Lowery Bretz**, Miami University, USA, **Hannah Sevian**, University of Massachusetts Boston, USA, **Daniel Tan**, Nanyang Technological University, Singapore, **Marcy Towns**, Purdue University, USA, **Georgios Tsaparlis**, University of Ioannina, Greece.

Titles in the Series:
1: Professional Development of Chemistry Teachers: Theory and Practice
2: Argumentation in Chemistry Education: Research, Policy and Practice
3: The Nature of the Chemical Concept: Re-constructing Chemical Knowledge in Teaching and Learning

How to obtain future titles on publication:
A standing order plan is available for this series. A standing order will bring delivery of each new volume immediately on publication.

For further information please contact:
Book Sales Department, Royal Society of Chemistry, Thomas Graham House, Science Park, Milton Road, Cambridge, CB4 0WF, UK
Telephone: +44 (0)1223 420066, Fax: +44 (0)1223 420247,
Email: booksales@rsc.org
Visit our website at books.rsc.org

The Nature of the Chemical Concept
Re-constructing Chemical Knowledge in Teaching and Learning

Keith S. Taber
University of Cambridge, UK
Email: kst24@cam.ac.uk

Advances in Chemistry Education Series No. 3

Hardback ISBN: 978-1-78262-460-8
PDF ISBN: 978-1-78801-361-1
EPUB ISBN: 978-1-78801-784-8
Print ISSN: 2056-9335
Electronic ISSN: 2056-9343
Paperback ISBN: 978-1-83916-745-4

A catalogue record for this book is available from the British Library

© Keith S. Taber 2022

All rights reserved

Apart from fair dealing for the purposes of research for non-commercial purposes or for private study, criticism or review, as permitted under the Copyright, Designs and Patents Act 1988 and the Copyright and Related Rights Regulations 2003, this publication may not be reproduced, stored or transmitted, in any form or by any means, without the prior permission in writing of The Royal Society of Chemistry or the copyright owner, or in the case of reproduction in accordance with the terms of licences issued by the Copyright Licensing Agency in the UK, or in accordance with the terms of the licences issued by the appropriate Reproduction Rights Organization outside the UK. Enquiries concerning reproduction outside the terms stated here should be sent to The Royal Society of Chemistry at the address printed on this page.

Whilst this material has been produced with all due care, The Royal Society of Chemistry cannot be held responsible or liable for its accuracy and completeness, nor for any consequences arising from any errors or the use of the information contained in this publication. The publication of advertisements does not constitute any endorsement by The Royal Society of Chemistry or Authors of any products advertised. The views and opinions advanced by contributors do not necessarily reflect those of The Royal Society of Chemistry which shall not be liable for any resulting loss or damage arising as a result of reliance upon this material.

The Royal Society of Chemistry is a charity, registered in England and Wales, Number 207890, and a company incorporated in England by Royal Charter (Registered No. RC000524), registered office: Burlington House, Piccadilly, London W1J 0BA, UK, Telephone: +44 (0) 20 7437 8656.

For further information see our website at www.rsc.org

For general enquiries, please contact books@rsc.org

For EU product safety enquiries, please email books@rsc.org or contact Royal Society of Chemistry Worldwide (Germany) GmbH, Römischer Hof, Unter den Linden 10, 10117 Berlin.

Preface

This is a book that is partially about cognition, partly about education, and partly about chemistry. Part of my motivation in writing this book was to do for the chemical concept what Linus Pauling did for the chemical bond in writing his '*The Nature of the Chemical Bond*': that is to provide some kind of coherent treatment of a topic that appears imprecise and confused in much of the literature. Concepts are central in teaching and learning, and I felt an attempt to address the topic for those who are working in chemistry education was sorely needed. Indeed, I considered that an authoritative account could support teaching – and that if the present attempt is judged inadequate in that respect, then this might at least provoke those who recognise its faults to attempt to do better.

In writing the book I have been very aware of my own limitations for the job. I appreciate that what I was writing was, in places, of its nature philosophy, and, in other places, I was attempting to offer insights from the history of chemistry. Yet I am not, and have no training as, either a philosopher of, nor as a historian of, science. Whilst I can certainly claim to be a chemist, I am not a research chemist – my research has only ever been in education.

So, if I have qualifications for writing this book, they must be found elsewhere. I have had, since my school days, a strong interest in science and how science comes to knowledge, and also in learning – and so how learners may come to knowledge. I have taught in secondary schools (mostly chemistry and physics), further education (mostly chemistry and physics, but also some science studies and research methods), and in university (mostly in science teacher preparation, and educational research methods). So over almost forty years I have spent a lot of time trying to communicate scientific concepts, and notions of how we can come to knowledge, to students.

So, this book is borne out of my teaching experiences as well as my research interests.

It seems (from my own experience) authors normally plan a book, and make a formal proposal to a publisher, at a point where they think they have a clear vision of what they want to achieve and how they will go about it. The actual writing process is often extended, and is where much of the detailed thinking is done. This often reveals that the vision needs a lot more work than the tidy proposal might have suggested. It may seem that once an author has the plan for a book it is just a matter of doing the writing, but that would both be naïve, and indeed probably make writing a book somewhat tedious. So, from the author's perspective, a good book proposal should reflect enough thinking to give confidence that the plan for the book is viable, but not so much that the writing seems a mechanical process of carrying out the plan.

Sometimes, during the writing, the author may need to revisit (or even reimagine) what the core of the book is – or was meant to be. During the writing process for this volume I asked myself why anyone should read this book with its 'enthusiastic amateur' treatments of topics from the nature of chemistry and the history of the subject, when there are plenty of scholarly accounts of such matters written by genuine experts in those fields. The answer is that although arguments in the book draw on a number of fields, its origin is in teaching chemistry, and the experience of teaching chemical concepts – and then of undertaking research into how students understood those concepts. In that sense, this book develops a theme from an earlier book (Taber, 2013) that although teaching is a professional activity, it often relies on what are little more than folk notions of key ideas ('concept', 'learning', 'understanding', 'knowing', *etc.*)

Teaching chemical concepts is a challenge. In part it is a challenge because once one has acquired a decent conceptual knowledge of the subject it becomes difficult to put oneself back in the position of a novice learner; in part it is a challenge because chemical concepts are abstract and often nuanced – and often highly inter-related (such that you never really understand some concepts until you also understand a number of others); in part it is a challenge because teachers of chemistry at all levels – through no fault of their own, it is in the nature of developing expertise – tend to operate with almost intuitive notions of the *nature of* the concepts they teach, and of what understanding those concepts actually means.

So, although this book is envisaged as a scholarly account suitable for graduate students and academics, I have tried to write in a way that addresses anyone who teaches chemistry, or works with students learning chemistry, as well as for those undertaking research into teaching and learning. It is a book that problematises the notion of the concept in chemistry, and sets out to show why teaching and learning chemical concepts can be so challenging. The book then offers an analysis of the background to much of what chemistry teachers and lecturers are doing in their classroom work. It is a book that comes from a privileged position that most chemistry teachers do not share: despite my very dubious qualifications in some of the key disciplines I draw upon, I am lucky to have had a career teaching across natural science and social science (education) that has given me opportunities to read a little across a range of fields and so consider how diverse perspectives might inform thinking about teaching chemistry. I have also benefitted from the generous participation of many students who were kind enough to share with me their own understandings of the concepts they have been taught. This offers a diversity of perspectives and experiences to make the present book viable (Taber, 2014).

I hope, then, that this book will be read not only by researchers and graduate students, but also by those teaching chemistry, at whatever level, and that it will resonate with teachers' experiences of the challenges they face in their classroom work. I would like to think it offers useful insights to help them better appreciate the nature and sources of these challenges, and so help teachers better support the learners they share our chemical concepts with.

Keith S. Taber

References

Taber K. S., (2013), *Modelling Learners and Learning in Science Education: Developing Representations of Concepts, Conceptual Structure and Conceptual Change to Inform Teaching and Research*, Dordrecht: Springer.

Taber K. S., (2014), *Student Thinking and Learning in Science: Perspectives on the Nature and Development of Learners' Ideas*, New York: Routledge.

Glossary

A number of key terms are discussed in this volume – being introduced and explained or developed at different points in the text. This glossary is provided for ready reference for the reader dipping into the book or requiring a quick reminder of how a term is being used.

Accredited concept   An accredited concept is a learner's personal concept (of some chemical notion, such as 'element' or 'acid') that has been judged as being sufficiently aligned with the target knowledge set out in a particular curriculum.

Alternative conception   An alternative conception is a way of conceptualising something that is judged to be inconsistent with the accepted ('canonical') way of thinking about that particular topic or phenomenon.

Authoritative concepts   This term is used in the book to refer to those personal concepts that experts in a field have developed that are sufficiently aligned with those of their peers (*i.e.*, other experts) such that they are widely judged canonical, and so facilitate ready communication within the community of experts.

Canonical concepts   A canonical concept is taken as the standard conceptualisation within a field. Canonical chemical concepts may be considered to have the authority of being *the* established ways of making sense of the discipline of chemistry.

Conception   A conception is a way of conceptualising some aspect of the world. A specific conception may be one aspect of someone personal concept (*e.g.*, 'the covalent bond is a shared pair of electrons' may be a conception that is part of, but does not exhaust, someone's {covalent bond} concept).

Glossary   ix

Conceptual content   This term conceptual content is used for the 'content' of a particular concept in the sense of the full set of meanings and associations it has – which could include such matters as examples of the concept, understandings of applications of the concept, etcetera.

Conceptual inductive effect   As concepts are embedded in conceptual networks, they are influenced by the *associations* of other concepts that they are directly linked to – or to put this another way, they are implicitly influenced by those concepts that they are *indirectly* linked to through other concepts.

Congenst   A hypothetical feature of a person's neural structure, representing aspects of previous experience, which, when activated, supports thinking characteristic of applying a concept: a concept-generating structure in someone's brain.

Creditable concept   A learner has a creditable concept when their personal versions of some chemical concept is sufficiently credit-worthy, at the level they are studying, to be evaluated as an accredited concept (if and when such an evaluation takes place).

Curricular models   It is seldom, if ever, possible to teach scientific concepts in their full sophistication and complexity including all known examples, properties, associations, etcetera; so, concepts are in effect modelled (*e.g.*, simplified, generalised, exemplified by prototypes) in the way they are represented in curriculum.

Historical concepts   These are concepts that once had wide currency in chemistry – and so might have once been considered canonical – but that are no longer part of the canon of chemical thinking.

Knowledge   A person's knowledge is understood here in the inclusive sense of ideas believed or being entertained as plausible (regardless of their objective correctness).

Mooted concepts   Not all scientific concepts become widely adopted and so can be considered canonical. The term mooted concepts is used to distinguish those scientific concepts that are currently under consideration, having been proposed in current scientific publications that have not yet, and may never, become widely used in the scientific discipline (be that chemistry generally, or a more specific research field).

Optimum level of simplification   A teaching presentation should simplify canonical concepts sufficiently for the material to be meaningful, and understood as intended, by the students; without being oversimplified so that it becomes an inauthentic representation of the canonical concept and/or a poor basis for further progression in learning.

Personal knowledge   Personal knowledge refers to the knowledge of an individual, and is therefore somewhat idiosyncratic.

Public knowledge   Public knowledge refers to what is widely taken for knowledge in society: in a science such as chemistry this is often understood in terms of what has been reported in the research literature – although it may be difficult to determine precisely what counts as public knowledge.

Scientific concepts Unlike everyday concepts that develop through informal cultural processes, scientific concepts are those that are mooted formally as part of scientific practice – understood here as an iterative process that seeks to conceptualise empirical experience theoretically, and test such theoretical inventions empirically, and so on.

Tacit knowledge Tacit knowledge refers to knowledge that a person would seem to have (as they appear to demonstrate behaviour informed by that knowledge) even though they are not consciously aware of what it is they know, and cannot readily deliberate on it. Tacit knowledge is related to the idea of intuition – an ability to make judgements without conscious reasoning.

Target knowledge The knowledge set out for a particular group of students to learn, as in a formal curriculum. For example, target knowledge about chemical reactions will vary between, say, middle school pupils; school leavers; and final-year undergraduates.

Typographical Conventions

Two particular typographical conventions are used in the text. One is to use curly brackets to put concept names in parentheses when referring to a concept rather than what the concept refers to: so, for example, to distinguish references to the concept {metals} from references to actual metals. Sometimes subscripted suffixes are used to distinguish distinct versions of concepts – such as the {acid$_{Arrhenius}$} concept and the {acid$_{Lewis}$} concept. The other convention is to denote the inverse of some group or category, so, for example, to use $\overline{acid}$ to denote anything that is not considered to be acid, and therefore to denote the concept of all those things that are not acid as {$\overline{acid}$}.

Acknowledgements

A text, even a single authored text, never has a single author.

Scholars are supposed to be careful about assigning credit, and to acknowledge all those who have informed their thinking. Yet learning is an incremental, interpretive, iterative, and, often, insidious, process, and so we are only ever vaguely aware of the sources informing many of our 'own' ideas. In a sense, then, any author is actually an inadvertent editor of a multitude of hidden, indirect, contributors to a work. I would like to acknowledge the role of this invisible college in the current work – all those who have influenced my own thinking through representing their own ideas in the public domain. I have cited some key influences in the text, but wish to acknowledge that my thoughts are inevitably the outcome of a complex blending of all the information I have made sense of during a lifetime of interpreting what I have heard and seen and read. The many misinterpretations are, of course, all the author's own work.

So, I readily acknowledge that this book is, in part, the product of my upbringing, my schooling, and all my other educative experiences over many years. In particular, I thank my parents for creating a supportive and loving home environment that provided the grounding for all that followed. I also thank my own teachers, and professional colleagues, and all those whose writing influenced my thinking (even when I have no explicit memory of that influence, or even the texts concerned, now). A teacher's thinking is developed in responding to the unexpected questions of those we teach. My own thinking has been developed in collaboration with those I have taught at school, college, or university level, as well as those students (whether I taught them, or not) who generously gave me some of their precious time to be interviewed about their understanding of concepts.

Advances in Chemistry Education Series No. 3
The Nature of the Chemical Concept: Re-constructing Chemical Knowledge in Teaching and Learning
By Keith S. Taber
© Keith S. Taber 2022
Published by the Royal Society of Chemistry, www.rsc.org

I would also like to thank the Royal Society of Chemistry for recognising the value of a book series focused on chemistry education as a scholarly field, and my editor there, Michelle Carey, for her patience and support.

To Philippa
always in my memory

Contents

INTRODUCTION

Chapter 1	The Challenge of Teaching and Learning Chemical Concepts		3
	1.1 Research is Underpinned by Theory		6
	1.2 Under-theorised Research		8
	1.2.1 The Mental Register		9
	1.3 Some Things that Should not be Taken for Granted		9
	1.4 What Do We Know, and What Do We Need to Know?		10
	1.4.1 Some Orientating Questions for the Reader		11
	1.5 The First-order Approximate Model of Conceptual Teaching		11
	References		12

CONCEPTS IN CHEMISTRY

Chapter 2	What Kind of Things Are Concepts?		17
	2.1 Making Sense of Curie and and and and and Meitner		17
	2.1.1 Concepts and Conceptions		19
	2.2 How to Best Understand Concepts		20
	2.2.1 Concepts as Categories		22
	2.2.2 The Abstract Nature of Concepts		24
	2.2.3 Concepts Are Mental Entities		24
	2.2.4 Concepts Are Tools Used in Thinking		26

Advances in Chemistry Education Series No. 3
The Nature of the Chemical Concept: Re-constructing Chemical Knowledge in Teaching and Learning
By Keith S. Taber
© Keith S. Taber 2022
Published by the Royal Society of Chemistry, www.rsc.org

		2.2.5	Concepts Are Only Apparent When Activated	27
		2.2.6	Concepts Act as Nodes in a Conceptual Network	31
	2.3	Representing and Exploring Conceptual Structures		32
		2.3.1	Representing Conceptual Structures	32
		2.3.2	The Natural Attitude – Talking, and Thinking, Like Mind Readers	36
	References			37
Chapter 3	**What Kinds of Concepts Are Important in Chemistry?**			**39**
	3.1	Four Types of Chemical Concept?		40
	3.2	Concepts Relating to Objects – Entities in Chemistry		40
	3.3	Concepts Relating to Events – Processes in Chemistry		42
	3.4	Concepts Relating to Qualities – Properties in Chemistry		45
	3.5	Meta-concepts – Concepts about Chemical Concepts		46
	References			49
Chapter 4	**Concepts as Knowledge**			**50**
	4.1	Concepts as Knowledge?		50
		4.1.1	Knowledge, Belief, and Truth	51
		4.1.2	We Should Not Believe in Scientific Knowledge	52
		4.1.3	A More Relevant Notion of Knowledge	54
		4.1.4	The Knower and the Known	54
	4.2	Conceptual, Procedural, and Episodic Knowledge		55
		4.2.1	Implicit Knowledge	56
		4.2.2	Mighty Oaks from Ignorant Acorns Grow	57
		4.2.3	Can Conceptual Knowledge Be Tacit?	58
	4.3	Fuzzy Concepts		59
	4.4	Manifold Concepts?		64
		4.4.1	Why Does It Matter?	66
	References			67
Chapter 5	**The Origin of a Chemical Concept: The Ongoing Discovery of Potassium**			**69**
	5.1	Chemistry as a Natural and Empirical Science		69
		5.1.1	The Nature of the Natural	69
		5.1.2	The Wider Context of Scientific Discovery	71

	5.2	Chemistry as an Empirical Science Depends on Imagination as well as Benchwork	72
	5.3	Thinking about Chemical 'Discoveries'	72
		5.3.1 Conceptualisation Is Shaped by the Cognitive Apparatus	73
		5.3.2 Constructivism	74
	5.4	The Discovery of Potassium: Imagining a New Substance	75
		5.4.1 The Creation of Potassium	76
		5.4.2 The Construction and Development of the {Potassium} Concept	78
		5.4.3 The Shifting {Potassium} Concept	80
		5.4.4 Potassium *sans* Isotopes	81
		5.4.5 Developing the {Metal} Concept	84
	References		85

Chapter 6   Conceptualising Acids: Reimagining a Class of Substances — 88

	6.1	Formation of a Concept of Acids	88
	6.2	Theoretical Elements of the {acid} Concept	89
	6.3	Lavoisier's Theoretical Account of Acids	90
	6.4	Acids and Hydrogen	92
	6.5	Current and Historical Scientific Concepts	93
		6.5.1 Inventing New Concepts	93
		6.5.2 Concepts Reaching Canonical Status	94
		6.5.3 Extending the Concept of Acid	96
		6.5.4 There Is Progression from the Lavoisier Concept to the Arrhenius Concept	97
		6.5.5 Have We Progressed Beyond Arrhenius in Discriminating Acid from **Acid** (Not Acid)?	98
		6.5.6 An Alternative History – Forming New Concepts Rather Than Expanding the Range of Existing Ones	99
		6.5.7 Lewis Revised the {acid} Concept	102
		6.5.8 Are Acids a Natural Kind, or a Chemical Convenience?	104
		6.5.9 Is There a Canonical {acid} Concept in Chemistry Today?	105
	6.6	When Are Chemical Discoveries Made?	108
	References		110

Chapter 7		**Concepts and Ontology: What Kind of Things Exist in the World of Chemistry?**	**112**
	7.1	Chemical Concepts and the Kinds of Things We Perceive in the World	112
	7.2	Concepts and Typologies	113
		7.2.1 Nesting of Concepts	113
	7.3	Do Chemical Concepts Represent Natural Kinds?	115
		7.3.1 The Importance of Natural Kinds in Science	115
		7.3.2 Induction, Bias, and Prejudice	117
		7.3.3 Conceptualising Kinds Assumes Some Essential Properties	119
	7.4	Natural Kinds in Chemistry	119
		7.4.1 Species as Natural Kinds? A Warning from Biology	121
		7.4.2 The Elements as Natural Kinds	123
		7.4.3 Are Natural Kinds Just the Operation of a Mental Bias?	124
	7.5	The Use of the Definite Article in Relation to Chemical Kinds	126
		7.5.1 Talk About Ideal Prototypes	128
		7.5.2 Which Methane Molecule Is Tetrahedral?	129
		7.5.3 No Methane Molecule Is Tetrahedral	130
	7.6	Constructing Typologies in Chemistry	131
	References		132
Chapter 8		**Chemical Meta-concepts: Imagining the Relationships Between Chemical Concepts**	**134**
	8.1	Concepts and Laws (and Law-like Principles)	135
		8.1.1 Avogadro's Law	136
		8.1.2 Conservation of Mass	139
		8.1.3 Chargaff's Rules	142
		8.1.4 Raoult's Law and Its Deviations	145
		8.1.5 Other Examples of Law-like Principles	147
	8.2	Concepts and Models	148
	8.3	Concepts and Theories?	151
		8.3.1 Particle Theories	153
		8.3.2 Micro-/Macro-distinctions	154
		8.3.3 Relating the Two Levels	156
		8.3.4 What Do We Mean By an Observable?	156
		8.3.5 Can We See Salt Dissolve in Water?	157
		8.3.6 Has Anyone Seen the Cat?	159

	8.3.7	Has Anyone Seen the Higgs Boson?	160
	8.3.8	Shifting from Observational Language	160
	8.3.9	Applying the Submicroscopic Concepts	161
8.4	Conclusion		163
References			164

TEACHING AND LEARNING CHEMISTRY CONCEPTS

Chapter 9   Accessing Chemical Concepts for Teaching and Learning   169

9.1	The Problem of Locating Canonical Concepts		169
9.2	Looking for Canonical Concepts in the Scientific Literature		171
	9.2.1	Lack of Coherence in the Literature	171
	9.2.2	Texts and the Nature of Knowledge	173
	9.2.3	Texts Contain Only Representations of Concepts	174
	9.2.4	Criticisms of This Position	175
9.3	Asking the Community of Chemists about Canonical Concepts		176
	9.3.1	Who Are the Chemists Who Know?	176
	9.3.2	Does Knowledge Need to Be Personal or Can It Be Distributed?	177
	9.3.3	The Human Knower – Information Resource System	178
	9.3.4	The Importance of Understanding	181
	9.3.5	Knowledge Distributed Across Networks of People	181
	9.3.6	Sharing Concepts Is a Process Involving Representation and Interpretation	183
	9.3.7	Can We Avoid the Need to Interpret Representations?	184
9.4	Can We Access Canonical Chemical Concepts from a Non-material Realm of Ideas?		185
	9.4.1	Do We Still Believe in Platonic Forms?	187
	9.4.2	Does World 3 Really Exist?	188
	9.4.3	Objectivity in Educational Questions	191
9.5	So Where Do We Find Chemical Concepts?		194
References			196

Chapter 10   How Are Chemical Concepts Represented in the Curriculum?   199

10.1	The Curriculum	199

10.2	Selection and Simplification		200
	10.2.1	Deciding What Counts as Chemistry	200
	10.2.2	Chemistry in Its Social Context	201
	10.2.3	Learning about the Nature of Chemistry	202
	10.2.4	Chemistry Promoting Intellectual Development	203
	10.2.5	Bounding the Chemistry Curriculum	205
10.3	Choosing the Chemistry		205
	10.3.1	Breadth or Width?	205
	10.3.2	Organising the Selection	206
	10.3.3	Teaching Concepts in Stages	207
10.4	Keeping It Simple		208
	10.4.1	What Shall We Teach about the {Metal} Concept?	209
	10.4.2	What Details Shall We Teach about the {Metal} Concept?	211
	10.4.3	The Spiral Curriculum	212
	10.4.4	The Optimum Level of Simplification	214
	10.4.5	Learning Progressions, Big Ideas and Threshold Concepts	216
10.5	Curricular Models		217
	10.5.1	How Do Chemical Reactions Take Place in England?	220
	10.5.2	When Is an Atom, Not an Atom? (When Is an Ion an Atom?)	222
	10.5.3	Learning Progression between Educational Stages	224
	References		226

Chapter 11 How Are Chemical Concepts Communicated? — **228**

11.1	Objectivity and Subjectivity		228
	11.1.1	Objectivity in Science	229
	11.1.2	Subjective Reports	230
11.2	Are Concepts Subjective or Objective?		232
	11.2.1	Personal Concepts and Canonical Concepts	232
	11.2.2	Canonical Concepts Are as Useful as Ideal Gases	234
	11.2.3	Chemical Education Without a Focus on Concepts	234
	11.2.4	Induction into a Community of Practice	235
	11.2.5	Learning as a Black Box	236

11.3	Information and Understanding		237
	11.3.1	Information in Chemistry	238
	11.3.2	Understanding and Meaning	239
	11.3.3	An Analogy with Data (and Evidence)	242
11.4	Understanding as Subjective		243
	11.4.1	Understanding and Explanation	244
	11.4.2	Meaningful and Rote Learning	245
11.5	So, How Are Chemical Concepts Communicated?		247
	11.5.1	Authoritative Concepts and the Impression of Canonical Concepts	248
11.6	Concluding Comments		250
References			251

Chapter 12 How Are Chemical Concepts Represented in Teaching? 253

12.1	Communicating and Teaching		254
	12.1.1	Teaching as Moving Students Towards Accredited Conceptualisations	255
	12.1.2	The Communication of Information Is Necessary but Not Sufficient for Teaching	257
	12.1.3	Lecturing	258
	12.1.4	The Non-linearity of Teaching and Learning	259
12.2	Framing Information to Allow Meaningful Understanding		260
	12.2.1	Modelling the Other Person's Sense-making	261
	12.2.2	Making Unfamiliar Concepts Familiar	264
References			272

Chapter 13 How Do Students Acquire Concepts? 274

13.1	A Personal Conception of Positive-ray Spectrum: Implicit Concept Formation		274
13.2	Three Levels of Description		275
	13.2.1	The Behavioural Level of Description	275
	13.2.2	The Physiological–Anatomical Level of Description	277
	13.2.3	The Information Processing Level of Description	279
	13.2.4	The Mental Register (Revisited)	279
13.3	Concept Formation		281
	13.3.1	Implicit Learning in Everyday Life	281
	13.3.2	Studying Concept Formation	283

	13.3.3	Inherent Pattern Recognition	283
	13.3.4	Phenomenological Primitives (P-prims)	284
13.4	Two Systems of Knowledge		288
	13.4.1	Spontaneous Concepts	289
	13.4.2	Learning Scientific Concepts	290
	13.4.3	Melded Concepts	292
	13.4.4	Engaging Implicit Knowledge in Building Deliberative Concepts	292
	13.4.5	Teaching Informed by the Knowledge-in-pieces Model	293
References			294

Chapter 14 What Is the Nature of Students' Conceptions? 297

14.1	The Person on a Learning Journey	297
14.2	Dimensions of Student Conceptions in Chemistry	298
14.3	Canonicity: More or Less Alternative Conceptions	298
14.4	Commonality: 'Popular' and Idiosyncratic Alternative Conceptions	299
14.5	Explicitness: Learners Are Aware of Some, but Not All, of Their Thinking about Chemical Topics	301
14.6	Commitment: Belief or Suspicion?	304
14.7	Multiplicity: Unitary and Manifold Conceptions	305
14.8	Connectivity: Discrete Conceptions and Conceptual Frameworks	307
14.9	Conclusion	310
References		310

Chapter 15 How Do Students' Concepts Develop? 313

15.1	A Personal Conception of Positive-ray Spectrum: Reflective Concept Development		313
15.2	Meaningful Learning Revisited		315
15.3	The Metaphor of the Conceptual Ecology		316
	15.3.1	Epistemological Sophistication	317
	15.3.2	Rich Ecologies	318
15.4	Two Types of Conceptual Change?		319
15.5	Accretion – Conceptual Addition		321
15.6	Conceptual Realignment – Correcting Discrete Conceptions		323

	15.7	Restructuring Conceptual Frameworks	328
		15.7.1 But Perhaps Copper Was a Magnetic Material after All?	328
		15.7.2 Integrating Conceptual Frameworks	330
	15.8	Studies of Conceptual Change	331
	References		335

CONCLUSION

Chapter 16 Lessons for Chemistry Education — 341

	16.1	Lessons for Chemistry Teachers	341
	16.2	Responsibilities of the Chemistry Teacher	342
	16.3	Understanding Differently	342
	16.4	Checking for Shared Understandings	343
	16.5	Always Imagine, but Never Assume	344
	16.6	Imagine (and Test Out) Alternative Possibilities	346
	16.7	Lessons for Chemistry Education Research	348
	16.8	The Research Literature	348
	16.9	Moving the Research Programme Forward	351
	References		353

Subject Index — 355

INTRODUCTION

CHAPTER 1

The Challenge of Teaching and Learning Chemical Concepts

Chemistry as a subject is conceptual. Students learning chemistry at school level, or in colleges and universities, are introduced to, and asked to master, a wide array of concepts. So, students at different levels are taught about acids, elements, oxidising agents, covalent bonds, d-level splitting, chemical shift, and so on. Each of these, and many other, foci of study can be considered concepts (the nature of concepts is discussed in Chapter 2). Concepts are central to understanding chemistry, and the understanding of chemical concepts is therefore a core concern in chemical education.

Chemistry is a science, and arguably one thing that characterises science is the interplay between empirical experience and theory. Chemistry is often said to be a 'practical' subject, but what makes such practical work a scientific activity is the way it is informed by, and feeds back into, the theoretical frameworks of the subject (this is explored further in Chapter 5). Those theoretical frameworks are populated and supported by the wide range of concepts chemists have developed to make sense of what has been seen (and heard, smelt, felt, and – more so in former times – sometimes tasted) in the laboratory. Reflecting this, an authentic chemistry education is rich with concepts that are set out in the curriculum, presented in textbooks, and taught in the laboratory and classroom or lecture hall.

Yet learning chemical concepts is not straightforward. Students – at all levels – often do not understand; or only partially understand; or indeed misunderstand; key concepts they meet in their studies of chemistry. Students in these situations are sometimes well aware they are confused or do not understand what is being taught: but that is by no means always the case. Indeed, it is not at all unusual for students to only partially understand, or indeed misunderstand, concepts that they think they do understand.

Advances in Chemistry Education Series No. 3
The Nature of the Chemical Concept: Re-constructing Chemical Knowledge in Teaching and Learning
By Keith S. Taber
© Keith S. Taber 2022
Published by the Royal Society of Chemistry, www.rsc.org

When students present with conceptions that are inconsistent with the target knowledge being taught, their ideas are often labelled using terms such as misconceptions, alternative conceptions or alternative frameworks. Such terms are justified because often (although not always) students' alternative ideas in chemistry are well established and strongly committed to (see Chapter 14). So, even when the teacher becomes aware that there is an issue, modifying student thinking may not be straightforward (see, in particular, Chapter 15). This is one of the core issues in chemistry education (Taber, 2002).

This is clearly not the only issue of importance in modern chemistry education. Another would be developing the most relevant curriculum for particular student groups (Eilks and Hofstein, 2015). This may be 'most relevant' in terms of preparation for further study, for professional life in chemistry, or for wider citizenship. This may also be relevance in terms of what students themselves feel is relevant to their lives, interests, and concerns. These two areas of concern may overlap, but need not.

What is most relevant for future chemistry PhD students may not be so relevant for most school students required to study the subject, or even for most undergraduates taking a general chemistry course to meet matriculation requirements (as happens in some national contexts). This is a very important issue: education uses valuable resources, so investment in education should be carefully targetted. Students, especially those in compulsory education, have a right to expect decisions influencing their lives (such as the curriculum they are expected to study) to reflect their best interests (Taber and Riga, 2016). Student motivation may be strongly influenced by learners' perceptions of course relevance, and this in turn influences course completion, achievement, and decisions about progression to further courses.

Perhaps future curriculum reviews and developments might reduce the conceptual load in high-school chemistry, or substitute some of the concepts presently included with new concepts perceived more relevant to the student group. Updating courses in further (post-compulsory) and higher education will over time lead to more recently developed chemical concepts replacing some of the traditional concepts as they come to be seen to be less important in chemical research and practice today. Scientific concepts themselves evolve (see Chapters 5 and 6), so, for example, the concept {acid} today is somewhat different from what was nominally the 'same' concept when it first appeared in a chemistry curriculum. (The use of parentheses (*i.e.*, {}) to mark concept labels is introduced in Chapter 2.)

So, the actual concepts met in chemistry courses will change. However, given the nature of chemistry as a subject, teaching chemistry will always involve teaching students a good deal of conceptual material, and much of that will remain challenging. That is, challenging for students to learn, and so challenging for teachers and lecturers to effectively teach. Moreover, it seems very unlikely that some of the core concepts of chemistry today, concepts known to often be found challenging, will ever come to be judged superfluous or anachronistic in chemistry courses. Certainly, concepts identified as core when I was a school-age student (Fensham, 1975), are still core today. For example, a serious study of chemistry as a science is likely to always need students to learn about elements and compounds; atoms and molecules; and periodicity.

A premise of this book, then, is that the very nature of the subject of chemistry means that understanding its conceptual content is challenging, and so many students will face learning difficulties with much of the material – and that, consequently, teachers will find that teaching these concepts effectively is not straightforward. If that seems pessimistic, then it is hopefully balanced by two other premises underpinning the book. One is the optimistic view that learning can be supported, and teaching can be informed, by an understanding of the nature of this challenge. Informing teachers at all levels about the nature of conceptual learning in chemistry, and why it often seems to go wrong, can allow them to develop more effective teaching strategies. A second optimistic assumption is that whilst, certainly, more research is needed (see Chapter 16), there is already a good deal of research that can help teachers understand in general terms the problems and potential solutions when teaching chemical ideas.

Some of this work derives from research in psychology, or in what are sometimes referred to as the learning sciences, and some originates in educational studies exploring general aspects of teaching and learning. Much of that work is generic in nature, and potentially as applicable to learning in, say, mathematics, or history, as in chemistry – but it nonetheless offers useful theoretical frameworks and perspectives for understanding the general nature of learning processes.

Complementing this, is research that is more specifically relevant to teaching chemistry in particular. Science education is now a very well-established field with a wide range of periodicals (including a number of highly regarded, well-established, research journals), and within science education there has been a vast amount of work exploring students' learning of science concepts, and the development of student thinking in response to teaching. Some of this work is specifically based on teaching and learning of chemical concepts.

In recent years, chemistry education research has become better established as a field in its own right, with its own journals, research groups, conferences, and the like. Again, one of the key areas of research within chemistry education has explored student thinking and learning in particular topic areas, and there has been much attention to the nature of student conceptions and the challenges of learning canonical chemical concepts (Taber, 2018a).

The present book is therefore an attempt to offer an account of the nature of chemical concepts, and the learning of those concepts, based on the current state of knowledge drawn from within and beyond chemistry education. The purpose of the book is two-fold. One goal relates to the area of research itself. Although there is a good deal of relevant research, we certainly do not know all that it would be useful to know about the teaching and learning of chemical concepts. This book also acts as a snap-shot of the current state of our knowledge in this important area, and by reviewing the current state of the field, the book offers indications of fruitful directions for further research. Hopefully, such an account will support those setting out on research, as well as those planning teacher education and development programmes. I hope also to offer a readable account that will be informative for all teachers of chemistry, to support teachers in developing greater understanding of the nature of the challenge of teaching chemistry concepts to students – and to therefore inform more effective teaching practice.

1.1 Research is Underpinned by Theory

It was suggested above that scientific research in chemistry depends upon interplay between theory and practical enquiry. This is so in any scientific field. Those of us who work in science education and/or chemistry education consider that our research should be, or at least aspire to be, scientific in nature. Educational research is essentially social science, but still science. The distinction is important, as clearly the assumptions and methods that apply in – say – enquiry into synthetic routes towards biologically active natural products, are not directly transferrable to enquiry into teaching and learning processes in a school classroom or an undergraduate chemistry laboratory.

However, there are expectations that would apply to those undertaking research in any field that we wish to consider as scientific. We expect chemistry research to be undertaken by those who have been prepared for chemical enquiry by induction into both the theoretical background and the experimental procedures and techniques needed to contribute to the particular specialist field. Someone who did not know their chemistry and had not been trained up in how to work in the laboratory is unlikely to produce empirical research that meets the requirements of peer review. For that matter, someone whose background and experience was (for example) in working with gases on vacuum lines would not be expected to suddenly switch to producing publishable work on cell biochemistry without a further period of specialist preparation.

The same kinds of expectations should apply just as much in chemistry education. Journals publishing research that makes recommendations for how teachers should go about their professional work practices need to subject that work to rigorous evaluation: to ensure the work is based on appropriate theoretical perspectives supported by relevant research literature, and that it has been carried out using a suitable research design that employs appropriate data collection and analysis techniques (Eybe and Schmidt, 2001; Taber, 2012b). This raises the question of the extent to which chemistry education, or science education more widely, might be considered research fields that deserve to be seen as scientific.

These fields certainly exist as well-established foci of scholarly and research activity (Gilbert, 1995; Fensham, 2004; Taber, 2012a), with periodicals, conferences, networks, book series, and the like. The more important journals certainly reject most submissions (which is one crude indicator of peer review rigour). There are, however, some distinctions between chemistry education as a field and most established areas of chemistry. Clearly methodologies used in educational research are often quite different from in the natural sciences, as is appropriate given the different foci of interest (Taber, 2016). Of more significance is the diversity of theoretical perspectives and methodological approaches that are employed within the field of chemistry education (Taber, 2014). Thomas Kuhn (1970) suggested that within a mature science there tend to be accepted norms relating to a range of considerations (such as terminology, key concepts, techniques, reporting formats and the like), whereas within chemistry education it is not unusual to see diverse terminology, alternative perspectives, and different methodologies applied to particular foci of interest.

In part, this could be argued to reflect chemistry education as a less mature field where a common 'disciplinary matrix' (Kuhn, 1974/1977) is still to be established, and where induction into the field is through a less formal training regime. Increasingly studies submitted for publication in chemistry education journals are led by scholars with PhDs in chemistry education – but many researchers in the field are primarily trained as chemists, or even in other disciplines such as psychology.

However, there is also a strong case to be made that the subject matter of chemistry education – where the primary foci are teaching and learning – need to be explored and understood from more diverse perspectives than the phenomena studied in most chemical research (Taber, 2014). Student learning difficulties can be primarily related to the intrinsic challenge of the conceptual material and the nature of human cognitive processes. Yet often pedagogical 'variables' (ordering of topics, choice of teaching models and analogies) and cultural considerations (how technical terms are used in everyday language; whether learners are encouraged to question the textbook and teacher presentations) are relevant, as well as institutional factors (for example a school regime that influences student aspirations and/or motivation). It is quite reasonable, therefore, to expect that a range of studies that are at one level considering the 'same' core issues (the teaching and learning of chemistry) may actually take very different forms, and offer complementary insights into a complex situation.

Enquiry that is scientific takes place within a research tradition, building upon existing research, and adopting core commitments that are shared among a community of researchers (Lakatos, 1970). A shared core commitment in chemical sciences might be to the value of conceptualising matter as particulate at a submicroscopic scale to build models that can be used to develop explanations of the observed behaviour of substances in the laboratory. That is, whilst chemists might reasonably disagree over the details of the most appropriate molecular models and how to apply them, there are few (if any) practising chemists who reject the use of explanations involving molecules, ions and electrons.

Such traditions, or research programmes (Lakatos, 1970), certainly exist in areas of science education. For example, a shared commitment among a community of researchers working on problems related to the present book would be that learners commonly form alternative conceptions of chemical ideas that significantly influence learning of the curriculum. This is one commitment of a perspective known as constructivism (or sometimes more specifically as pedagogic or psychological or personal constructivism). The present book will draw heavily on the constructivist programme (Taber, 2009) that has been extremely influential in science education for thinking about learning (Tobin, 1993; Jenkins, 2000). However, where Kuhn suggested that in the natural sciences the norm was for one such tradition to dominate within a research specialism at any one time, and Lakatos suggested coexisting programmes were necessarily rivals; alternative complementary approaches that examine different aspects of the complex natures of teaching and learning may augment each other productively within educational work, without necessarily being in direct competition.

1.2 Under-theorised Research

So, the argument here is that research in chemistry education certainly can be scientific and fit within the kinds of research programmes or traditions (sometimes labelled paradigms) said to exist in the natural sciences, but that the less orderly disciplinary structure seen in chemistry education could reflect a less mature scientific field or the greater complexity of what is being studied. It is likely both factors contribute (Taber, 2014). The less established nature of work in chemistry education is certainly detectable in some of the published research in the area reviewed by this book. Here, I will refer to two examples, one relating to terminology, and one to methodology.

Learners' ideas have been described in the literature as misconceptions, alternative conceptions, intuitive theories, alternative frameworks, and indeed in a whole raft of other ways. Some scholars have offered careful analyses of particular terms, and arguments for particular preferences. Careful reading of some of these studies suggests the diverse terminology reflects significant differences in the conceptualisations of student ideas and thinking (see Chapter 14). However, as has long been recognised (Abimbola, 1988), usage of terminology is not consistent. That creates a challenge for researchers – but sadly one that is not always faced. So it is not unusual for studies submitted for publication, and indeed published, to simply adopt a term (commonly 'misconceptions' or 'alternative conceptions') as a catch-all, and completely sidestep the issue. Whilst the problem of terminology creates a challenge for authors and referees, and whilst writing intended primarily to inform teachers may well benefit from being uncomplicated, it seems questionable that those researching the issue should write research papers that do not acknowledge and engage with what is recognised as a central problem in the research field.

1.2.1 The Mental Register

Perhaps even more significant is how some researchers into learners' ideas and learning seem to adopt (at least in their research reports, and so one assumes in their research) oversimplistic ideas about how we might find out what others are thinking. This relates to what has been described as *the mental register* – the set of terms we use in everyday life to discuss mental events and processes (Taber, 2013). These terms are common in social discourse – and for the purpose of informal conversations we all know what we mean by such terms as 'ideas' and 'thinking', and we all know that usually the best way to find out what somebody else thinks, is to ask them.

However, if one is doing research into what students think, then one needs to take a more sophisticated approach that problematises the nature of thought and the epistemological question of how we can ever be sure we understand what someone else is telling us they are thinking. As with any area of research, an analysis of key ontological (*e.g.*, what counts as an idea?) and epistemological (*e.g.*, how do we get trustworthy knowledge about someone else's ideas?) assumptions is essential to developing a methodological approach one can be confident produces valid and reliable data. A careful consideration of such matters is a topic for a book in itself (Taber, 2013) – but that does not excuse how so many research papers in chemistry education say virtually nothing about these key research issues, and treat the 'collection' and reporting of thinking and ideas as unproblematic.

1.3 Some Things that Should not be Taken for Granted

This book is then informed by a premise that professionals in chemistry education – such as those who are teachers or researchers (or both) – need to problematise things that we can too easily take for granted. Indeed, this is part of the rationale for research and scholarship. We all know, at some level, that we teach concepts – and can readily list some examples – but it may not be so easy to define what we mean by a concept, or explain what kind of entity a concept is, such that we can take that into account in our work. This might be compared with the period when atomic theory was gaining acceptance, such that chemists started to consider that matter was made up from atoms, but before there was a strongly evidenced model of what kind of entities atoms were best considered to be.

A core focus of education is learning, and again those working in chemistry education know – in general terms – what they mean by learning, and may be quite expert in applying formal techniques supposed to assess that learning. However, professionals will also be aware that such formal assessments often produce simple output measures (78%, grade C+, a 2(i) classification, *etc.*) that put aside questions of what learning really is, and rely on measuring it in terms of behaviours: getting the right answers in tests. Yet research has shown that slight changes in context or question phrasing that seem irrelevant to the expert often lead to students answering test questions quite differently (Palmer, 1997; Taber, 1997): what does it mean to say a student has, or has not, learnt something in these circumstances?

Research looking for learning gains during educational innovations often use a pre- to post-test design that incorporates delayed as well as immediate post-tests. If, as can be common, students get the correct answers immediately after being taught the module, but then give wrong answers some weeks later (Gauld, 1986), we need to consider what is a sensible criterion for judging the learning to have occurred (and if we make the judgement on an immediate post-test, do we then consider then some kind of 'unlearning' to have then occurred?) Sometimes learning gains are actually greater in delayed measures – as something happens in the learner's brain in the weeks and months *after teaching is completed* that means their test performance improves after some kind of incubation period. So how can we understand the influence of teaching if it may have effects that are not apparent till long after the teaching episode happened?

1.4 What Do We Know, and What Do We Need to Know?

This book then seeks to offer an examination of conceptual learning in chemistry that attempts to move beyond what we might readily take for granted: to explore what we know (and what we do not yet know) about the nature of chemical concepts, how learning of such concepts occurs, and how teaching can be informed by such considerations to bring about more effective learning.

At one level the treatment here may be considered philosophical, if not in a formal sense (there are much better qualified candidates than the present author for undertaking formal philosophical analysis). However, this book is centrally concerned with questioning what we know (or, sometimes, assume we know) about a core feature of the chemistry curriculum, chemistry concepts, and about the processes of teaching and learning those concepts.

The book is intended to inform those working in chemistry education, whether as researchers or teachers or both, and has been written with such colleagues in mind. I very much hope that this book will be read and understood by many chemistry teachers (whether teaching in a middle school or on a post-graduate programme or somewhere in between) and will substantially inform professional thinking and so teaching practice. However, driven by the view expressed above that sometimes research in chemistry education has been under-theorised and so is less rigorous and

The Challenge of Teaching and Learning Chemical Concepts

useful than it might potentially be, there has been no attempt to simplify technical language or avoid complications that are relevant to the core issues discussed. Rather, to support readers, the author has liberally employed pedagogic devices such as analogies and metaphors where these seemed useful (*e.g.*, "This might be compared with the period when atomic theory was gaining acceptance..." above). This has been done in the spirit that teachers are advised to adopt (see Chapter 12), as offering starting points for thinking about ideas that may be unfamiliar to some readers.

Chemistry teaching is a highly skilled job that draws upon a body of specialised knowledge. The chemistry teacher needs good subject (*i.e.*, chemical) knowledge, a good understanding of the students they are teaching, and what has been termed pedagogical content knowledge or PCK (Kind, 2009) – knowledge of pedagogical ideas worked through within the specific teaching topic. Just as chemical knowledge is specialised knowledge where the teacher needs to keep up with recent developments, PCK is a specialised body of knowledge that is being continuously developed by reflective teachers (Taber, 2018b). That is what would be expected of any area that seeks to be considered scientific, whether within chemistry itself or the increasingly active area of chemistry education. This volume can contribute to teacher PCK as well as offering a survey of an important area of chemistry education research for those who look to take this field of research forward.

1.4.1 Some Orientating Questions for the Reader

If a book of this kind is seen as a scholarly examination of its topic, then useful questions in the context of the present enquiry might be:

- Are there canonical chemical concepts?
- Do experts (chemists, teachers) share the same concepts?
- Can novices (students) be considered to acquire canonical chemical concepts?

1.5 The First-order Approximate Model of Conceptual Teaching

These questions are motivated by a naïve model of teaching that often appears to be grounded on assumptions that:

1. There are canonical chemical concepts that are agreed by experts, presented in curriculum, and held by competent teachers.
2. Students generally lack chemical concepts prior to formal instruction, but through effective pedagogy can be supported in acquiring canonical chemical concepts.
3. Teaching is successful to the extent that students acquire canonical chemical concepts.

This book will certainly not suggest that such a model has no value, as it is rather difficult to see how chemistry teaching can proceed without some such general scheme. Yet it will be argued that each of these three points are problematic. At best, 1 and 2 are gross simplifications (as will be illustrated in this volume), and, once that is acknowledged, then care is indicated in how 3 is applied in evaluating teaching.

I am going to suggest that this model, points 1–3 above, should be considered a 'first-order approximate model of conceptual teaching'. It would be disrespectful to generations of teachers who have operated under such working assumptions to describe this model as being zeroth order, as it clearly works to some extent, as least most of the time, with some students – but it includes a high level of simplification that may retard high-quality, refined work. One purpose of this volume is to suggest what a more sophisticated model, drawing upon research and scholarship in chemistry education and other fields, would be like.

References

Abimbola I. O., (1988), The problem of terminology in the study of student conceptions in science, *Sci. Educ.*, **72**(2), 175–184.

Eilks, I. and Hofstein, A. (ed.), (2015), *Relevant Chemistry Education: From Theory to Practice*, Rotterdam: Sense Publishers.

Eybe H. and Schmidt H.-J., (2001), Quality criteria and exemplary papers in chemistry education research, *Int. J. Sci. Educ.*, **23**(2), 209–225.

Fensham P. J., (1975), Concept formation, in Daniels D. J. (ed.), *New Movements in the Study and Teaching of Chemistry* (Vol. 199–217). London: Temple Smith.

Fensham P. J., (2004), *Defining an Identity: The Evolution of Science Education as a Field of Research*, Dordrecht: Kluwer Academic Publishers.

Gauld C., (1986), Models, meters and memory, *Res. Sci. Educ.*, **16**(1), 49–54.

Gilbert J. K., (1995), Studies and Fields: Directions of Research in Science Education, *Stud. Sci. Educ.*, **25**, 173–197.

Jenkins E. W., (2000), Constructivism in School Science Education: Powerful Model or the Most Dangerous Intellectual Tendency?, *Sci. Educ.*, **9**, 599–610.

Kind V., (2009), Pedagogical content knowledge in science education: perspectives and potential for progress, *Stud. Sci. Educ.*, **45**(2), 169–204.

Kuhn T. S., (1970), *The Structure of Scientific Revolutions*, 2nd edn, Chicago: University of Chicago.

Kuhn T. S., (1974/1977), Second thoughts on paradigms, in Kuhn T. S. (ed.), *The Essential Tension: Selected Studies in Scientific Tradition and Change*, Chicago: University of Chicago Press, pp. 293–319.

Lakatos I., (1970), Falsification and the methodology of scientific research programmes, in Lakatos I. and Musgrove A. (ed.), *Criticism and the Growth of Knowledge*, Cambridge: Cambridge University Press, pp. 91–196.

Palmer D., (1997), The effect of context on students' reasoning about forces, *Int. J Sci. Educ.*, **19**(16), 681–696.

Taber K. S., (1997), Student understanding of ionic bonding: molecular versus electrostatic thinking?, *Sch. Sci. Rev.*, **78**(285), 85–95.

Taber K. S., (2002), *Chemical Misconceptions – Prevention, Diagnosis and Cure: Theoretical Background* (Vol. 1), London: Royal Society of Chemistry.

Taber K. S., (2009), *Progressing Science Education: Constructing the Scientific Research Programme into the Contingent Nature of Learning Science*, Dordrecht: Springer.

Taber K. S., (2012a), The nature and scope of chemistry education as a field, *Chem. Educ. Res. Pract.*, **13**(3), 159–160.

Taber K. S., (2012b), Recognising quality in reports of chemistry education research and practice, *Chem. Educ. Res. Pract.*, **13**(1), 4–7.

Taber K. S., (2013), *Modelling Learners and Learning in Science Education: Developing Representations of Concepts, Conceptual Structure and Conceptual Change to Inform Teaching and Research*, Dordrecht: Springer.

Taber K. S., (2014), Methodological issues in science education research: a perspective from the philosophy of science, in Matthews M. R. (ed.), *International Handbook of Research in History, Philosophy and Science Teaching* (Vol. 3), Dordrecht, Netherlands: Springer, pp. 1839–1893.

Taber K. S., (2016), Methodological Issues in Science Education Research, in Peters A. M. (ed.), *Encyclopedia of Educational Philosophy and Theory*, Singapore: Springer, pp. 1–6.

Taber K. S., (2018a), Alternative Conceptions and the Learning of Chemistry, *Isr. J. Chem.*, DOI: 10.1002/ijch.201800046.

Taber K. S., (2018b), *Masterclass in Science Education: Transforming Teaching and Learning*, London: Bloomsbury.

Taber K. S. and Riga F., (2016), From Each According to Her Capabilities; to Each According to Her Needs: Fully Including the Gifted in School Science Education, in Markic S. and Abels S. (ed.), *Science Education Towards Inclusion*, New York: Nova Publishers, pp. 195–219.

Tobin, K. (ed.) (1993), *The Practice of Constructivism in Science Education*, Hilsdale, New Jersey: Lawrence Erlbaum Associates.

CONCEPTS IN CHEMISTRY

CHAPTER 2

What Kind of Things Are Concepts?

If conceptual learning is central to chemistry education, then those working in chemistry education should have a good understanding of the nature of conceptual learning. A good place to start would be in considering the nature of chemical concepts, or indeed more generally, concepts themselves.

This chapter sets out an account of what we mean by concepts. The following two chapters consider the main kinds of concept drawn upon in chemistry, and how the notion of concepts link to the more general idea of knowledge. These chapters offer a discussion of important issues that will provide a background to thinking about concepts in chemistry. This touches upon the nature of science (both natural science such as chemistry, and the social sciences, where academic work in education is often located), questions of ontology (the nature of things) and epistemology (how we can come to knowledge, and what status such knowledge can have) in science, and suggests why the nature of concepts is worthy of such attention in the context of chemistry education.

2.1 Making Sense of Curie and and and and and Meitner

At this point it is useful to introduce a typographical device to help distinguish between when I refer a concept itself rather than the application or referent of the concept. That is, for example, when I refer to someone's {acid} concept, rather than to acids themselves. In other words, in a book that will be referring a lot to concepts *qua* concepts, it will be useful to have means to distinguish in the text when a word is being used to refer to a concept in abstract rather than using the concept label (*e.g.* 'acid') to refer to what is understood by that concept (as a term such as acid would normally be used in chemical texts). An example may make this clearer. Consider the sentences:

- Metals have lustre.
- Metals is used as part of a dichotomous classification.

The first sentence suggests a property of certain substances, those considered to be metals. The second sentence, however, would not make sense if it was referring to substances. (Moreover, there is something of a clash between the use of the singular 'is' and the apparently plural 'metals'. "Metals is..." does not seem to make for good English.) However, if the word metals is being used in the second sentence to refer to the concept 'metals', for example the set of someone's conceptions of 'metals', then the sentence has a coherent meaning: "[The concept] 'Metals' is used as part of a dichotomous classification." Then what is apparently the same word – 'metals' – is actually acting as a homonym – one of two words with different meanings, but spelt (and pronounced) the same.

A similar issue arises in linguistics all the time, where there is a need to distinguish the word itself as the focus of interest, rather than what the word normally represents. A riddle I recall being posed in my childhood was to produce an authentic sentence that had five consecutive 'and's. A solution would be the hypothetical complaint of the manager of the company Curie and Meitner to a sign-writer about a sign which had been commissioned (see Figure 2.1). The manager bemoans "there is a different size gap between Curie and and and and Meitner".

The second and fourth 'and's refer to the painted word considered as an object in its own right, whereas the other three instances are normal uses of the word as a conjunction. The sentence might be easier to decode if some sort of indication was used of where the writing on the sign was being referenced:

There is a different size gap between

'Curie' and 'and'

and

'and' and 'Meitner'

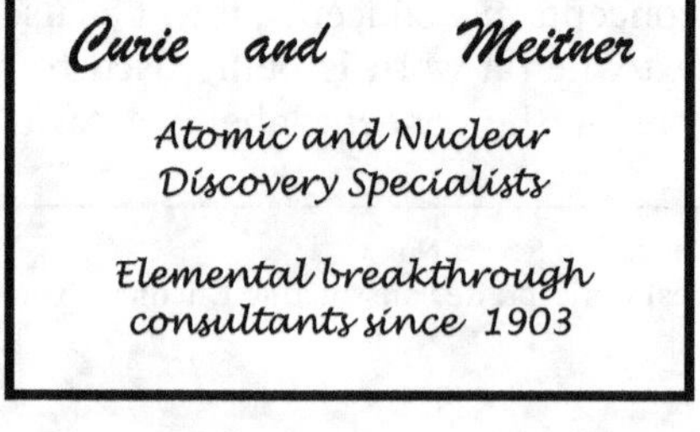

Curie and   Meitner

*Atomic and Nuclear
Discovery Specialists*

*Elemental breakthrough
consultants since 1903*

Figure 2.1 An asymmetrical sign.

or even

> There is a different size gap between
> ('Curie' and 'and') and ('and' and 'Meitner')

To support readability in the present volume, curly brackets – {} – will be placed around concept words when they are being used to refer to the concept as an entity in itself, *e.g.*, "...the concept {knowledge} is itself a rather problematic one in some ways...". So, in terms of the example used earlier

- Metals have lustre.
- {Metals} is used as part of a dichotomous classification.

This implies that in the second sentence the word metals is not being used to refer to the usual referent (substances considered metals), but rather to refer to the concept {metals} itself. Hopefully, this simple typographical device will aid readability, and the placing of words in parentheses in this way will not be distracting once the convention becomes familiar. In sentences that are phrased to refer to, for example, "...the concept element..." the brackets are not strictly needed to clarify meaning, but the convention (*i.e.*, "...the concept {element}...") is used for consistency.

An argument can be made that the concept {element} and the concept {elements} should strictly be considered related rather than identical as the singular and plural discriminate differently – that the {elements} concept is a compound built from the {more than one} concept in association with the {element} concept. So, the concepts {element} and {elements} stand in subtly different relationships to other concepts. Whilst this may be strictly true, the relationship between singular and plural concepts is consistent, and where the distinction is important it is generally clear which is applicable. I have therefore in places in the book slipped between singular and plural concepts without highlighting this shift as if there is no distinction. I think that simplifies the use of language in places and should not lead to confusion.

2.1.1 Concepts and Conceptions

The terms concept and conception are sometimes used as synonymous. So, in the literature, an alternative conception may in effect often be taken to mean a person's concept when it is inconsistent with canonical science. However, there is commonly a difference in usage that will inform how the terms are used in this volume.

A person's 'conception of' something may refer to their conceptualisation of any aspect of reality. So, for example, the phrase used above 'metals have lustre' that refers to a single proposition can be understood to be 'a conception'. Someone using this phrase in a meaningful way (that is, rather than simply parroting something read in a book, rote learned without understanding – see Chapter 11) presumably has an understanding of what is meant by 'metal' and 'lustre'. This being so, we would consider that they have both a {metals} concept and also a {lustre} concept. Probably 'lustre' is only one of the associations the learner has for 'metal' (*e.g.*, they may also consider 'metals are ductile' or 'metals conduct electricity'), and indeed they could have other associations for 'lustre'. In this case, then, this specific conception is an association between two concepts: {metal} and {lustre} and so can be considered one aspect of the learner's {metal} concept, and also one aspect of their {lustre} concept. This is true even if it at some stage in their learning this is the only association, and so in effect the full 'content', of that concept. (Perhaps, for many learners, the 'content' of their concept {lustre} may indeed comprise one direct association: as indeed may also be the case with their concepts {sonorous}, {ductile} and {malleable}!)

So, concepts have a particular focus on one type of entity, event, *etcetera*. The alternative term conception is seen as related, but more general, and in particular may relate to the associations understood to link certain concepts. If a teacher were to refer to "Julie's conception of metals" then this would refer to how Julie broadly conceptualises metals, so Julie's {metals} concept: so conception and concept become synonymous in this particular usage. If, instead, the teacher referred to "Julie's conception of how metals react with acids" then this conception would be considered to relate to *one aspect of* Julie's {metals} concept (as well as relating to her {acids} and {reaction} concepts) – so only part of the 'content' of her {metals} concept.

2.2 How to Best Understand Concepts

It will be argued that we can best understand concepts in terms of a number of basic features:

- Concepts act as categories.
- Concepts are abstractions from experience.
- Concepts are mental entities.
- Concepts are tools used in thinking.
- Concepts are only apparent when activated.
- Concepts act as nodes in a conceptual network.

Each of these points will be developed further later in the chapter, but in outline:

Concepts act as categories in that they are the basis of discriminations we make. When we have a particular concept, then we know when that concept applies, or somewhat applies. If we have a concept of {acid}, or {classroom}, or {gifted learner}, then we can make judgements about whether something is an acid; is a classroom; or is a gifted learner.

This is leaving aside, for the moment, such issues as: whether we make what others would consider correct judgements; whether our use of concepts is consistent; and the level of confidence we may have in using them. (These

What Kind of Things Are Concepts?

are issues that will arise later in the book.) Someone who had no concept {metalloid}, or no concept {hybrid orbital}, would have no basis for discriminating experience on that basis. If a student has no concept of what is meant by the label 'metalloid', then should a teacher ask them whether something was a metalloid, the best they can do is guess – so perhaps a metalloid is an android fabricated from metal: a metal robot in human form?

Concepts are abstractions from experience in the sense that concepts are formed based on some experience that is being made sense of. Human beings have a natural tendency to make sense of experience (Brock and Taber, 2017), and forming concepts is part of this process (see Chapter 13). Again, there are complications. Young children form concepts from their direct experience of the physical world, but education seeks in some ways to short-cut this concept formation process by helping people form concepts based on second-hand experience – what other people tell us (Vygotsky, 1934/1994). Nonetheless, concepts are abstract in nature.

The concept of {conical flask} can be used to refer to a very great many specific objects that are classed as conical flasks: some are much larger than others; some have thicker glass; they may have different markings on their side; they may vary somewhat in the 'steepness' of their sides, and so their overall geometry; some may have small chips of glass missing on the lip, or be cracked and residing in a disposal bin; some may contain materials, whilst others are 'empty' (apart from air); some may be damp from being washed, and sitting 'bottom-up' as they dry; or they could be in locked cupboards, or still in the supplier's boxes in the store room, or on lorries in transit, or in sinks, or on the bench; they may be bunged or corked, and may be connected to other glassware.... All of this seems obvious because we are so used to our human ability to abstract some critical features that discriminate kinds of objects, here conical flasks, and so can make such discriminations despite the diversity of flasks and the quite different contexts in which they occur. When asked about 'concepts', we are more likely to think of examples such as {hybridisation} and {entropy} rather than {conical flask} – but a person who can distinguish conical flasks from other things has a {conical flask} concept.

This links to the next point, that concepts are mental entities. Concepts do not exist in the physical world (even if they refer to things that do) – but we make sense of our experiences, of what does seem to exist (things, processes) by conceptualising them. That is, we form mental representations. Each of us has our own mental representation (image) of 'the' conical flask, which is part of our own concept {conical flask}, and that is not limited to any particular flask but concerns the general idea. Unlike real flasks, that exist in the public space of the physical world, our flask concepts are private, mental features. This raises the issue of how we know we 'share' concepts. This is clearly a key question in education, where teachers seek to share canonical scientific concepts with learners – such as {hybridisation} and {entropy} (as well as {conical flask}) – and this will be a major concern later in the book (see, in particular, Chapters 11 and 12).

The final point discussed here is that although we cannot find concepts out in the environment (only entities that we might apply our concepts to), that does not mean concepts exist in some kind of splendid isolation. Concepts do not exist in some mental solitary confinement, but rather are linked to other concepts in ways that are essential to their nature. Concepts form networks or webs (to use some common analogies) and the meaning a person has for a particular concept depends in part on how that concept is linked to their other concepts.

2.2.1 Concepts as Categories

Concepts are used in making discriminations when we make sense of experiences, and so are the basis of how we categorise, as we perceive the world. As I write this, I can see my computer screens; a few twigs and leaves from my garden hedge appearing just above the screen directly in front of me; the house on the opposite side of the street; as well as a lamp-post and a tree; and a clear blue sky in the background. That is a fairly everyday experience, but it represents a good deal of conceptualisation. The sensory data my eyes relay to my brain is based on differential illumination of areas of my retinas – that is various patches of colour. However, what reaches my consciousness is not abstract information about patches of colour, but awareness of objects.

We all occasionally find we are looking at, or listening to, something and thinking 'what is THAT?' – but this experience is very much the exception, so much so that it may even be troubling, as we are so used to seeing and hearing recognisable *things* or *situations* (doors, and buses, and conical flasks; running water, a passing jet, children playing) rather than trying to make sense of raw sense data. That the surface of the house opposite has different colours and textures (the dark brown roof tiles, yellowish brickwork, glass windows, white window surrounds, *etc.*) is only noticed when I deliberately pay attention, and is otherwise just part of an experienced Gestalt of an integral 'house'. (As I re-read this description, I realise that despite intending to describe sensory information, I have, without noticing, described the windows as 'glass' which is an automatic interpretation – it is what I see because my nervous system makes preconscious inferences in perception.) Moreover, I tend to see the same colours under varied lighting conditions, and have never actually felt the texture of that particular brickwork or the windows or door, even though I sense how they feel as part of my perception of the house.

Only parts of the house, the tree, the lamp-post, and (hardly any of) the hedge, are visible from my position, but my sense-making system applies concepts {hedge}, {house}, {tree}, and {post}, that both identifies more extensive and coherent objects than I can currently actually see (the house is taken to have depth, although from my current viewpoint it could be a facade, as sometimes used in films), and also requires each of them to be attached to the ground rather than floating in the air – although based on the sensory information currently available from light incident upon my retinas there is no immediate evidential basis for preferring full, grounded objects rather than partial, floating ones.

Our brains do all this making-sense work so automatically that it may seem trivial to make these points, but in classrooms the teacher's brain is often doing similar automatic work using concepts that the students do not (yet) fully share. Teacher and student see the classroom window equally well, but the teacher may 'see' a reaction or an oxidation or a displacement or evidence of a potassium salt or of an unsaturated compound – when the students do not. What is seen in terms of the concept by the expert may just be patches of colour or undifferentiated noise to the novice. Learning to see the chemistry being taught at the learner's resolution (Taber, 2002), recognising how the evidence for the chemistry seems to a novice, is a key teaching skill.

Some examples of the concepts met in chemistry learning include {element}, {molecule}, {acid}, {formula}, {distillation}, and, as suggested above {conical flask}. This list includes a range of types of entity (see Chapter 3), but *as concepts* they are all abstractions (see below). That is, the concept refers to something general – not one of the chemical elements, or a specific molecule, or a particular conical flask, but rather the idea of an element or of a molecule or conical flask. At one level, then, concepts act as categories.

The concept {element} is a type that can be considered to have membership. Hydrogen can be considered a member of the set of elements, so is covered by the concept, but ammonia is not. Of course, hydrogen can only be considered a member of the set of elements by someone with a {hydrogen} concept (as well as an {element} concept), and {hydrogen} is also an abstraction that potentially relates to many actual samples of materials.

Some, but not all, articles found in the laboratory would be considered as conical flasks. Some laboratory activities would be considered as distillation, and not others. So even if a concept (*i.e.*, {distillation}) describes a process (*i.e.*, distillation) it can still be considered to act as a category, as it allows the conceiver to distinguish examples and non-examples.

In writing about concepts in this way, I have inevitably labelled concepts with names. This raises an issue of whether it is possible to have a concept of {element} without using the term 'element'. In principle, it is possible to have a concept without such a label. A student who was not familiar with the term 'conical flask' could be asked to sort laboratory glassware into different sorts, and may separate out all the conical flasks, from all the beakers, and all the watch glasses, *etc.* This will indicate the student was operating with concepts, including a concept of {conical flasks}, even if they had no verbal labels for the different classes of glassware they were forming. Labels such as 'conical flask' make communication much easier. However, they may also seem to suggest that two people using the same concept label have *the same* concept, and, as will be discussed later in the book, this is not necessarily the case. As we can easily point to conical flaks, students are more likely to share much the same {conical flask} concept than much the same {oxidation} or {resonance} concept. It may also be easier to find out if people share much the same {conical flask} concept than much the same {oxidation} or {resonance} concept.

2.2.2 The Abstract Nature of Concepts

First, however, it is useful to consider the abstract nature of concepts. Concepts such as {molecule}, {reaction rate}, and {entropy}, refer to abstract ideas, and so may easily be accepted as abstractions – but concepts more generally are abstractions, even when they are applied to very concrete objects. Conical flasks are material objects, things, that can be washed out, put away in cupboards, and purchased from laboratory suppliers – or dropped and broken in undergraduate laboratories and charged against a student's laboratory deposit. The actual objects are real enough, but *the idea of* a conical flask, the concept {conical flask} (that allows us to distinguish a conical flask from, say, a beaker), is an abstraction from the real objects: a conical flask is a physical object: but {conical flask} is an abstraction. (How such abstractions are formed is discussed in Chapter 13.)

The concept {distillation} is also an abstraction, if not referring to a set of objects, but rather to a set sequence of processes. The abstract concept {distillation} includes some processes, but not others – for example heating a melting point tube would not come under the abstraction of the {distillation} concept, although heating a sand bath might – depending on the wider context of that specific process.

The concept {molecule} is similarly an abstraction, but in this case not relating to a class of directly and readily manipulatable objects (like conical flasks) or observable processes (like distillation), but rather something that is already inherently abstract. A molecule of ammonia is in effect conjectural in the sense that it cannot be shown to, or readily manipulated by, students, but can only be inferred as a theoretical object used as a component of explanations in chemistry (see Chapter 3). Nonetheless the concept is still applied in making discriminations: a conjectured NH_3 molecule is admitted as a molecule, but an Na^+-Cl^- ion pair is not usually covered by the concept. That many students *are* likely to consider an Na^+-Cl^- ion pair *is* a molecule (Taber, 1997) is an important complication that exemplifies an important theme of this book (see Chapter 14) – people often develop non-canonical concepts.

2.2.3 Concepts Are Mental Entities

If we ask where we might find concepts, where they exist, we need to consider them as mental entities. If concepts are the basis of discriminations, then they must be found where discriminations are made – inside the heads of people. If concepts are abstractions from experience, then those abstractions are undertaken within and with brains.

There is a problem in identifying minds with brains. A brain is a physical object that can in principle be observed, weighed, *etcetera*. Mind is more tenuous, and is a construct used to frame our conscious experiences – what we think, dream, imagine, plan, decide, believe, *etcetera*. That distinction need not be problematic in itself – mind is widely considered to be emergent from brain, and so offers a different level of analysis – similar to how the chemist can think about a reaction in terms of the change in the actual material observed at the bench, and how this can be understood in terms of the interactions of molecules and ions and electrons (or, collectively, quanticles for brevity). In the same way, we might understand mental experience as the observable outcome of activity at a very different level – currents flowing in tiny neural circuits. Thinking, a subjective experience, is considered to correlate with physical processes occurring in brain structures (see Chapter 13).

The complication, however, is that there is much brain activity that contributes to our conscious thinking but that is not itself directly experienced. There is much *preconscious* 'thinking' (that is processing of information, which we might consider should count as thinking) which we cannot consciously access, but which informs our conscious thinking. An intuition (Brock, 2015) is the conscious outcome of preconscious processing of information – so we may, for example, judge someone untrustworthy, but without really knowing why. We are conscious of our distrust, which is based on processes in the brain that draw on cues we may not be consciously aware of. This is an important topic, as preconscious processes are believed to be very important in concept formation and learning (and indeed in doing science).

To stretch an analogy used above, the chemist observing a reaction moving to equilibrium sees the gross changes (in colour, volume, *etc.*), but cannot observe the underlying activity at the level of quanticles (*e.g.*, molecules, ions). Once the system reaches equilibrium, then, assuming no external drivers for change are applied, it appears to be fairly inert – in a static state. However, the quanticles remain highly active, with the forward and reverse reactions continuing apace – albeit at the same pace so as to cancel each other on the molar scale. If we could directly observe the reaction at the level of quanticles, the equilibrium would be a very dynamic equilibrium indeed. There is much activity, but (like preconscious thinking) it is not readily detected.

So, in suggesting that concepts are mental entities we potentially invite a degree of ambiguity. We might ask whether this means that for something to count as a concept it should be explicitly accessible to conscious introspection, or whether we should include as concepts the basis for making discriminations that seem to be undertaken entirely preconsciously, such that the person concerned cannot explain their use of the 'concept'? This is a reasonable question, as either position could be supported, and it is perhaps a matter of semantics (how we might choose to define concepts).

If we are interested in how people make discriminations – when a student decides something is an acid or an example of oxidation for instance – the student's preconscious processing may be as important as their deliberate thinking processes. The view taken here is there is an important distinction between these two categories – it has consequences in teaching and learning as it is easier for teachers (and learners themselves) to engage with students' conscious thinking – but both are important, and within the remit of concepts for the purposes of this book. This perspective will fit with the discussion of concepts-in-action, below.

The exact relationship between brain and mind is not fully understood, and it is even sometimes suggested that consciousness is not functional, but rather something of an epiphenomenon (Pockett, 2004) – that is, it provides an awareness of some brain activities but makes absolutely no difference to those activities (so, for example, what we experience as *thinking that leads to a decision*, might actually be simply *the awareness of* that decision-making process).

Regardless of the functional role (or otherwise) of consciousness, it does provide us with access to some of our thinking, including some evaluations and decisions we make. The qualifier 'some' is important, as to borrow a rather overused metaphor, the thinking we are aware of is like the tip of an iceberg: most of our cognition takes place beneath the surface and without conscious awareness. That is not necessarily a bad thing, and it may protect us from information overload – like the head of a large organisation who needs to preserve her time and energy to deal with strategic matters, the limited scope of conscious awareness may need to be protected from most routine cognition and only alerted on a need-to-know basis. For present purposes, the critical point is that we apply concepts in carrying out our thinking processes, so they are considered as mental (rather than physical) entities.

2.2.4 Concepts Are Tools Used in Thinking

If concepts are mental entities, which are used to make discriminations – so linked to understanding what kind of things exist in the world and the characteristics of those things – then concepts are part of a 'mental toolkit'. Scientific work relies on thinking – on the processing of information in brains. In part, it relies on imaginative thinking to make creative leaps, that is to imagine possible ways the world might be, so as to make sense of observations, or to imagine ways we might investigate the nature of things. It also relies on logical reasoning – being able to see what follows from what; and what excludes what. For example, the use of syllogism:

Sodium is a metal.
Metals are good thermals conductors.
Therefore, sodium will be a good thermal conductor.

This might be part of an argument for considering sodium as a coolant in the design of a nuclear reactor: along with other considerations such as the high liquid range of sodium (as it remains a liquid to over a thousand Kelvin), and its relatively weak behaviour as a neutron absorber. It is clear that even in this short vignette, the reasoning used relies upon a range of concepts: {sodium}, {metal}, {thermal conductivity}, {coolant}, {nuclear reactor}, {liquid range}, {temperature}, {neutron}, and {neutron absorber}.

A wider consideration of this possibility would include sodium's high chemical reactivity, and how it becomes the radioactive isotope ^{24}Na when it does absorb neutrons. In determining the significance of the latter complication one consideration will be the relatively short half-life (less than a

What Kind of Things Are Concepts? 27

day) of ^{24}Na. Here, a number of additional concepts are drawn upon: {chemical reactivity}, {radioactivity}, {isotopes}, {half-life}.... These concepts are not all of the same type (see Chapter 3) – some relate to substances, some to processes – but they are all the basis of making discriminations.

We can distinguish sodium from non-sodium, or perhaps (using a typological convention white text on black background to represent the complementary set of everything that is not sodium) {sodium}; coolants from non-coolants ({coolants}); isotopes from non-isotopes ({isotopes}). Concepts have ranges of convenience (Kelly, 1963) where they can readily be applied. So, we can use the concept {isotope} sensibly in making discriminations in relation to thinking about sodium and other elements, but it has less direct application to thinking about temperature for example. We can distinguish ^{24}Na from {^{24}Na} (non-^{24}Na) – which in some contexts where we would use the {^{24}Na} concept might primarily mean ^{23}Na, but in general refers to *anything* that would be discriminated as not being ^{24}Na.

Some such discriminations are binary, and some are more nuanced. We can use the concept {sodium} as the basis of a dichotomy: we can sort everything in the world into two categories – examples/instances/samples of sodium and ... anything else. We may not tend to think about the {sodium} (*i.e.*, not sodium) category in that way – that is, we seldom explicitly operate with the concept {not sodium} or {sodium}. Yet, whenever we use the {sodium} concept to make discriminations those things excluded from the category defined by the concept {sodium} are implicitly being categorised as not sodium and so members of the category defined by an implicit {not sodium} or {sodium} concept.

The concept {coolant} might be said to be fuzzier. Virtually any fluid could be considered to act as coolant in some circumstances, even if it would be unlikely to be deliberately chosen to do so. The category defined by the concept {coolant} is therefore more open – to degrees of membership, and to be applied in context-sensitive ways. In making discriminations with the concept {coolant} we are likely to be considering other factors than whether something actually can act as a coolant: perhaps, is the substance *used commonly* as a coolant? is it an *effective* coolant? is it a *feasible* coolant in a specific context in relation to boiling temperature, cost, reactivity, toxicity, *etcetera*. In the chemistry laboratory, we are unlikely to consider human sweat as a candidate coolant to use to prevent a laboratory set-up overheating, although in everyday contexts it is an example of a coolant that we are all familiar with, and indeed rely upon for our own well-being.

2.2.5 Concepts Are Only Apparent When Activated

If concepts are not physical entities, then they cannot be seen, touched, weighed, *etcetera*. They do not have unique emission spectra or give rise to characteristic absorptions in the infra-red range. We might question whether they really exist at all. The fields of psychology and education have found the notion of concepts (that is, the {concept} concept) very useful as an explanatory device for making sense of much experience, so this book follows that tradition and treats concepts as real things – albeit mental entities. We might consider concepts to be akin to atoms, in that they have a useful explanatory value, and can help us make sense of a good deal of data. This means, however, that we never see our own concepts, and even less anyone else's, but can only characterise them through indirect means.

Most chemists treat atoms as real objects. Under certain conditions, for example in a scanning-tunnelling microscope (STM), nature provides certain phenomena, say patterns of tiny currents, that allow us to produce an image that we interpret as the arrangement of atoms on a highly magnified surface. When we turn off the power to the instrument, the image is gone, as is the phenomenon that was produced by the experimental set-up (the currents), and so we no longer have evidence of those atoms. Perhaps they only exist when we turn on the machine – perhaps they are an artefact of the experimental set-up and the rest of the time the surface is perfectly smooth with no bumps we might conceptualise as atoms? I do not make that as a serious suggestion, but if a student were to raise this possibility, we might wonder how we could counter it. One argument is that nature is regular and would not play tricks on us by being different only when we look.

Yet in the double-slit experiment nature seems to play just such tricks. A single electron will act as a wave (demonstrate its wave-like qualities), and when its wave front meets the two slits will seem to produce secondary wave sources that duly interfere (as indicated by where electrons can be detected after they have passed through such an arrangement). It does this until we set up a means to 'find' the electron before it diffracts at the slits, after which the electron is duly found to be localised and demonstrates particle-like qualities. It behaves like a particle that can only pass through a single slit (or miss the slits altogether and be absorbed). Again, I am not suggesting atoms are behaving like that in the STM, but rather I am asking how readily we might persuade the awkward or genuinely inquisitive student otherwise.

Concepts are only detectable when they are activated – used to think with. If we consider our own concepts, then we can introspect on them to some extent – either directly or indirectly. We can produce 'maps' of our concepts (Novak, 1990). We can consider whether we have definitions, or can list any characteristics, give examples, relate them to other concepts, and apply them in various ways. We can 'introspect' on

- How do we recognise conical flasks?
- Why is water not an element?
- What are the common characteristics of metals?

We can do this deliberately. Also, someone who is sufficiently reflective may notice how they are thinking with a particular concept: "I notice that I am unsure if I should refer to van der Waals' forces as chemical bonding – so how do I define 'chemical bonding' and decide the range of convenience

What Kind of Things Are Concepts?

of this term?" This then relies on a level of metacognitive awareness (Whitebread and Pino-Pasternak, 2010).

Similarly, we may infer something about another's concepts (Taber, 2013) in a similar way – by 'testing' them through direct interrogation ("what is a transition element?") or noticing how they seem to apply concepts (thinks: "Hm. So you are telling me that there are equal amounts of reactants and products in this reaction – I wonder if that is something you assume must be true for any equilibrium?") and perhaps following up on points of interest ("you said that a strong acid has a pH of 1; do you think that always applies?")

Consciousness has a limited focus – we can only think about a modest number of things at any one time. This is related to working memory, an attribute that allows us to 'keep in mind' and to work with, mentipulate, a modest array of verbal and/or imagistic information (Baddeley, 2003). So, for example, we may consider a hypothetical chemist who is thinking about the ideal gas equation. Whilst she is thinking about the gas equation she is unlikely to also be thinking about displacement reactions and the halogens or the catalytic specificity of enzymes. At the point when she is thinking about the ideal gas equation there is evidence available to her, and perhaps others if she is 'thinking out loud', that she has a concept {ideal gas equation}, but probably not that she also has a concept of {halogens} or {catalytic specificity of enzymes}. On another occasion she may be thinking, and talking, about the halogens, or about the catalytic specify of enzymes, but probably not, at the same time, the ideal gas equation.

This suggests that the evidence that someone 'has' these concepts has to be collected at different times, when they are actively being used in thinking. Yet we would normally assume the concepts remain at other times, and do not only appear when we think about them. That is, it seems common sense that the concepts continue to exist, and so may be activated, at different times. Like atoms when the STM is powered-off, they are assumed present even when not apparent.

However, the wave–particle duality found in situations such as the double-slit experiment offers an alternative analogy. Under some observing conditions electrons seem to behave like waves, but under other conditions they appear to be particles and do not show their wave-like properties. Concepts could be more like the particle behaviour of electrons, a kind of behaviour that appears under some conditions. In this scenario, there is some underlying feature (*cf.* the electron) that when activated in a particular way – thinking – appears as what we identify as a concept (*cf.* the measurement of a particle position), but which is actually distinct. That is, perhaps the concept is best understood as the activation of something else, which for the sake of argument I will label as a 'congenst'. In this model, the congenst continues to exist when we are not activating it to produce the concept.

This may seem unnecessarily fanciful, after all one widely adopted scientific heuristic considered to derive from William of Ockham (or Occam) suggests that when there are several potential explanations, then, all other things being equal, we should prefer the simplest. That is of course just a rule of thumb, and gives no assurance of us making the right choice. Often in science simple models and explanations evolve into more complex ones as greater evidence offers additional explananda that the original versions do not fully explain. 'All other things' seldom really are equal.

My reason for complicating matters here relates to apparent qualities of people's conceptualising. People acquire concepts, which develop over time. Their observed concepts may also be erratic – not only applied selectively in what would seem to be equivalent contexts, but with shifting patterns of inconsistency over time (see Chapter 14). People may on some occasions not use appropriate scientific concepts that they had previously seemed to have mastered, only to later be able to apply them again. If we consider chemical concepts to be 'stored' somehow in brains, then the underlying brain structures seem to often be very labile and inconsistent – at least in many students. No doubt brain structure does change (our experiences are constantly leading to modifications of synaptic connections – see Chapter 13) but we might doubt these changes are so capricious.

There are good reasons therefore to use a working assumption that concepts are generated at a particular time, in a particular context, drawing upon an underlying resource that is related to, but not identical to, the concept itself. At one level this is surely so – the underlying brain structures are believed to be neural circuits and the activity is electrical patterns that are ontologically completely distinct from the subjective experience of thinking with concepts. However, that is not quite the distinction I was making: the issue is not whether the underlying anatomical structure is the concept (clearly not – that involves an ontological category error: they are different types of entity) but whether it 'stores' the concept. If we answer 'yes' then we have to explain whether inconsistent responses that are often seen from students reflect ongoing shifts in the storing of concepts such that the stored concepts are ebbing and flowing between activations, or whether the student has stored a good many different versions of nominally the same concept that are activated on different occasions.

However, if we assume instead that there is something more stable (certainly modified over time, but not in such an apparently haphazard manner) that acts as a kind of substrate from which concepts are generated, then it is less difficult to see how the same underlying concept generating substrate or structure (or perhaps, 'congenst') might be activated differentially at different times under different conditions. A concept is then more like a hand drawn from a pack of cards in the sense that a different hand might be drawn from the same deck at another time. Or we might consider someone who on one occasion takes a Swiss army tool from her pocket to tighten a loose screw, and, on another occasion, takes the same tool from her pocket to cut an apple into slices: the same resource is activated and applied differently in different contexts.

A term that is sometimes applied to concepts when people seem to apply them inconsistently is 'multifacetted' (see Chapter 4). Perhaps a congenst

What Kind of Things Are Concepts?

can be multifacetted, and the concept activated reflects a particular facet that is illuminated by a particular context or question. Clearly, we are here dealing with metaphor and analogy, as neither concepts nor their representations in brains are directly observable.

For the sake of readability, especially for readers who may not read through the book in chapter order, in most places in the text I will not seek to distinguish between the concept as activated and the conjectured congenst assumed to be underpin this – but it is worth readers bearing in mind that notions such as conceptual growth, conceptual development, and conceptual change should sensibly relate to the structures underpinning conceptual thinking, rather than the conceptions people are aware of at some particular moment. Thinking is by its nature in a state of flux – as in the so-called 'stream of consciousness' – and education seeks to provide relatively stable resources to channel that flux in preferred directions.

2.2.6 Concepts Act as Nodes in a Conceptual Network

Another important quality of concepts is that they tend not to exist in isolation, but rather are related to other concepts. Earlier in the chapter the term 'content' was used in relation to concepts, such that the content of a concept would be the aggregation of all its associations. So the concept is not the tag it may be given (a label such as 'oxidation'), but the full range of meanings due to its associations within the wider web or net of concepts. It is therefore a simplification to equate concepts with simple verbal definitions and characterisations. We can appreciate the importance of conceptual relationships by examining how we understand specific concepts. A few examples will suffice.

The canonical concept of {chemical bonding} is related to other concepts such as {molecules}, {ions}, {electrons} and so forth. Someone who had no {atoms}, {electrons}, {nuclei}, {ions} or {molecules} concepts could not reasonably be said to have acquired the scientific {chemical bonding} concept.

Similar points can be made about other canonical concepts:

- The concept {element} is inherently linked to the concept {substance} that is in turn inherently linked to the concept {matter}.
- The concept {transition element} is inherently linked to the concepts {element} and {periodic table}.
- The concept {melting} is inherently linked to the concepts {solid}, {liquid} and {energy}.
- The concept {distillation} is inherently linked to the concepts {mixture} and {separation}.
- The concept {hybridisation} is inherently linked to the concept {atomic orbital}

Many more examples could be provided.

This is not to say that someone could not have some conception of transition element – perhaps linked to concepts such as {metal}, {coloured salts}, {variable oxidation state} – without linking it to the concept {periodic table}, but such a conceptualisation would be deficient in an important sense. It would be linked to some other concepts but not all those that might be considered to be essential for the full chemical concept. Someone who does not hold a scientific notion of what is meant by substance in chemistry surely cannot be considered to have acquired the canonical chemical concept {element}, even if they demonstrate appropriate applications of their {element} concept in some contexts. The notion of alternative conceptualisations of chemical concepts will be considered in Chapter 14.

Such distinctions (between what is canonical, and what is alternative) are matters of degree. What is considered essential for someone's concept to be credit-worthy can vary at different educational levels. A reader seeing "The concept {melting} is inherently linked to the concepts {solid}, {liquid} and {energy}" may have thought that {entropy} and/or {order} should be added to the list. A {melting} concept that has no associations to entropy and/or order is surely deficient – yet in early secondary school education such a (deficient) concept would be a 'creditable' concept. It may be evaluated to be sufficiently canonical to match a curricular model of the scientific concept at that grade level: that is, to be an 'accredited' concept (see Chapter 12) in that educational context.

As concepts are used as tools in thinking, when we engage in particular reasoning processes we might draw upon an extended range of conceptual links from the vast conceptual net. For example, Figure 2.2 offers a schematic showing a large set of concepts (represented as nodes – some shown with concept labels), with some of the connections that might be activated in a context used as an example above: considering whether sodium might be suitable for a coolant in a nuclear reactor.

2.3 Representing and Exploring Conceptual Structures

Figure 2.2 is only meant as a representation of how we might understand networks of concepts, and should not be taken as some kind of realistic model. If, as suggested above, concepts are mental entities then they cannot be directly observed in others (and sometimes not even in ourselves) – yet we normally talk in everyday life as if we are mind-readers, and researchers sometimes report results from enquiry into learner conceptions as if people's concepts can be readily characterised (Taber, 2013).

2.3.1 Representing Conceptual Structures

The way we commonly define concepts in verbal terms (an element is...; an acid is...; *etc.*) may over-emphasise the verbal aspects of concepts – for example against imagistic aspects. Some concepts may have imagistic tags rather than verbal labels (that is, we picture something), and some concepts may have no labels (see the discussion of implicit knowledge in Chapter 4).

What Kind of Things Are Concepts?

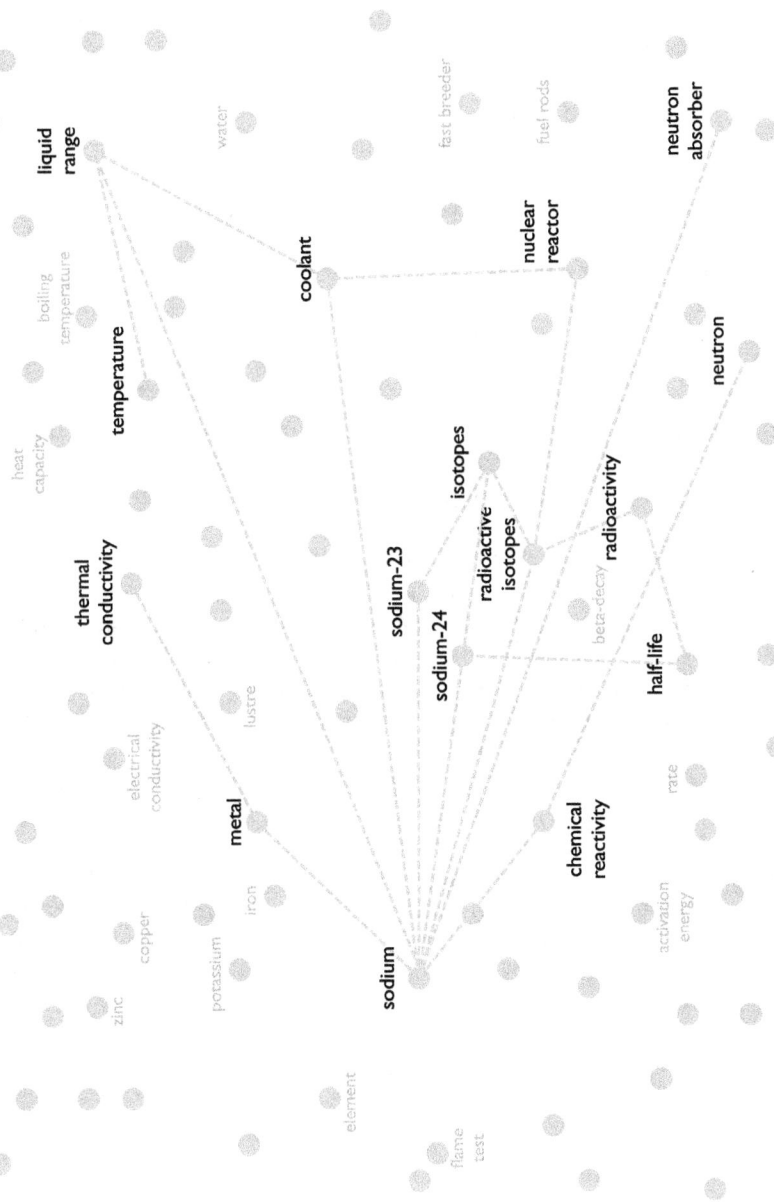

Figure 2.2 Schematic representing the hypothetical activation of a net of concepts when reasoning about the potential of sodium as a coolant in a nuclear power station.

Representations, such as Figure 2.2, should only be seen as schematics, and to reflect a kind of hypothetical model of the nature of concepts that should not be mistaken for a realistic representation. Although some of the inactivated nodes are labelled with concept labels, if we understand concepts as depending upon their inter-relationships, then it would make no sense for an isolated node with no relationships with other nodes to be considered as a concept. (The nodes in a spider's web cease being nodes when there are no strands to connect them into a network.) A concept such as {chemical reactivity} can be understood as the sum total of all the connections between the node showing the verbal label of the concept {chemical reactivity} and any other nodes that will be activated in different contexts: even if there are likely few situations where all those links will be active at once. Each connection (association between concepts) that might potentially be activated represents one conception that is part of the total 'content' of the concept.

Figure 2.2 maps out the concepts in a two-dimensional space where a multidimensional 'conceptual space' may be more appropriate, and the figure shows connections as activated or not as if this were a binary matter, which is another simplification. Certainly, neural connections, which presumably underpin conceptualisation, have variable and changeable connection strengths, and indeed adjustments in these strengths may well underpin learning (see Chapter 13).

Figure 2.2 is not meant to show all the viable connections that might be activated, but only to reflect activation in one particular context (here sodium as possible coolant in a nuclear power station). If the context were the question of how sodium might be extracted from its salts, a different set of associations would be activated. So, the {sodium} concept-in-action will be different in these two cases, as different parts of the associative net will be primed in different contexts.

To use a term introduced earlier, the net of associations that can potentially be activated for, say, sodium is the concept generating structure, the congenst; and what might be termed the concept-in-action is generated by this in response to particular contextualised stimuli (see Figure 2.3). So, the following questions might all activate a person's {sodium} concept, but in each case the concept as activated in that person's mind, and subsequently represented in their response to the questions, involves a limited sub-set of the associations they have available.

- What is the crystal structure of sodium?
- Is sodium a transition element?
- Is it true that there is sodium in my toothpaste?
- Does this flame suggest this is a sodium compound?

If we wanted to study the full range of associations, we would need to probe a person's understanding in a sufficient way to trigger all their available associations such that they become represented in observable behaviour (*i.e.*, what they say, or write, or draw, *etc.*). If we were able to

What Kind of Things Are Concepts?

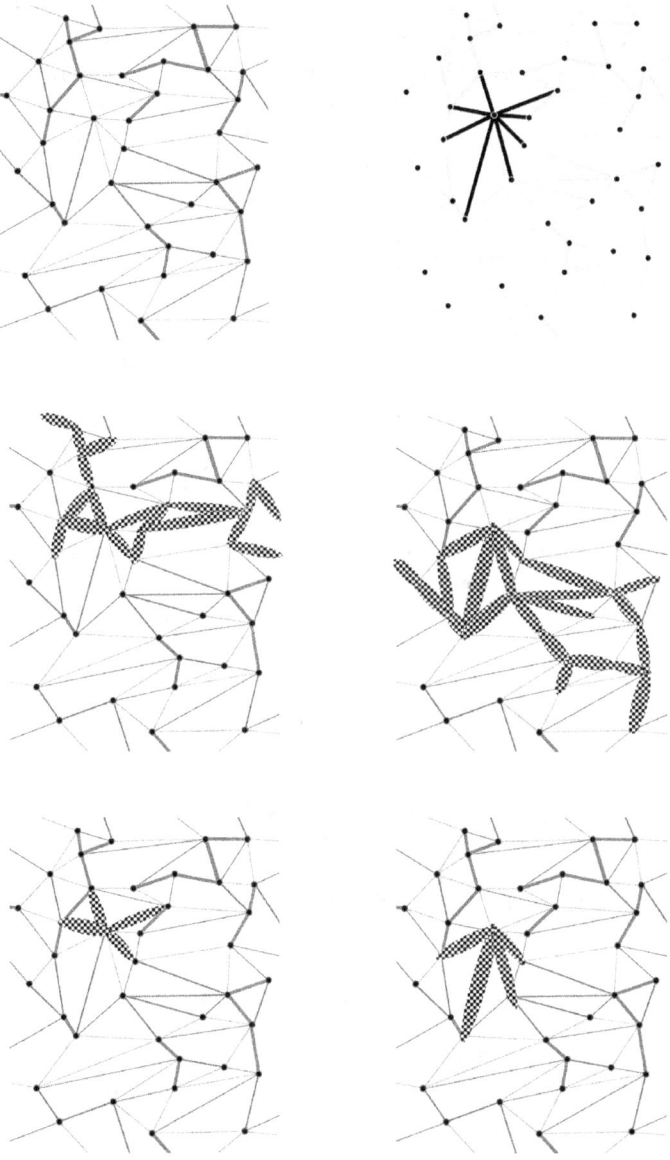

Figure 2.3 Not all the associations of one concept with other concepts will necessarily be activated together. If some region of conceptual structure is modelled as associations of different strengths (top left), then the content of a particular concept can be understood as a set of associations (top right). When the concept is part of a pattern of activation in a particular context some, but not all, of these associations may be recruited and brought to mind (mid-left); and in another context (mid-right) different associations may be recruited. The concept-in-action may therefore seem quite different (bottom left, bottom right) in different contexts.

undertake such a comprehensive probing, and to do so within a small enough window of time that their underlying set of associations did not change before we had finished, then we might feel we have fully characterised their concept; that we had identified its total 'content'. The proviso about time-scale is significant as there is evidence to suggest that activating knowledge often modifies it. This is the basis of an area of research known as microgenetic studies, where researchers look to explore learning by probing learners at a high enough frequency that they can detect changes as they occur (Brock and Taber, 2016).

Another issue that a representation such as Figure 2.2 raises is how, if each node has potential associations to others, and activation of one node may lead to activating some of these links, there is potential for activation to spread through the network (*cf.* the middle row of images in Figure 2.3). This also raises the question of what has been called a conceptual inductive effect (Taber, 2015). This term was suggested by analogy to the chemical inductive effect, where the characteristics of a bond does not only depend upon the atoms connected by that bond (*e.g.* a carbon–carbon bond) but on the other atoms or groups bonded to those atoms: the carbon–carbon bond in ethane, for example, does not have precisely the same properties as the carbon–carbon bond in ethanol or ethanoic acid. Something similar applies with associated concepts (see also Chapter 14).

2.3.2 The Natural Attitude – Talking, and Thinking, Like Mind Readers

There are some issues here that are critical for thinking about chemical concepts in the scholarly way necessary to (a) support research, and (b) inform improving teaching practice. It is easy for all of us to fall into adopting a kind of folk psychology when talking about mental lives (Taber, 2013). This seems likely to be related to it being part of normal human development to acquire what is described as 'theory of mind' (Wellman, 2011). Theory of mind allows us to recognise that others have mental experiences much as we do, and that they may believe different things from us; and to appreciate that others may be wondering about what we are thinking, and that we my deceive others so that they misjudge what we are thinking; and so forth.

These abilities are automatic (and may seem trivial) to most adults, but like many aspects of cognition have to develop during childhood. It has been suggested that when young children realise that people's beliefs do not necessarily reflect reality (that they can be wrong, or even lie), this provides an important starting point for developing scientific and critical thinking (Kuhn and Pearsall, 2000) which requires the ability to appreciate the conjectural nature of statements about the world, and the relationship between knowledge claims and evidence.

With developing theory of mind comes the appreciation that we may all act to some degree as mind readers – that is, that we can have some kind of access to the contents of others' minds by interpreting our observations of their behaviour. The development of a theory of mind is important in order for people to be able to operate in social contexts, but in everyday life it does seem to lead to people largely taking for granted the ability to infer others' thoughts. Within the context of educational scholarship, however, it is important to problematise this process as many learning difficulties in a subject such as chemistry may be seen as, in part, the failure of students to read teachers' minds, and indeed *vice versa* (Taber, 2013). If teaching can be considered the task of making the unfamiliar familiar (Taber, 2002), research and scholarly enquiry often involves making the familiar unfamiliar again: that is, challenging and problematising what might be taken for granted.

As an analogy, we might find a young child discovering that a magnet will pick up several small pieces of metal and drawing the conclusion that metals are magnetic (an example used again in Chapter 15). This is the kind of generalisation that is at the basis of much informal learning, but from the perspective of formal science there has been insufficient testing of the idea, and a lack of attention to the nature of the evidence and argument, so that the conclusion is based on no more than anecdote. We should not be too hard on the young child, but we would not expect an adult scientist to be so cavalier in reaching conclusions.

In the same way, in everyday life, we often draw inferences about people's moods, views, ideas, attitudes, and so forth, based on a few snatched data points in the form of odd comments, expressions, shrugs of shoulders, and so forth. That is fine as it goes, but is like the child noticing her magnet picks up a couple of paper clips and a staple. The inferences we draw may indeed often be correct, but they cannot be considered to be robustly based on sound evidence. So in educational research that aspires to be scientific we need to do more than collect and seek to interpret a few isolated observations, if we are to make claims about student thinking, and draw inferences about their conceptual understanding.

References

Baddeley A. D., (2003), Working memory: looking back and looking forward, *Nat. Rev. Neurosci.*, **4**(10), 829–839.

Brock R., (2015), Intuition and insight: two concepts that illuminate the tacit in science education, *Stud. Sci. Educ.*, **51**(2), 127–167.

Brock R. and Taber K. S., (2016), The application of the microgenetic method to studies of learning in science education: characteristics of published studies, methodological issues and recommendations for future research, *Stud. Sci. Educ.*, 1–29.

Brock R. and Taber K. S., (2017), Making sense of 'making sense' in physics education: a microgenetic collective case study, in Hahl K., Juuti K., Lampiselkä J., Lavonen J., and Uitto A. (ed.), *Cognitive and Affective Aspects in Science Education Research – Selected Papers from the ESERA 2015 Conference*, Dordrecht: Springer, pp. 167–178.

Kelly G., (1963), *A Theory of Personality: The Psychology of Personal Constructs*, New York: W. W. Norton & Company.

Kuhn D. and Pearsall S., (2000), Developmental Origins of Scientific Thinking, *J. Cognit. Dev.*, **1**(1), 113–129.

Novak J. D., (1990), Concept mapping: a useful tool for science education, *J. Res. Sci. Teach.*, **27**(10), 937–949.

Pockett S., (2004), Does Consciousness Cause Behaviour?, *J. Conscious. Stud.*, **11**(2), 23–40.

Taber K. S., (1997), Student understanding of ionic bonding: molecular versus electrostatic thinking?, *Sch. Sci. Rev.*, **78**(285), 85–95.

Taber K. S., (2002), *Chemical Misconceptions – Prevention, Diagnosis and Cure: Theoretical Background* (Vol. 1), London: Royal Society of Chemistry.

Taber K. S., (2013), *Modelling Learners and Learning in Science Education: Developing Representations of Concepts, Conceptual Structure and Conceptual Change to Inform Teaching and Research*, Dordrecht: Springer.

Taber K. S., (2015), The Role of Conceptual Integration in Understanding and Learning Chemistry, in García-Martínez J. and Serrano-Torregrosa E. (ed.), *Chemistry Education: Best Practices, Opportunities and Trends*, Wiley-VCH Verlag GmbH & Co. KgaA, pp. 375–394.

Vygotsky L. S., (1934/1994), The development of academic concepts in school aged children, in van der Veer R. and Valsiner J. (ed.), *The Vygotsky Reader*, Oxford: Blackwell, pp. 355–370.

Wellman H. M., (2011), Developing a theory of mind, in Goswami U. (ed.), *The Wiley-Blackwell Handbook of Childhood Cognitive Development*, 2nd edn, Chichester, West Sussex: Wiley-Blackwell, pp. 258–284.

Whitebread D. and Pino-Pasternak D., (2010), Metacognition, self-regulation & meta-knowing, in Littleton K., Wood C., and Kleine-Staarman J. (ed.), *International Handbook of Psychology in Education*, Bingley, UK: Emerald, pp. 673–711.

CHAPTER 3
What Kinds of Concepts Are Important in Chemistry?

The previous chapter characterised concepts as mental entities that are formed as abstractions from experience. This chapter considers the different kinds of abstractions we make in chemistry when forming chemical concepts, to offer some background for some of the ideas that will be developed later in the book. In particular, it will suggest four rather general classes of concept that are needed to do, or successfully study, chemistry. There is some permeability in these classifications, but collectively they seem to generally offer a useful typology. There are also some significant differences between the typical concepts falling within these general types.

The chapter will also introduce an important distinction related to how concepts are formed, which will be developed later in the book, relating to the extent to which concepts are automatically or deliberately formed. Some concepts seem to be presented to us in perception. These concepts are abstractions from our experiences, but the abstraction occurs preconsciously, at a level in the brain that operates automatically, without us making any deliberate effort. Other concepts formed in chemistry require conscious deliberation.

Some of those concepts formed automatically are a direct basis for useful formal chemical conceptions, although often the abstractions people 'naturally' make do not usually match precisely with formal chemical concepts. So, some chemical concepts are in effect the formalisation of concepts that people naturally acquire directly from their perceptions of the world, and other chemical concepts require more work to reconstruct them from direct perception. Other concepts are what might be called 'meta-concepts' that are formed by reflecting on concepts themselves and how they are related, or how those concepts can be operationalised for particular purposes.

This will be illustrated through various examples. As a preview, we might consider the differences between:

- The {sulfur} concept refers to something that can be directly handled and perceived.
- The {sulfur molecule} concept refers to a theoretical entity considered to exist based upon indirect observations.
- The {concentric shells (or solar system) model of the atom} concept concerns a thinking tool sometimes used in introductory chemistry teaching that is understood to be a simplified representation of (theoretical) objects – that is, the concept concerns a model of something rather than a physical entity.
- The {ideal gas law} concept concerns a theoretical relationship between other concepts ({pressure}, {volume}, {temperature}, {amount of substance}) that applies to a theoretical type of substance (an ideal gas).

3.1 Four Types of Chemical Concept?

I am presenting a model of types of abstractions made and used in chemistry that, although it oversimplifies the diversity of chemical concepts, can be a useful framework for thinking about the concepts we use and teach. In this system, chemical concepts may concern (i) objects; (ii) events; (iii) qualities; or (iv) may be what I will call meta-concepts – concepts that draw upon and relate other chemical concepts.

Concepts in the last category, meta-concepts, are clearly especially abstract as their formation depends upon engaging in a kind of deliberate mentipulation of other concepts. However, although categories (i)–(iii) include some concepts that are much closer to direct perceptual experience, these categories also include examples of entities, processes, and properties, that are actually theoretical and not directly observable.

3.2 Concepts Relating to Objects – Entities in Chemistry

The first category of concept concerns the kinds/types of things that are conceptualised to exist in the physical world. Imagine a student who is handling a lump of sulfur. This is an object. The student recognises the piece of sulfur as an object that has material form. They recognise it as having some degree of permanence. If they place it on the bench in front of them to make some notes, and then look back they will recognise the same lump of material, rather than imagine some new material has appeared to replace the lump they placed there. This is obvious – but requires having a concept. So, we might say they conceptualise the piece as a regularity in their environment – they discriminate the lump of material from its surroundings, they recognise it is the same piece of material as they turn it over (even if it has an irregular shape such that it has a different appearance when

What Kinds of Concepts Are Important in Chemistry?

looked at from different directions), and they recognise it as being the same object at different times. This level of conceptualisation occurs automatically, and develops early in life.

That is not a *chemical* concept, but clearly it is a necessary basis for forming chemical concepts. If the student saw and handled several pieces of sulfur of similar form they would almost certainly recognise certain perceptual similarities – the feel, the colour, the smell – and it is very likely they would (automatically) abstract that this is a 'type' or 'kind' of thing. They would form a concept that covered different instances of this type of object. Likely this would also cover quite different scales: a piece of sulfur the size of a small ball bearing and a piece the size of a fist might be readily recognised as being specimens of this same kind.

The student would not inherently know this is what other people call sulfur. They might conceptualise it as a kind of material, but would not be able to know from immediate experience that it was a chemical substance or indeed an element. Moreover, if this was plastic sulfur they may not include in the same concept an object that was a lump of crystalline sulfur, and which had different appearance; or a sample of flowers of sulfur, which looks quite different.

So, such a concept that is formed naturally is helpful to chemistry, but is only a starting point for developing the chemical concept. A student who is handed a piece of sulfur and responds, 'ah yes, element 16, oxygen group, S_8 molecules', is clearly working with a theoretical concept that is not simply the outcome of abstracting perceptual similarities from different specimens that they have come across.

We can consider a wide range of chemical concepts within this category of types of object or entity – some of these are built directly on abstractions from perceptual data, and others rely on abstraction that involves considerably more deliberation. So an indicative list of examples might include:

- {metals};
- {alkali metals};
- {potassium};
- {acids};
- {molecules};
- {electrons};
- {p orbitals}.

These are all entities used in chemistry. In a sense, they all refer to the types of things chemists consider exist in the world. They are part of a chemical ontology. However, some of these types can be directly linked with directly experienced entities (like our lump of sulfur) and some are theoretical entities (such as p orbitals). Clearly this list could be extended considerably.

Two of these examples are discussed in more detail later in the book, where it is considered how the chemical concepts {potassium} (see Chapter 5) and {acids} (see Chapter 6) became established, and have been developed, in chemistry itself. These types, and especially potassium, may be considered natural kinds. The term natural kind refers to a class of object that is found to exist in nature. That is, nature seems to present to us different examples that appear to be the *same type* of entity. There is assumed to be something about a natural kind that makes its members essentially, necessarily, members of that kind, rather than contingently so. The notion of natural kinds will be explored more in Chapter 7.

3.3 Concepts Relating to Events – Processes in Chemistry

A different type of concept a person might form concerns types of event they might perceive, and abstract as having some common nature. So, rather than concepts referring to objects, these concepts may be better considered as being of processes. Some examples of chemical concepts relating to processes would be:

- {melting};
- {combustion};
- {neutralisation};
- {ionisation};
- {homolytic bond fission};
- {rehybridisation}.

Again, this is a list that could be extended considerably.

Perhaps an obvious example of a concept of event relevant to chemistry might be {fires}. A person who observes a number of instances of fire is likely to – automatically – recognise similarity and so form a concept to cover these similar but distinct events. They will also be introduced to the everyday notion of burning as a process occurring in fires. They may perceive how, for example, a piece of wood, hisses and crackles and becomes blackened and smaller during a fire, and how paper changes colour (sometimes as if a wave of change is moving across the sheet) and shape and breaks up into small pieces of material that may be carried up on the flames. Fires produce changes in materials, and usually damage or destroy them, and often leave charred or no remains.

The everyday concept of {burning} is usually not simply directly abstracted from seeing fires, but is normally – partially at least – acquired from the culture: from how others describe the process. This concept offers a basis for constructing the chemical concept of {combustion} – which of course is directly connected to the {chemical reaction} concept and so to the concept of {substance} in chemistry.

Neutralisation is another class of chemical reaction, but there does not seem to be an everyday concept that is naturally developed (*cf.* {burning} for

What Kinds of Concepts Are Important in Chemistry?

{combustion}) that can be used as the starting point for developing the chemical concept {neutralisation}. There is not necessarily any direct perceptual similarity between observed neutralisations that allow people to automatically abstract some kind of commonality as the basic for a concept to discriminate neutralisation from neutralisation (*i.e.*, not-neutralisation). That is, not all neutralisations 'look' the same, and some neutralisations may inherently appear as much, or more, like some non-neutralisation events, than some other neutralisation events.

The {neutralisation} concept is useful in chemistry, but has to be largely developed from theoretical considerations – using other chemical concepts. In teaching we can bias learners' perceptions by presenting examples of neutralisation using an indicator or pH meter to provide perceptual similarity and so support discriminations. Strictly, however, we are here offering students something more complex – a system of neutralisation reaction + indicator, or a system of neutralisation reaction + pH meter – and asking them to associate the similarity between particular aspects of these wider systems (discriminating when indicators change colour and when they do not; or when a pH meter reading shifts significantly, from when it does not) with the chemical reaction as it would occur without those embellishments. I am not here suggesting this is inappropriate or cheating (even if it seems reminiscent of Pavlovian conditioning), just highlighting that it requires introducing additional complexity and so asks students to do something more complicated than just notice a perceptual similarity in the event itself – as when they might see flames as a common feature associated with different fires they obverse.

Some of the other examples here are even more abstract, as they are not processes that are readily observed at all. Where two neutralisations can be observed and may not be obviously the same kind of event, the breaking of a bond in a chemical reaction cannot be observed at all. This type of event is a theoretical notion that students are asked to learn about and add to their ontology of the kinds of things that happen in the world. It clearly cannot be taught through directly showing examples, as combustion might be.

Rehybridisation is an even more abstract example. Most chemists imagine bonds really do break during chemical reactions, even if they might grudgingly admit that bonding is a theoretical notion. Some chemists may think of rehybridisation as something that really happens, but arguably it is purely a change in conceptualisation. If we consider orbitals to exist as real entities (which is itself open to debate), and if we accept that the assignments of orbitals found in hydrogen can be strictly applied to non-hydrogenic species, that is atoms and ions with more than one electron – which is contended (Scerri, 1989) – then we might consider that rehybridisation is not a physical process that occurs to orbitals at all, but a choice between different permitted ways to visualise something abstract in familiar physical space. That is, rehybridisation may be considered more a mathematical operation than a change in the material world.

So perhaps this is less like a Hindu deity choosing to appear on earth as one avatar rather than another, and more like the apocryphal blind men describing the elephant, but where we make a conscious choice to close our eyes and then feel and describe what the trunk is like rather than the ear. Rehydridisation may perhaps be considered a process, but it is some way from the kind of process (*e.g.*, burning, melting) that can be observed and pointed out to another.

There are other examples of chemical concepts that seem to be processes, but where we might question such a characterisation. We might think of keto–enol tautomerism. Some substances can exist as a ketone or alcohol form, and naturally exist as a mixture of the two forms (so raising the question of whether this is indeed a 'substance' in chemical terms). There is a process by which the two forms interconvert. We might conceptualise the mixture as containing molecules of ketone, molecules of the isomers that are alcohols, and molecules in intermediate stages during the transition. However, if one molecule was observed over an extended period of time (and of course molecules are not readily observed, and what might be considered an extended period of time for a molecule might seem a fleeting moment for human observers) there is not a ketone molecule or an alcohol molecule, but a molecule that shifts at various points on a structural continuum between extremes that are ketone and alcohol molecules. So, if we imagine a freeze fame (*i.e.*, a frozen moment in time), we might have some molecules of ketone and molecules of alcohol and some molecules we could not readily put in either camp; but, if instead we imagine a sample of the substance sitting on the bench in a beaker over some time then none of the molecules 'are' molecules of ketone or alcohol. These molecules are – rather – dynamic systems spreading over and between both possibilities.

So, we could think of this shifting between forms as a process that involves the disappearance of one entity and the appearance of another, but we could also conceptualise this as the dynamic structure of a single entity – the keto–enol molecule. Arguably, it is not a matter of asking what the correct conceptualisation is, that is, asking what reality is actually like. There are choices here in how we decide it is most useful to conceptualise nature – and students have to form their {keto–enol tautomerism} concept without ever seeing an actual molecule.

Moving one step beyond this example, we might consider delocalisation of the kind that occurs in the π ring in the benzene structure. Again, the entity concerned is a molecule, a theoretical entity. Here, the structure may sometimes be conceptualised as a resonance between different canonical forms (the Kekulé structures, perhaps also the Dewar structures). However, we would probably not want students to imagine the resonance process as reflecting some kind of ongoing reaction mechanism (as described above for the keto–enol tautomerism) but rather as a purely theoretical scheme to justify aromaticity and the non-integral C–C bond order in the structure.

This is not a rare example. Similar arguments apply to conceptualising the structures of electron-deficient compounds such as the boranes or an ion as

What Kinds of Concepts Are Important in Chemistry?

ubiquitous as the sulfate ion (*i.e.*, the tetraoxosulfate(VI) ion). These are indeed abstractions of abstractions, of things that cannot be directly observed: even if we consider they 'really' exist. Again, my purpose here is not to tell readers something they are not aware of, but rather to remind readers who are very familiar with the concepts of how far chemical thinking sometimes is from what students can directly observe at the bench.

3.4 Concepts Relating to Qualities – Properties in Chemistry

A third category of concepts used heavily in chemistry concerns qualities that are perceived in chemical phenomena. These are usually physical and chemical properties. A far-from-exhaustive, indicative, list of examples would be:

- {hardness};
- {lustre};
- {ductility};
- {melting temperature};
- {electrical conductivity};
- {electronegativity};
- {isomerism};
- {electron donor};
- {tetrahedral geometry}.

Again, some of these qualities are directly open to sensory perception. The student with the lump of sulfur can dig her fingernails into it (readily, if it is the plastic form), and we can appreciate that clean metal surfaces are often shiny. Other properties depend upon setting up appropriate observational conditions. We can directly perceive the ductility of metals – but not by simply picking up an iron rod and trying to stretch it with our bare hands.

We form concepts related to properties, such as temperature, pressure, volume that can be given numerical values. A temperature of 300 K could refer to a wide range of different substances – but only applies at specific times (some copper may be at a temperature of 300 K now but 320 K soon after). The concept of {a temperature of 300 K} (which can discriminate instances of this temperature form non-instances) is clearly related to, but more specific than the concept {temperature}. Knowing that something is at a temperature of 300 K also relies on some form of instrumentation: what allows us to abstract a commonality from a lump of iron at 300 K, a beaker of solution at 300 K, and a flask of air at 300 K, are the similar thermometer readings in the different cases. Of course, temperature is a theoretical concept as well as an empirical one – it is linked to theoretical models such that a chemist's concept of {a sample at 300 K}, that applies across the iron, solution, and air, incorporates associations relating to particle motion.

Other qualities here are much less directly observable. So, electronegativity is a theoretical concept. Sometimes it is quantified using measurements, providing students with a sense of something definitive and absolute – but not measurements an introductory student can make in a school laboratory. Indeed, the property of electronegativity that is sometimes presented with numerical values clearly derives from a mathematical model of electronegativity.

So, although chemistry uses concepts based on observable properties ({hardness}), or readily measurable properties ({melting temperature}), it also works with concepts that are some way from being based on direct perceptions. Observing a sample of methane does not suggest a tetrahedral geometry; that a species acts as an electron donor or a good leaving group is not apparent from observing a reagent in its bottle, nor indeed can it be directly inferred from watching a reaction where the imagined process (*e.g.*, donating electrons, leaving a molecule) is considered to be occurring. So assigning these types of properties is dependent upon thinking with other, already established, theoretical concepts.

3.5 Meta-concepts – Concepts about Chemical Concepts

It seems then that even in the categories of concept concerning entities, processes, and properties there are graduations in how readily the chemical concepts can be developed from direct observation. Some chemical concepts do relate to what we can directly perceive: we can make discriminations using a {hardness} concept when feeling samples of materials, or a {lustre} concept when looking at substances. Others require some setting up of condition for observation. We can make discriminations using a {sonorous} concept but need to hit our test specimen with a suitable object. As clean metal surfaces are assumed to have lustre, then before applying the {lustre} concept to help make discriminations of metals from non-metals, we should clean the surface of the sample of material. We can apply the process concept {melting} by observation, but to apply the property concept of {melting temperature} we need to set up the conditions for melting, and also the means to make a measurement.

Other concepts in these categories are further from direct observations, which is not to say chemistry does not make relevant observations, but that the connection between the concept and the observation may be much more indirect and rely on an extensive theoretical framework that itself depends upon on a range of other formal concepts. Indeed, even when using a melting point apparatus, taking a reading from the length of a thin column of mercury in a glass tube to determine meting temperature relies on a chain of logic drawing on theoretical considerations. In this case it is easy to forget this, as even novice students are likely to take it for granted (because of familiarity, as much as theoretical knowledge) that a

What Kinds of Concepts Are Important in Chemistry? 47

liquid-in-glass thermometer measures temperature and that a lens magnifies, without distorting, the process occurring in the tiny melting point tube.

Some chemical concepts that refer to entities (*e.g.*, the sulfate ion), to processes (*e.g.*, the second ionisation of sodium), or properties (*e.g.*, the tetrahedral geometry of the methane molecule; the electronegativity of fluorine) are by their nature theoretical, referring to non-observables that are deduced or inferred (or simply constructed) through arguments drawing indirectly on observations that are mediated through other concepts. In a sense, these could be seen as meta-conceptual, but I am here reserving that sense for concepts that refer to ideas that are more abstract than entities, processes, or properties.

I would suggest that a non-exhaustive indicative list of examples of what I am here labelling meta-concepts would be:

- {ideal gas law};
- {Lewis acid model};
- {VSEPRT (valence shell electron pair repulsion theory)};
- {ligand field theory};
- {Le Chetalier's principle};
- {Aufbau principle}.

There are basically two types of meta-concept in this list (although, again, this is not an absolute distinction). The list includes concepts of models, and concepts of laws, principles and theories. As an example of a model, we can consider something like the Bohr model of the atom. Learners of chemistry are expected to learn the concept {atom} that refers to a (not directly observable, so theoretical) entity. They are also introduced to different models of the atom, and expected to understand these not as realistic representations of atoms, but as models – as simplifications with particular ranges of convenience or application. So the {Bohr atom model} concept is linked to the {atom} concept but also relies on the epistemological concept {model}. It is a meta-concept in the sense that it is a conceptualisation formed not of something considered to exist in material form – but only as an idea. So atoms may exist in the physical universe, but the Bohr atomic model only exists in minds. Although it is purely imaginary, that has not prevented it being a concept that has done real work in the discipline of chemistry.

My statement that the Bohr atom model does not exist in material form but only as an idea might seem to ignore all the many representations of the model in textbooks and so forth. Physical models can certainly exist as objects, but many models we use in chemistry are purely theoretical or mathematical models, and *representations of them* are not identical to the model itself (perhaps they are models of the models, emphasising certain aspects, making certain compromises). So, for example, many representations of the Bohr atom model are two-dimensional. This point is explained in more detail in Chapter 8 in relation to the example of 'the' model of DNA structure.

Laws and principles are considered meta-concepts as again they tend to be about operations made on concepts. So, forming a concept of Boyle's law means appreciating a relationship between the concepts of {pressure} and {volume} (and {amount of substance} or {mass}, and {temperature}) applied to substances discriminated by the concept {gas}. In a particular case, Boyle's law can be applied to actual measurements made at the bench, but the concept {Boyle's law} is an abstraction of all the possible instances of such measurements using these concepts. As another example, the {Aufbau principle} concept involves using a set of abstract rules when operating on a model. No one ever applies the Aufbau principle to an atom (that surely involves a category error), only to a model of an atom. So, again, the concept is not applied directly to actual entities that are observed and manipulated, but rather to a conceptual abstraction that itself ultimately derives from making sense of observations and measurements.

Finally, theories are frameworks of relationships between different concepts. Theories supply predictions and explanations that can certainly be applied to real situations, but they are framed in terms of concepts, abstractions that will apply across a wide range of cases. So, ligand field theory is relevant to real chemical systems, but concerns theoretical entities like orbitals and energy levels, and properties such as electron spin and bond formation.

A reader may wonder why I class these theoretical 'entities' – laws, principles, models, theories – as meta-concepts, and so a kind of concept. A law is a law; a theory is a theory; and surely, it could be suggested, these are something *other than* concepts. Indeed, this is so. However, chemistry students are not asked to simply apply models, laws, principles and theories, but to recognise them as such. This requires forming a concept. A student therefore needs to form a concept of Le Chetalier's principle, so that she can, not only apply the idea, but also recognise when it is being discussed, and can write or talk about it. So, students are expected to work at both levels – both working with the concepts such a meta-concept draws upon (so, perhaps, correctly applying Hund's rules in an appropriate context) and working with the meta-concept itself (so being able to report what Hund's rules are, and where they can be applied). This theme is developed in Chapter 8.

This is especially important if we believe that teaching a science involves teaching about the nature of the science (Duschl, 2000; Allchin, 2013) as well as some of its content. So, we want students to not only learn about key models, but appreciate their nature (including limitations) as models. We need them to understand the idealised nature of some principles and laws that may only strictly apply in ideal cases. We want them to understand that theoretical knowledge is a human construction that is open to review and improvement in the light of further scientific work.

References

Allchin D., (2013), *Teaching the Nature of Science: Perspectives and Resources*, Saint Paul, Minnesota: SHiPS Educational Press.

Duschl R. A., (2000), Making the nature of science explicit, in Millar R., Leach J. and Osborne J. (ed.) *Improving Science Education: The Contribution of Research*, Buckingham: Open University Press, pp. 187-206.

Scerri E. R., (1989), Transition metal configurations and limitations of the orbital approximation, *J. Chem. Educ.*, **66**(6), 481-483.

CHAPTER 4

Concepts as Knowledge

An examination of the nature of peoples' concepts needs to engage with the idea of knowledge – what is meant by knowledge, and how concepts feature in someone's knowledge.

4.1 Concepts as Knowledge?

It can be suggested that our concepts make up part of our knowledge of the world. This is certainly a position taken in this book, but begs the question of what 'knowledge' is. The *idea of knowledge* may initially seem fairly straightforward, even if *establishing knowledge* is not always so simple. People generally consider they know what is meant by 'knowledge' and the notion is used widely both in everyday talk, and in the more specialist professional discourse of teachers and others working in education.

That is, the terms 'knowledge', and 'knowing', 'knew', 'knows', *etcetera*, are part of the lexicon of everyday conversation – what was referred to in Chapter 1 as the mental register (Taber, 2013). However, it was also suggested there that if we either wish to undertake research, or ensure our practice is informed by research, then *we need to use such words as technical terms that have a specified meaning*, and we should not (in professional discourse) be content with the more fluid meanings of much everyday conversation. That is, we have to problematise notions such as knowledge. If, as suggested earlier in Chapter 2, concepts are the basis of the discriminations we make, then we need to interrogate the concept {knowledge} and ask what our criteria are for deciding something is (or is not) knowledge.

This chapter will explore some themes that will underpin discussion later in the book. It will examine what we mean by 'knowledge', and how concepts relate to knowledge. It will also consider how knowledge is related to the

Concepts as Knowledge 51

person who is said to know, and what this means for ideas like canonical knowledge or public knowledge. The chapter also considers whether we can have knowledge that we do not know we have, and so concepts we cannot explicitly report. The chapter then returns to the theme of Chapter 2, on the nature of concepts, and explores some of the complications in characterising the nature of concepts used in chemistry.

4.1.1 Knowledge, Belief, and Truth

A traditional notion of knowledge, that was discussed at least as far back as the Ancient Greeks, is that knowledge consists of true, justified (or reasoned), beliefs (Bhaskar, 1981). A belief is something that a person thinks to be the case – perhaps, for example, a hypothetical student Lewis believes that vinegar is an acid. A belief is justified if it is based on sufficient, reasonable grounds. Of course, what counts as sufficient and reasonable is open to question. Our hypothetical student, Lewis, argues that he knows vinegar is acidic because (a) his chemistry teacher, Miss Lowry, told him so; (b) he read it in a textbook; and (c) he tested some vinegar with universal indicator solution and the paper changed colour, going red. Most readers might find this a fair basis for Lewis's belief, even if not totally beyond challenge.

That is, we could imagine ways these sources of evidence could be unsound. We can imagine that it is quite possible that Lewis may have misheard or misremembered what Miss Lowry said. Similarly, it is possible Lewis misremembered or misread his textbook. It is also possible that Lewis is incorrectly recalling testing the solution with indicator – perhaps remembering testing a different solution. It is possible something labelled as vinegar was actually something else, or Lewis (not realising there are other types of indicator available) used a different indicator assuming it was Universal indicator. So, perhaps Miss Lowry or the textbook said orange juice, not vinegar, was an acid, and Lewis is mis-recalling. Perhaps Lewis remembers testing a different acidic solution. Perhaps he used phenol red, not universal indicator. These are all possibilities, although as we consider that vinegar is acidic (so have no particular reason to doubt the grounds of Lewis's belief), we are likely to take his reasons at face value.

However, consider instead that Lewis actually believes that ammonia solution is an acid, and that Lewis argues that he knows ammonia solution is acidic because (a) his chemistry teacher, Miss Lowry, told him so; (b) he read it in a textbook; and (c) he tested some ammonia solution with universal indicator solution, and it changed colour, going red. This is an entirely parallel example, open to many of the same potential errors as before. This time we consider the belief false, and so are more likely to assume there are (and so look for) problems with the grounds on which he came to that belief. As we are able to take a 'God's eye view' (*i.e.*, we think we know what is true) or at least an expert view (our knowledge can be considered authoritative in this domain) we are more likely to suspect the grounds for believing ammonia solution (rather than vinegar) is acidic to be erroneous.

Perhaps we can find other potential reasons for a false belief. Perhaps Miss Lowry did say that the ammonia solution was an acid: it is not impossible she has flawed beliefs herself, even if it seems unlikely that a chemistry teacher would be mistaken on something so basic (if readers will excuse the pun). Miss Lowry might have been deliberately misleading Lewis (again we would suspect that is not a likely scenario, but it is not out of the bounds of possibility – perhaps provoking him to challenge or test her claim). Or Miss Lowry may have made a 'slip of the tongue', or may have misheard a question Lewis was asking, or may have actually been addressing another student when she said "that's an acid" although Lewis thought she was talking to him. The textbook might have contained a typographic error (or have been written by someone with a flawed belief, or who made a slip of the typing fingers). There are all kinds of reasons why a belief might be poorly founded, and this may apply to a correct belief as well as a false one. Lewis's friend Bronsted may have deduced that ammonia solution would be acidic because acids contain hydrogen ions that become dissociated in water, and ammonia contains hydrogen. That is a false argument, but based on some reasonable (reasoned), if incomplete, chemical thinking.

To be counted as knowledge, using this traditional notion, it is not enough that a belief is justified, but it also needs to be true. Using this criterion, we may accept that Lewis has knowledge that vinegar is acidic (if convinced about the soundness of his reasoning, and that he was strongly committed to the belief), but reject the suggestion that Lewis has knowledge that ammonia solution is acidic – because we believe that we know that is not the case. So, to apply the 'true, reasoned, belief' notion of knowledge, we have to know what is true. In this case, that might seem unproblematic – we know vinegar is acidic and ammonia is not – but as a general principle we cannot always assume we know which beliefs are true. Indeed, an important scientific principle is to always reserve the possibility of admitting new evidence, or new arguments, that may give us cause to reconsider our ideas.

4.1.2 We Should Not Believe in Scientific Knowledge

Arguably, then, the search for scientific knowledge is not a matter of beliefs. Scientists seek to develop models and theories that have value in making predictions and explanations, and that they can therefore treat as provisional accounts of aspects of the natural world. These accounts are often partial, and may have limits to their range of ready application, yet they may still be our current best ways of making sense of important phenomena. But we should not 'believe' scientific ideas, as then we lose the critical capacity to keep them in 'probation' and to be prepared to question them when appropriate.

It follows from this that scientific education is not (or should not be) about persuading students of the truth of scientific ideas, but rather of their utility and value as (provisional) ways of understanding that can be

Concepts as Knowledge 53

productively adopted until such time as they may be improved upon (Taber, 2017). Many students struggle to appreciate how much of the chemistry they are taught is theoretical and based on models (with their approximations, or questionable assumptions, or limited ranges of application) – and so adopt such ideas as if 'truths' – which is unhelpful (Taber, 2010). Students who 'know' (*i.e.*, 'believe') that an acid is (*i.e.*, always, necessarily) something that releases hydrogen ions in solution, or who 'know' (*i.e.*, 'believe') that 'the' atom (sic, every atom) has electrons arranged in concentric shells, often then struggle when they are asked to learn about a new more sophisticated definition, or model, or theory, inconsistent with what they have taken as an absolute truth about some aspect of the natural world.

If we adopt the traditional notion of knowledge (as true, justified, beliefs) we should strictly avoid making claims such as:

The student *knows* the universe was created in the big bang.
He *knew* that heavy elements were created in nuclear reactions in stars.
She *knew* that Kekulé had a dream that gave him insight into the structure of benzene.

Or even

She *knew* that Kekulé claimed he had a dream that gave him insight into the structure of benzene.

The first two examples concern ideas that currently have scientific credibility (and may well be 'true' – accurate accounts of some aspect of nature) but, as with all science, have to be open to being revised or rejected in the light of new information. As Popper (1945/2011) noted, "if we can never *know* – that is, know for certain, there are no grounds here for smugness". No one but Kekulé could have known (for sure) if he had a dream (and even he could have been later suffering from a delusion about the matter). A student may read accounts of the dream, but she is not able to meet Kekulé and hear the claim first hand. A published account said to be written by Kekulé is available in the literature (Rothenberg, 1995), and whilst I have no particular reason to doubt that the account was produced by Kekulé, no one today can be absolutely certain (beyond possible doubt) these are his own, unadulterated words.

The level of scepticism adopted here reflects scientific criticality, and in the case of Kekule's dream in particular (Ball, 2002), the Royal Society motto that we should take no one's word for things. So, if we adopt a notion of knowledge as true, reasoned, belief, we either need to set ourselves up as knowing for sure what is true (a stance that seems inconsistent with scientific values) or knowledge becomes an empty set (the set of reasoned beliefs we can authoritatively say are definitely true). Scientific knowledge is better considered to be the set of ideas about the natural world considered to be based on sufficiently strong evidence and arguments to be *provisionally* adopted until such time as we may have reason to revisit those provisional evaluations.

4.1.3 A More Relevant Notion of Knowledge

Equating knowledge with true beliefs would then effectively exclude references to knowledge and knowing from educational discourse, so it would seem to make more sense to continue to use the term knowledge but ac-*knowledge* that the knowledge people have can be flawed (it does not consist of true, justified, beliefs). From an educational point of view then, the traditional notion of knowledge is not very useful. People (including scientists, teachers and students) have a wide range of ideas about aspects of the material and wider natural world, to which they are committed to different degrees (see Chapter 14). These ideas may be based on very modest – or alternatively sometimes substantial – warrants: with evidence that may, or may not, be reliable, and reasoning that may, or may not, be sound. The resulting ideas may, or may not, be true reflections of the natural world (something that we mere mortals cannot definitively judge), and – more to the point – will match current accepted scientific ideas to different degrees.

It is very important in education to know what ideas students have about chemical phenomena, principles and theories, and how they understand and explain them. The educator wants to know what ideas a student has under active consideration for potentially making sense of chemistry: the range of (presumed) facts, their versions of chemical concepts, and so forth. As this is not what was traditionally meant by knowledge, we could find an alternative term for this – but we would have little use for the term 'knowledge' if we adopted the classical meaning.

The alternative approach taken here then is to refer to *the ideas a student has under active consideration as ways of making sense of chemistry* as knowledge – and to understand a person's chemical knowledge to not be limited to true, justified, beliefs, but rather to encompass the range of relevant thinking the person entertains. There are here then two distinct {knowledge} concepts: {knowledge$_{\text{true justified belief}}$} and {knowledge$_{\text{currently entertained notions}}$} but as we are not presented with opportunities to use the former there is normally no need to highlight the distinction in discussing students' knowledge: so I am adopting the latter concept.

4.1.4 The Knower and the Known

If we understand knowledge as {knowledge$_{\text{currently entertained notions}}$} then it is clearly linked to a particular individual who is entertaining the particular notions: on this understanding, knowledge is personal knowledge – in the sense of someone-in-particular's knowledge. Some students' knowledge will include the idea that sulfur exhibits multiple valencies and some students' knowledge (in the sense used here) will include the idea that solutions of table salt contain solvated sodium chloride molecules. However, we might be more comfortable in that latter case suggesting a person 'thinks', rather than 'knows', that solutions of table salt contain solvated sodium chloride molecules.

There is also a common discourse concerning what can be referred to as public knowledge. Public knowledge concerns the knowledge that is put into the public domain, which in a science (Ziman, 1968) such as chemistry usually means knowledge reported in the research journals and similar literature. Here, knowledge is treated as though it can be disconnected from any particular knower, and becomes a public resource that has some kind of independent existence. It will be suggested later (see also Chapter 11) that whilst this notion of public knowledge is useful, it is also problematic. This is important for the theme of this volume, as we often consider that we evaluate an individual's knowledge by comparison with the canonical disciplinary knowledge of chemistry, which is taken to be a form of public knowledge.

4.2 Conceptual, Procedural, and Episodic Knowledge

Personal knowledge is sometimes divided into different kinds, such as procedural knowledge, episodic knowledge, and conceptual knowledge. An example of procedural knowledge would be knowing how to ride a bicycle without falling off, or being able to set up the kit to carry out a distillation. Episodic knowledge might involve recalling one's marriage ceremony, or graduation in chemistry, or some such event in one's life. Some people seem to represent in their brains something akin to a vault of mental 'videos' of their entire life (in extreme cases they can be successfully tested on what happened on some randomly selected date), but many of us only have clear, discrete, recollections of special or otherwise noticeable episodes.

I cannot now remember many details of the practical work I undertook as an undergraduate, let alone at school. This failing is at least consistent with the findings of research that asked school students to recall laboratory practicals they had carried out in science, and what the practical demonstrated (Abrahams, 2011). Generally, learners, even those who claim to enjoy science laboratory work, can only report a small number of the many school practicals that they have carried out and then may not be able to explain the point of the activity. Those sessions I can remember tend to be memorable for reasons other than the scientific outcomes.

I recall a first-year undergraduate physical chemistry laboratory circus activity that involved using a radioactive source. There was no obvious source provided at the station set up for the practical. To find the source I dismantled the interlocking bricks of the lead castle to find a large metal box (like a cubic cake tin); opened the metal box to find a divided block of wax; took out the top half of the wax block to find a cylindrical metal tin; and unscrewed the lid from the tin to find a radioactive source, that when carefully removed with tongs was physically too large to fit into the apparatus provided. When the demonstrator was asked for assistance, he helpfully commented that he had not seen anyone else dismantle the lead castle – which contained a stronger source used to irradiate the source that I was actually meant to use in the practical. I have a clear memory of this episode, though sadly no recollection of the actual practical activity I carried out once the correct source was located.

Conceptual knowledge would include such matters as knowing transition-metal salts are often vividly coloured – in contrast to the episodic memories of four different afternoons when I (thought that I) followed the 'recipe' supposed to produce an isomer that gave purple crystals only to end up on the first three afternoons with beautiful, but definitely green, crystals. (Or, indeed it may have been the other way around – even that detail may now be unreliable.) Conceptual knowledge comprises the set of concepts one has acquired, including the various relationships between them (see the discussion of concepts as nodes in a conceptual network in Chapter 2).

4.2.1 Implicit Knowledge

As well as dividing knowledge into such categories as procedural, episodic and conceptual; knowledge can also be classified as explicit or implicit (or tacit). Episodic knowledge would seem by definition to be explicit: we can recall the episode. (We may have also acquired implicit knowledge about certain types of situation by abstracting patterns in various episodes we have experienced, but if we do not recall the specific episodes, this is not considered episodic knowledge.) Much procedural knowledge may be implicit.

If a bicycle-riding chemist was asked to explain (a) how they would separate a mixture of liquids by fractional distillation, and (b) how they rode a bicycle without falling off, they would likely be able to offer different levels of detail of these two procedures. The chemist could likely provide a fairly detailed account of various steps taken in the distillation – when to add anti-bumping granules; how to adjust the level of heating, *etcetera*. However, this can be understood as actually drawing upon conceptual schema that have been learned. There is also the procedural knowledge involved in how the Bunsen burner is handled, and how it is recognised that it is time to switch flasks to collect a new fraction – things that can be described to some extent but may be, partially at least, implicit.

The importance of implicit knowledge in science was first highlighted by a chemist (Michael Polanyi) – although the phenomenon had previously been noted by the potter and entrepreneur (and grandfather of Charles Darwin) Josiah Wedgwood, in the context of seeking to determine reproducible kiln conditions (Chang, 2004). Polanyi (1962) was a research chemist who came to be known as a philosopher. He recognised, from working in a research laboratory, that science depended upon developing a good deal of what he termed tacit knowledge. Bench scientists develop the equivalent of a gardener's so-called 'green fingers': they acquire a level of expertise in handling materials and apparatus that is more like the knowledge needed to ride a bicycle than, say, applying valence-shell electron-pair repulsion theory to

Concepts as Knowledge 57

predict the geometry of molecules. The experienced research scientist can get procedures and equipment that they use regularly to work when a novice who is carefully following the same instructions, and seems to an observer to be doing the 'same things', fails (and so perhaps produces the isomer with green crystals when they were supposed to synthesise the isomer with purple crystals).

This has been found to be a major issue with some new techniques and apparatus – such that despite the supposed objectivity of science, requiring that details in papers should suffice to reproduce results, novelty often only gets transferred between laboratories when expertise is shared by face-to-face contact: that is, actually spending time in the laboratory where the technique has been developed (Collins, 2010). The importance of tacit knowledge in science complicates a simple account of how science develops by experimental tests and replication (Collins, 1992).

So, for example, in Millikan's famous oil-drop experiment, that first established a reasonable (by modern standards) estimate for the electronic charge, an examination of Millikan's laboratory notebooks shows that he did not use all his results in coming to the published figure (Gauld, 1989). Millikan selected some runs from his experiments, and decided to reject others when he considered the results were unreliable. This seemed to be often a matter of tacit knowledge – results being rejected without any obvious objective reason, but based on a 'feel' for the experiment and based on close familiarly with the apparatus. As Millikan got a result reasonably close to the current values, hindsight seems to suggest his judgement was sound. Had he got a result completely different from the currently accepted value of 'e' then historical judgement might have been that his selective reporting of results was poor scientific practice. Perhaps the lack of reproducibility of cold fusion experiments (Close, 1990) results from tacit knowledge that Fleischmann and Pons developed, that other scientists lacked – although the general view today is that their work was flawed.

4.2.2 Mighty Oaks from Ignorant Acorns Grow

Although we may use procedural knowledge to ride a bicycle, without being able to interrogate that knowledge and specify it to others, we are able to reflect upon the activity, and conceptualise our experience. However, if applying procedural knowledge does not require conscious awareness, then some would suggest that this kind of 'knowledge' does not require a (conscious) knower. It could be said that a person in a coma still has the procedural knowledge to keep pumping blood around their body. It could also be argued that an acorn has the knowledge needed to grow into an oak tree, something that, given supportive conditions, it can achieve without any reflection or deliberation. Indeed, it has been argued that plants do indeed behave in ways that should be considered "active, purposeful and intentional" based on "assessment of the future" supported by "the acquisition and processing of information" (Trewavas, 2009, p. 606). The plant is – from such a perspective – applying (procedural) knowledge, but this does not require conscious experience and reflection.

4.2.3 Can Conceptual Knowledge Be Tacit?

Yet conceptual knowledge would seem to be associated with the ability to have conscious experience. We use concepts as thinking tools (see Chapter 2), to reflect on experience; to deliberate on possible courses of action; to theorise about the world. Yet even when we are actively problem solving, we rely on automatic thinking processes outside of conscious awareness.

We all sometimes exercise judgements where we can make evaluations (and sometimes be confident in them) without being able to logically explain the basis of our decisions. We have intuitions (Brock, 2015), and these reflect the ability of our brains to recognise patterns in the environment that are useful for predicting future outcomes. These 'expectations' become built into the functioning of our perception/cognition and used to evaluate situations and make decisions. We use this implicit knowledge to make discriminations, to select courses of action, to evaluate options, and so forth, in situations where we are engaged in conscious and deliberate mental activity, even if we only have awareness of some of the cognition involved. (In addition, we are only ever aware of some aspects of our thinking, whatever the context.)

There is clear evidence then that some of the knowledge we rely upon in our conscious thinking is of the form of implicit elements that we access as intuitions rather than reasoned arguments. There is a genuine question of whether this implicit knowledge should be labelled conceptual – whether we can have 'tacit concepts' – or if we should describe this type of knowledge differently. In traditional tests of concept acquisition, 'subjects' (participants) are considered to have acquired an artificial concept (perhaps the target concept is shapes with less than five sides that are blue or green) when they can make discriminations according to the concept (for example dividing figures of various shapes, numbers of sides, colour, into examples or non-examples of the concept). Initially the test subject can only guess whether a shape is an example of a concept that has not been explained to them ("Is this shape an example of a 'pauloid'?"), but given feedback on their performance (whether or not the discriminations were correct, but not why) they gradually learn to make the correct discriminations. They may eventually spot the rules that define the concept, but may well demonstrate a high level of correct discrimination before reaching this point – so they have attained the concept (at a preconscious level), even before they can explain it.

We can imagine many parallel examples in chemistry. We can imagine that a learner might be able to do well if shown samples of different substances and are asked 'is this a metal?' or 'is this material crystalline', even if they could not give an explicit account of the criteria used. If a student was able to given canonical responses when asked to suggest the shapes of a wide range of simple molecules, including some unfamiliar, but could not explain

Concepts as Knowledge 59

how they came to make the discriminations we would likely consider some conceptual learning had occurred (even if we would not be totally satisfied with a high level of performance that was not supported by a clear explanation of the principles involved). In everyday life, it is much easier to be able to correctly discriminate between buses and trucks, or between coots and moorhens, than to provide a set of clear criteria that would allow someone else or a machine to accurately and reliably make those discriminations.

There is not a simple dichotomy here – and indeed initially tacit knowledge elements may become more explicit over time (this is discussed in Chapter 13). One psychologist used the alternative term 'personal constructs' (Kelly, 1963) to encompass both explicitly accessible concepts (those we can reflect on) and the tacit elements of cognition (those where we can only reflect on the outcome). For the purpose of this volume I am including implicit knowledge elements that support those discriminations we are aware of making, as kinds of concepts. So the concept {conceptual knowledge} is taken to be the knowledge that can be used in decisions we can deliberate on, even when some of that knowledge is itself tacit. Implicit knowledge of this kind is important in conceptual learning and will be discussed later in the book.

4.3 Fuzzy Concepts

It seems then that that concepts are not all of the same kind, and indeed that the {concept} concept may itself be open to degrees of membership. Concepts of this kind are known as fuzzy concepts (Leung and Lam, 1988). A fuzzy concept does not have a clear and definite boundary. We can best understand this in contrast to an equivocal or well-defined concept. An example of the latter might be the concept {element}. In chemistry, the concept {element} is well defined, such that there is likely to be very little disagreement among experts when making discriminations. Manganese is an element, but water is not; rhodium is an element, but an sp^3 hybrid orbital is not; nitrogen is an element, but the air is not.

There is always potential to think of ways even experts might quibble – should ozone be considered an element as such, or an allotrope of an element? Even if some chemists disagreed over that, they would be arguing over semantics rather than disagreeing about the fundamental nature of elements. That is, the experts would have a common core to their {element} concepts, even if they defined the concept such that their individual conceptual boundaries were located slightly differently.

In the wider culture, outside of science, we can find some concepts that seem much less well defined. An obvious example is musical genres: where people may well disagree if a particular recording should be considered as falling under different labels: jazz and/or fusion? When does rock music become 'heavy'? What actually counts as 'progressive' rock? What makes a recording count as 'soul'? How do we establish a boundary line between compositions that are to be considered as part of the baroque and classical periods? Deciding what counts in such matters is often a matter of opinion, or taste, whereas we expect that discriminations about what is a chemical compound, or a reaction intermediate, or a conical flask, should be standardised within the chemical community.

Some concepts used in chemistry may, however, be considered to be less equivocal. For example, we might consider the concept {acid} (see Chapter 6) or perhaps the concept {oxidising agent}. The ways that these concepts have been used to discriminate examples from non-examples have changed historically, and different definitions (which include and exclude different substances) are still sometimes used today. We might counter this by suggesting that the chemist's concept {acid} needs to be understood as subsuming several other concepts (such as {Lewis acid}, *etc.*), and that therefore any ambiguity is a matter of laziness in not clearly specifying which specific {acid} concept was being considered. Perhaps, but if so (and Chapter 6 suggests it may not be so clear that is the case) we may have to concede that sometimes concepts (which have been referred to as atoms of thought) can have complex substructure.

However, in other cases, it seems that concepts may define sets that offer different degrees of membership, or even have different memberships at different times. Consider the concept {covalent bond}. There are some examples of bonds that experts will all agree fit the concept {covalent bond}: the bonds in H_2, N_2, O_2, the C–C bonds in ethane, or cyclohexane, are surely covalent bonds for all chemists. Similarly, it seems unlikely any chemists are seriously going to suggest the bonding in KF, or that in potassium metal, is covalent. Yet, it may not be quite so clear whether the bonds in NH_3 or H_2O are covalent. Or, perhaps, even whether the C–C bond in the ethanol molecule '*is*' covalent.

Imagine a chemist who has two children at school. One, Pierre, is a fourteen-year-old taking a secondary science course, and the other, Marie, is a 17-year-old and is taking an advanced elective chemistry course leading to university entrance examinations. Both children ask mum to check their chemistry homework.

In the context of his homework Pierre has used his concept {covalent bond} to discriminate examples from non-examples, and has judged NH_3 and H_2O to have covalent bonding (Figure 4.1). His mother checks his work and tells him she agrees. In the context of her homework Marie has used her concept {covalent bond} to discriminate examples from non-examples, and has judged that neither NH_3 nor H_2O have covalent bonding (Figure 4.2). Her mother checks her work and tells her she agrees.

It appears that in this scenario the concept {covalent bonding} has some flexibility – in one context the application of the mother's {covalent bonding} concept discriminates to include NH_3 and H_2O, but in another context the 'same' concept discriminates to exclude these same candidates. The bonds in NH_3 and H_2O have not changed, of course, but the judgement has been contextualised. If we are using a dichotomous classification of substances between ionic and covalent, we make one judgement; it we are making a

Concepts as Knowledge

Substance	Covalent bonding
H_2	✓
KF	✗
NaCl	✗
NH_3	✓
H_2O	✓

Figure 4.1 Homework question given to a hypothetical chemist's 14-year old son, Pierre: Which of the substances will have covalent bonding rather than ionic bonding?

Substance	Covalent bonding
H_2	✓
N_2	✓
O_2	✓
NH_3	✗
H_2O	✗

Figure 4.2 Homework question given to a hypothetical chemist's 17-year old daughter, Marie: Which of the substances will have covalent bonding, rather than polar bonding?

more nuanced judgement about whether a bond is polar or non-polar, then covalent comes to mean something more specific.

We can imagine other examples where the concept might apply in different ways depending upon nuance. The C–C bond in ethanol would normally be considered covalent, but if the question was whether it was a 'pure' covalent bond then the inductive effect that leads to a non-symmetrical charge distribution in the C–C bond may lead us to judge otherwise. A chemist is unlikely to consider the bonding in copper to be covalent, rather than metallic – yet might suggest the bonding in copper had significant covalent character. So according to the precise context we might consider the C–C bond in ethanol as covalent, or not, and the bonding in copper to be completely, or only partially, excluded from being covalent.

{Covalent bonding} is not an isolated or special example. Consider the concept {reversible reaction}. In schools, students are sometimes taught that most chemical reactions that proceed are irreversible – so a mixture of hydrogen and oxygen may readily be provoked into changing into water, but one can wait an awfully long time for water to spontaneously decompose back into hydrogen and oxygen.

$$\text{hydrogen} + \text{oxygen} \Rightarrow \text{water}$$

On the other hand, the Haber process, for example, relies on a reaction that is reversible: between hydrogen and nitrogen. The reaction

$$\text{hydrogen} + \text{nitrogen} \Leftrightarrow \text{ammonia}$$

is said to be reversible as the direction in which the reaction 'goes' can be varied by changing the conditions – the equilibrium position can be shifted. This is sometimes a useful distinction in chemistry. However, whether a reaction can be considered reversible is a matter of degree: all reactions are potentially reversible, depending upon conditions. Most school children have seen the van't Hoff voltameter used to electrolyse water into hydrogen and oxygen. There is usually a 'cheat', in that the water used is not pure, but in principle one could create the conditions for a pure sample of water to decompose:

$$\text{water} \Rightarrow \text{hydrogen} + \text{oxygen}$$

So, a more advanced chemistry student learns that all reactions can be considered as equilibria, even if for many reactions under common conditions the equilibrium position is very much towards the products (or reactants, for example, if waiting for water to spontaneously decompose).

A more mundane example would be the concept {high melting temperature}. This can clearly be considered a kind of composite, or sub-concept, relying on the concept {melting temperature}. We can imagine this concept applied to discriminations:

- Iron has a high melting temperature (>1500 °C), but sodium (98 °C) does not.

Or

- Water has a high melting temperature (0 °C) but ammonia (–78 °C) does not.

These discriminations are each sensible enough, although when juxtaposed it seems that 0 °C is a high melting temperature, and 98 °C is not! What counts as a high melting temperature depends upon a particular context. This is commonplace enough, but highlights how the application of concepts may not be straightforward as we usually apply concepts to make discriminations relative to specific contexts.

Another fuzzy concept is {chemical stability}. Students tend to suggest that the sodium ion (Na^+) is more stable than the atom (Taber, 2003). It is difficult to make such a judgement in absolute terms. The most neutral situation we can probably imagine to compare 'like with like' is an isolated sodium atom, compared with an isolated sodium ion separated from a lone electron:

$$Na \leftrightarrow Na^+ + e^-$$

This of course represents the first ionisation of sodium, which is endothermic. Energy is required to ionise the atom. In corollary, energy is released when the atom forms from the ion plus an electron. So, the latter process, but not the former, is spontaneous – in contrast to the expectations of many students who consider the presence of an octet as a reliable sign of stability (Taber, 2009). So, in these terms, students commonly get this question wrong and assume Na^+ is more stable than Na. Yet one must have sympathy.

In the element, under familiar conditions, there is a metallic lattice that may be considered to contain (to a first approximation) Na^+ ions. Something similar tends to be true in its alloys. In its compounds, sodium tends to exist as salts containing (to a first approximation) Na^+ ions. In solution, sodium compounds tend to give hydrated sodium ions. So, in most (or even all) chemical contexts that students learn about, Na^+ is stabilised (in a metallic lattice; in an ionic lattice; by solvating species) and it would appear to be stable. So, such judgements need to be contextualised. There are many examples of species that might be considered stable or unstable in solution, depending on the pH.

As a final example of a fuzzy concept, consider the distinction sometimes taught in school science classes between chemical changes and physical changes (Taber, 2002). The concept {chemical change} is here often associated with large energy changes (sometimes bangs and flashes), irreversibility, and bond breaking and/or formation, as well as the production of a new substance. Yet, as just discussed, many chemical changes are usually considered reversible (and all are in principle); many chemical changes occur with modest energy changes; and many changes normally considered 'physical' involve changes in bonding. Dissolving NaCl requires overcoming ionic bonds and the solvation of ions; even melting ice involves the breaking of many of the hydrogen bonds in the solid structure. The concepts {chemical change} and {physical change} certainly seem very fuzzy.

4.4 Manifold Concepts?

One way of making sense of some of these complexities is in terms of manifold or multifacetted concepts – that is, concepts that appear to have different facets. The example of 'acid' might be seen in this light (see Chapter 6). So perhaps here are a range of more specific acid concepts – {Lewis acid}, {Brønsted–Lowry acid}, *etc.*, which are subsumed under the more general superordinate concept {acid}. If this is so, we might say that the concept acid is not itself fuzzy, but rather a superordinate concept with a number of different facets that will be activated and applied in different contexts. So, some of those concepts that could seem fuzzy might be better characterised as being manifold, or multifacetted.

This might seem convincing for {acid} (but see Chapter 6 for a more nuanced discussion), but it may be harder to persuade readers to treat the concept {covalent bond} in a similar way. In the case of covalent bonds, it does not seem that there is a superordinate concept {covalent bond} that encompasses several subordinate concepts (perhaps such as {pure covalent bond}, {slightly polar covalent bond}, {primarily metallic bond with substantial covalent character}, {dative covalent bond}, *etc.*), but rather {covalent bond} is the kind of concept that seems fuzzy as it has meaning that shifts somewhat depending upon context of application. (Despite this, as soon as I suggest, say, {primarily metallic bond with substantial covalent character} then this becomes a viable basis for conceptualising and discriminating examples and non-examples: so, it can be a viable concept, even if I suspect it is not one commonly adopted by most chemists.)

Again, the concepts {chemical change} and {physical change} do not seem multifacetted in the sense of having multiple meanings, but fuzzy in that the defining characteristics do not usefully and cleanly map onto the world to provide us with clear discriminations. There are cases, such as solvation, where it may be difficult to persuade a student that this is an example of a physical change and not a chemical change – new bonds are formed, and new chemical species, complex ions (which may sometimes be relatively stable) appear. Precipitation may seem the reverse of dissolving, yet it could be considered either a chemical change or a physical change depending upon the context. If two solutions are mixed and there is spontaneous precipitation then this is normally considered a chemical change (but perhaps not according to the English National Curriculum – see Chapter 10), but if a supersaturated solution was suddenly cooled and crystallisation occurred this would usually be considered a physical change – a change of state.

Given that concepts are mental entities, applied in thinking, only indirectly observable by inference from how people use them (see Chapter 2),

Concepts as Knowledge

it may not be possible, or indeed sensible, to seek to characterise them precisely and to claim definitively that one concept is best seen as having fuzzy boundaries ({physical change}, say), and another as being multi-facetted (perhaps {acid}). (This could be a valid characterisation for one chemist or student's thinking, yet not for another.)

It is worth considering another concept here, though – that of 'element'. If a chemistry lecturer was visiting a friend who was a historian of science, she might be surprised to find the friend rummaging around his desk mumbling about having lost the elements. The response 'they are there in essence, but now form part of the compounds in those objects on your desk' would be unlikely to help. If the friend explained he was looking for his copy of Euclid's 'Elements' then any confusion should abate. Euclid's work uses the term element in a sense of fundamental constituent parts. This is clearly a more general meaning of element, distinct from, but related to, the more specific chemical concept of the {elements}.

But what if the chemist's friend wanted to talk about some work he was doing on classical texts about the elements – air, earth, water, fire and ether? The phrase 'the elements' here would seem to refer (in principle) to {the elements} as the concept is used by the chemist herself: the basic building blocks from which all material substances are comprised in various proportions. In one sense, this is the 'same' concept – yet when applied would lead to very different discriminations. The classical elements are no longer seen as elements, and the modern elements would have been seen as compounds of the classical elements. This raises the question of how we understand the relationship between these different meanings of the word 'element': do they relate to different concepts (are the words homonyms), or the same concept, or something else?

The general everyday use of the word 'element' surely refers to a different (if related) {element} concept to the more specific chemist's {element} concept. The elements of, say, embroidery, does not refer to fundamental substances. Perhaps this general {elements} concept subsumes the more specific chemical {elements} concept? Indeed, perhaps our disciplinary concept should only be referred to as the {chemical elements} concept and our use of the shorthand version (both here in referring to the {elements} concept, and generally when referring to the elements themselves) is just a liberty we take in the context of discourse with other chemists? That is, it is specialised usage that should be reserved for use (i) within a specialised discourse community (chemists), and (ii) when we seek to induct novices into that community through using authentic disciplinary practices in teaching (*e.g.*, talking like chemists).

Yet the distinction between air, earth, fire, and water, on the one hand and hydrogen, helium, lithium, *etc.*, on the other appears to reflect different discriminations made according to the same basic principle. Both uses of 'the' concept refer to the types of substance that might be considered fundamental, but the decision about what to include or exclude has changed over time. We might consider this a manifold concept – one concept with different aspects – but we could alternatively argue that these are distinct concepts unfortunately having the same name (but perhaps better distinguished as {modern chemical elements} and {classical elements}). We have here a more extreme case of the situation found with the concepts {acid} and {oxidation}.

As concepts are mental entities, we might even wonder if the answer depends on *which particular mind* is thinking about the issue and activating a concept. If we can only find out about concepts by seeing how people behave when using them, we might find (assuming we have suitable methods) that one chemist has a superordinate concept {element} and in different contexts uses this to think in terms of the modern chemical or the classical discriminations; but another chemist in effect two discrete {elements} concepts that are both labelled 'elements' but activated in different contexts.

4.4.1 Why Does It Matter?

The suggestions that it may be difficult to determine the difference between a fuzzy concept and a manifold concept, or a manifold concept and several concepts with the same label (name) may seem to focus on a rather esoteric issue. We might ask: do these distinctions matter?

I would remind readers of the model represented in Figures 2.1 and 2.2 (in Chapter 2). It was suggested earlier that concepts as enacted, and as indirectly inferred from what people say and do, exist in the moment, but are dependent upon a largely stable (on a short time scale) underlying concept generating structure (the 'congenst'). In view of how this must ultimately be encoded in networks of neural connections (see Chapter 13), it was envisaged as a great many nodes, which are heavily interconnected with different connection strengths (making activation of some parts of the system more likely to lead to activation of other parts). This is, of course, only a model, as no one knows precisely how concepts are represented in brains, but seems well motivated by what we know about neural structure.

Such a model suggests there may not be any *absolute* distinction between a system of related but distinct concepts and a concept with different facets: as in both cases everything is inter-weaved through links that may be more or less direct, and may have different strengths. This means that our descriptions of concepts (as observed to be enacted) as unitary, multifacetted, and so forth may only be matters of degree. Such a model allows us to understand conceptual change (*e.g.*, learning) by gradual changes in the system (see Chapter 13), through developing, strengthening, and atrophying, of connections. This means there may be evolution in the underlying structure that allows shifts relating to the kinds of distinction being made here. Two concepts that seem distinct, might – over time – come to be one; a single concept may get differentiated; weakly associated concepts may become more strongly associated; *etcetera*. The underlying structure is always to some extent in flux, so the concepts themselves, as revealed when the person uses them, may change their characteristics. Concepts-in-action (as actually applied, when the underlying neutral structure is activated) may become less, or more, fuzzy, for example.

This all matters in education because chemistry teachers are expected to teach canonical chemical concepts, and students on chemistry courses are asked to learn canonical chemical concepts. In addition, of course, chemistry teachers are often expected to not just improve their students' knowledge and understanding – but also to assess it. Yet this chapter suggests that concepts can be 'slippery' entities, and it should be clear from the treatment here that assessing someone's conceptual knowledge is far from straightforward (Taber, 2013).

A widely recognised feature of good teaching practice is that it employs formative assessment – that teachers should be constantly monitoring student thinking to inform their own teaching (Black and Wiliam, 2003). If student concepts were unitary, with clear ranges of application, and applied uniformly across all pertinent contexts, then undertaking formative assessment during teaching would be a good deal more straightforward.

If however, as seems to be the case, students' (indeed, people's) underlying conceptual structures are actually much more dynamic, and nuanced, and their thinking is often very specifically context dependent (as is argued here, and as is suggested by much research evidence) then it becomes so much more challenging for a teacher or lecturer to "really know what the students are thinking" (rather than what they reported thinking at that moment, in response to that particular question, asked in some unique wider context). The issues raised in this chapter, then, do have repercussions for teaching – and themes raised here will resonate through the rest of the book.

References

Abrahams I., (2011), *Practical Work in School Science: A Minds-on Approach*. London: Continuum.
Ball P., (2002), Chemistry in soft focus, *Chem. Br.*, **38**(9), 32.
Bhaskar R., (1981), Epistemology, in Bynum W. F., Browne E. J. and Porter R. (ed.), *Macmillan Dictionary of the History of Science*. London: The Macmillan Press, pp. 128.
Black P. and Wiliam D., (2003), In praise of educational research: formative assessment, *Br. Educ. Res. J.*, **29**(5), 623–637.
Brock R., (2015), Intuition and insight: two concepts that illuminate the tacit in science education, *Stud. Sci. Educ.*, **51**(2), 127–167.
Chang H., (2004), *Inventing Temperature: Measurement and Scientific Progress*. Oxford: Oxford University Press.
Close F., (1990), *Too Hot to Handle: The Story of the Race for Cold Fusion*. London: Allen and Unwin.
Collins H., (1992), *Changing Order: Replication and Induction in Scientific Practice*. Chicago: University of Chicago Press.

Collins H., (2010), *Tacit and Explicit Knowledge*. Chicago: The University of Chicago Press.

Gauld C., (1989), A study of pupils' responses to empirical evidence, in Millar R. (ed.) *Doing Science: Images of Science in Science Education*. London: The Falmer Press, pp. 62–82.

Kelly G., (1963), *A Theory of Personality: The Psychology of Personal Constructs*. New York: W. W. Norton & Company.

Leung K. S. and Lam W., (1988), Fuzzy concepts in expert systems, *Computer*, **21**(9), 43–56.

Polanyi M., (1962), *Personal Knowledge: Towards a Post-critical Philosophy (Corrected version ed.)*. Chicago: University of Chicago Press.

Popper K. R., (1945/2011), *The Open Society and Its Enemies*. London: Routledge.

Rothenberg A., (1995), Creative cognitive processes in Kekulé's discovery of the structure of the benzene molecule, *Am. J. Psychol.*, **108**(3), 419–438.

Taber K. S., (2002), *Chemical Misconceptions – Prevention, Diagnosis and Cure: Theoretical Background* (Vol. 1). London: Royal Society of Chemistry.

Taber K. S., (2003), Understanding ionisation energy: physical, chemical and alternative conceptions, *Chem. Educ. Res. Pract.*, **4**(2), 149–169.

Taber K. S., (2009), College students' conceptions of chemical stability: The widespread adoption of a heuristic rule out of context and beyond its range of application, *Int. J. Sci. Educ.*, **31**(10), 1333–1358.

Taber K. S., (2010), Straw men and false dichotomies: Overcoming philosophical confusion in chemical education, *J. Chem. Educ.*, **87**(5), 552–558.

Taber K. S., (2013), *Modelling Learners and Learning in Science Education: Developing Representations of Concepts, Conceptual Structure and Conceptual Change to Inform Teaching and Research*. Dordrecht: Springer.

Taber K. S., (2017), Knowledge, beliefs and pedagogy: how the nature of science should inform the aims of science education (and not just when teaching evolution), *Cult. Stud. Sci. Educ.*, **12**(1), 81–91.

Trewavas, (2009), What is plant behaviour? *Plant, Cell Environ.*, **32**(6), 606–616.

Ziman J., (1968), *Public Knowledge: An Essay Concerning the Social Dimension of Science*. Cambridge: Cambridge University Press.

CHAPTER 5

The Origin of a Chemical Concept: The Ongoing Discovery of Potassium

This chapter, and the next, consider discoveries in chemistry, and how new chemical concepts are formed. In particular, as examples, two cases will be looked at in some detail – the discovery of potassium (and so the formation of the chemical concept {potassium}) and, in the next chapter, the development of ideas about acids (and the formation and refining of the chemical concept {acid}).

5.1 Chemistry as a Natural and Empirical Science

Chemistry is one of the 'natural' sciences, concerned with understanding and explaining natural phenomena, rather than a social science that concerns itself with social phenomena that only arise from human culture and institutions. Like most such conceptual distinctions this is in part a matter of convenience and deliberately constructing definitions, rather than absolutes. Some non-human animals exhibit behaviour that might be judged as evidence of a primitive type of culture (Laland and Hoppitt, 2003), but we generally still consider such phenomena as part of the natural world and within the remit of natural, rather than social, science.

5.1.1 The Nature of the Natural

It is not uncommon for a distinction to be presumed in everyday discourse between what is natural and what is man-made: yet those things made by people working in science are certainly within the natural realm, rather than the result of some magical or supernatural activity. We might consider the example of transuranic elements, that are 'man-made' and so 'artificial' (they are artefacts of human activity) – but certainly natural.

A notion of what is natural tends to be strongly held by most people, even if they may not always have a clear rationale for their intuitions. The term 'chemicals' may imply, for some, artificial, manufactured substances. So, for example, some people consider that they will be healthier if they eat foods that do not contain any chemicals. The cynical chemist might respond that they would not be healthy, but simply starve – as all foods are comprised of chemical substances.

There has been a related consumer move (among some in wealthier nations who can afford to be choosy) to buy 'organic' foods. The rationale behind this may not be entirely spurious, as it is often concerned with avoiding foods that may contain residues potentially damaging to health, and/or encouraging sustainable farming by avoiding foods produced by highly intensive farming methods that may be considered irresponsible. Yet, for many of those buying such foods, there is an assumption that they are avoiding 'chemicals' in their food.

There are many toxins in natural products, and whilst chemistry is not innocent of having developed pesticides and other agents that have done a good deal of harm, neither 'natural' substances nor 'artificial' ones can be collectively considered either safe or dangerous – nor 'good' or 'bad'. Some of the greatest achievements of applied chemistry have involved the synthesis of useful natural substances (for example, for medicinal purposes) that can have great advantages over using natural products: chemical manufacture may reduce costs where a compound is only known in species found in remote places, offers ready quality assurance, and often less waste, and may avoid the need to kill organisms to meet human needs.

Countering this, we should avoid hubris – the wide-scale prescription of the synthetic drug thalidomide in a number of countries in the late 1950s and early 1960s led to tens of thousands of incidence of either deaths of unborn children or serious birth defects. Chemistry as a science does neither social harm nor good, but provides knowledge that can be applied – to make profits, to give consumer choice, to fight wars, to save lives and improve heath, and so forth. Chemists are moral agents, and need to consider potential implications of their work, but the responsibility for how science is applied is a broader societal matter.

The social world presumably arises 'naturally' as part of the evolution of ('natural') human beings, certainly rather than being supernatural in the sense of invoking causes and entities that are not part of the physical universe (and outside the remit of natural science). The distinction between natural and social science might be considered in similar terms to the argument for distinguishing between chemistry and physics. It may be that there is nothing in chemistry that could not *in principle* be understood and explained in terms of physical concepts. Yet, even if that were assumed, it is sensible, given the complexity of chemical phenomena, to employ concepts at the chemical level and not seek to reduce phenomena described by chemists in terms of concepts such as {oxidation}, {acid}, {resonance}, {transition state}, and so forth, to purely physical descriptions.

Similarly, social phenomena may ultimately be complexities emerging from the physical world, but no one is going to make sense of immigration policy, or differences in average school examination results between different ethnic groups in some national context, or decide the best way to educate the most gifted children, by solving the Schrödinger equation any time soon.

Chemistry includes the production and characterisation of synthetic compounds that would be extremely unlikely to occur 'in nature' (if, again, human activity is stood outside of nature for the purposes of such a description). Chemistry might be considered the science that studies substances through their characteristics and interactions. Synthetic compounds are clearly substances rather than social phenomena, even if their synthesis may rely upon not only human activity, but activity that is framed within human cultural institutions (disciplines, fields, chemical education, laboratories, *etc.*).

5.1.2 The Wider Context of Scientific Discovery

Despite the existence of some synthetic compounds on earth being entirely contingent upon these human institutions, we have no trouble as chemists seeing them as chemical entities that fall within the discipline of chemistry, and a field of research within that *natural* science discipline. There are scholars who are interested in scientific discovery who are less interested in the natural phenomena themselves, than in the institutional structures and particular cultural conditions within which the science can take place. Sociology of science has its own focus that exists alongside the science and offers complementary levels of explanation. Arguably, a complete explanation of the synthesis of a new compound would describe the chemistry, but also the conceptual, cultural and social context in which it occurred. But then a *complete* explanation for how some new synthetic compound appeared upon earth would need to address how there came to be intelligent living beings able to do the synthesis – and how the earth formed, and indeed the origin of the universe itself! Thankfully, we are rarely interested in 'complete' explanations of that kind.

However, sociologists and psychologists of science offer insights, which remind us that chemical 'discoveries' are not inevitable. The word discoveries is placed in scare-quotes there because of the implication that something that can be discovered must have already existed, and has just been stumbled across and found. This may not be the most helpful way of thinking about some of the things that might be referred to as discoveries in chemistry. It is likely that some compounds that have been synthesised by chemists never existed in the universe before they were prepared in a laboratory. If so, these compounds were not discovered in nature so much as discovered in the imagination (as a possibility) and only then realised, constructed, *in vitro*. The possibility is 'discovered' in the mind of a chemist, and then the compound as a natural substance is in effect the confirmation of a hypothesis. Some discoveries may be serendipitous (mauveine comes to mind), but this description still applies: the chemist imagines what the unexpected product could be, and then carries out analytical procedures to test the idea.

5.2 Chemistry as an Empirical Science Depends on Imagination as well as Benchwork

Chemistry is also sometimes described as one of the empirical sciences, a term that seems most useful in distinguishing sciences from philosophy – a distinction that evolved historically (with natural science being at one time called natural philosophy). This is certainly a notion that is often adhered to in chemistry education where it may be claimed that chemistry is a practical subject, or a laboratory subject – for example, when arguing the case that authentic chemistry teaching requires access to specialised learning spaces (teaching laboratories) and resources (glassware, chemicals, *etc.*).

This is an important argument – an authentic chemistry education must include laboratory work. As a science, however, the focus on the empirical must necessarily require the practical to be considered in relation to the theoretical. Chemistry is a science because it develops through the interplay between observation and experiment on the one hand and conceptualisation and theorising on the other. This is represented schematically in Figure 5.1.

Figure 5.1 suggests that science is an iterative process where observation motivates the development of theory, which in turn motivates new observations to test hypotheses and refine concepts. In chemistry, observations may well be framed in terms of experiments – but this is more the case in some areas of science than others that may have to rely on 'natural experiments'; that is, observing from within the diversity of naturally occurring conditions rather than manipulating conditions. It is worth noting that the close association of 'conceptual' with 'theoretical' here does not imply these terms are being used synonymously: concepts are seen as being essential components of theory, but not as being sufficient to be considered theory by themselves (see Chapter 8).

5.3 Thinking about Chemical 'Discoveries'

It is then naïve, indeed too simplistic, to see scientific discovery as a process of noticing what already exists in nature (whether that noticing is serendipitous or the outcome of a designed experiment). Rather it will be argued that what is 'discovered' must often be understood to be as much a creative act of imagination that occurs inside the head of a scientist, as a process of finding something in the external world. This is not to suggest that there is no real physical world where chemistry happens that is being observed, nor that the concepts developed by scientists necessarily refer to imaginary

The Origin of a Chemical Concept: The Ongoing Discovery of Potassium

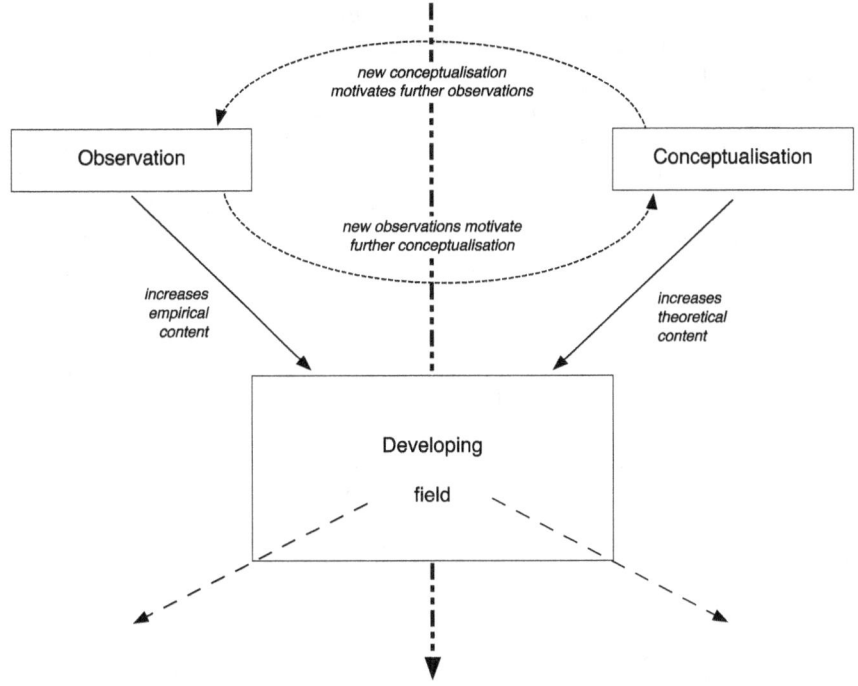

Figure 5.1 Scientific fields develop through the iterative development of empirical and theoretical content.

entities. Concepts are a form of mental representation of what is perceived and noticed in experience (see Chapter 13). The concepts are the products of imagination being applied to make sense of what is being observed – but they are not perfect representations of natural entities that exist 'in nature'.

5.3.1 Conceptualisation Is Shaped by the Cognitive Apparatus

That is, that conceptualisation is necessarily a mental activity, something that occurs in human imagination, and so it cannot be assumed to provide perfect representations of the natural world, even though it is constrained and informed by empirical work. This is because the human perceptual apparatus is not a perfect instrument for observing nature, and it is heavily dependent upon existing conceptualisations of the world to interpret sense data (Taber, 2014).

The human perceptive–conceptual system develops early in life and becomes highly effective: our bodies may sense patches of light or vibrations in the air, but what is presented to adult consciousness has already been interpreted as objects and activity in our surroundings (this is discussed in Chapter 2). We experience sights and sounds, not light and noise. We see people and retorts and flasks and we hear questions and retorts and explanations – even though such perceptions represent the outcome of a great deal of 'computing' (processing) activity to make sense of sensory data (see Chapter 13).

This is essential for normal adult functioning (imagine it was otherwise), but the cost of readily making sense of the world is that the system we have developed to interrogate and interpret sensory data is not perfect, and indeed has become biased in terms of perceptions that 'worked' for us previously (both in terms of evolved genetic predispositions reflecting the experiences of our ancestors, and our personal experiential histories).

Our brains seem to be set up (*i.e.*, to have evolved) to offer type II rather than type I errors: that is, although the system sometimes throws up false negatives (we may take some time to realise that the object moving by the bushes at twilight is just a paper bag being blown about by the breeze), false positives are more likely. We recognise these false positives when the friend we approach in the street turns out to be someone we do not know, or when the tune we have recognised resolves into an unfamiliar composition, but probably most instances are never noticed. People see faces everywhere – in butter melting on toast, on craters on Mars, in the most primitive emoticons :-) to such an extent that this seems to be part of our common genetic inheritance. When we come across something novel, we may initially perceive it as something familiar – the brain works to match the incoming information to a familiar pattern (see Chapter 14).

5.3.2 Constructivism

This is one message of a perspective on cognition and learning known as constructivism (Glasersfeld, 1989). In science education, constructivism is a major perspective on learning and teaching that highlights the challenges for students learning canonical science (Gilbert, 1995; Matthews, 2000; Fensham, 2004; Taber, 2009). The psychological processes that support conceptual development in learners and professional research scientists are *fundamentally* the same – an eleven-year-old, or a 16-year-old, or a 21-year-old student have the same *basic apparatus* for making sense of the world as the Nobel prize winning chemist. Of course, that apparatus matures (Piaget, 1970/1972), and is more efficient in some people than others, and the knowledge base it draws upon will be vastly different in different cases. Typically, the professional chemist knows more chemistry than the school pupil, indeed knows more complex and advanced chemistry, and has a more integrated and coherent structure for chemical knowledge, and has many fewer 'misconceptions': yet the fundamental processes of perception and cognition are the same.

It used to be widely believed that brain structures, and so the thinking that depends on them, became fairly fixed in adolescence, but it is now recognised that the brain retains a high level of plasticity throughout life (The Royal Society, 2011). The brain continues to modify its structure in response to experience in adults. However, experience is interpreted in terms of existing understandings (and so the interpretations made of previous experiences), and in a sense the system is biased to be conservative – people tend to appreciate having 'comfort zones' and may continue in their habits of thought, as in their other habits. The system is also believed to be inherently conservative in terms of how working memory (the brain 'module' that we consciously use when thinking deliberately about something) operates. Working memory has limited capacity (Miller, 1968), but is able to treat complex material that is highly integrated (being well established in memory) as single 'units' of information (Gobet *et al.*, 2001). Teacher frustrations when students seem to fail to follow something relatively straightforward are often due to something that is indeed straightforward *relative to the existing conceptual structures of an expert* appearing much more complex from the perspective of the less well established and integrated conceptual understanding of a novice.

These points are often made in relation to student learning in chemistry classes, but it is worth also bearing them in mind when considering how chemistry itself has developed. Whilst historical chemists may certainly not have been novices, in terms of modern chemical ideas their conceptual structures may have offered no more support when now-familiar concepts were first mooted than those of a learner today meeting those concepts in their first modern chemistry course.

5.4 The Discovery of Potassium: Imagining a New Substance

Potassium was first isolated as an element in 1807 when Humphry Davy electrolysed potash (potassium hydroxide). Here is a description of this event:

> In October 1807, when Humphry Davy passed an electric current through some moist potash, minute globules of potassium burst through the crust of potash and took fire. At the sight of this, Davy "bounded about the room in ecstatic delight". He had good reason to be joyous, for the successful decomposition of the fixed alkalies [*i.e.*, alkalis] seemed to justify his recently expressed belief that the power of the voltaic apparatus gave hope of discovering "the true elements of bodies". Whether the new metals were "true elements" or not, became the unspoken question in the voluminous discussion which followed their discovery. The nature of the alkalies from which the new metals were made had been a point of uncertainty ever since Lavoisier had omitted them from his list of simple substances. Though there had been much conjecture concerning their possible composition, "that metals existed in the fixed alkalies seems...never to have been suspected". Some of Davy's delight must have come from surprise as well a success. (Siegfried, 1963, p. 248)

There are a number of interesting points that arise from this account. For one thing, it is clear this was not some completely serendipitous discovery. Davy was carrying out a programme of research motivated by theoretical considerations. The principle of some substances being elements was well established, but it was not entirely clear which substances should be considered 'true elements'. Some elements were known. Some other substances were understood to be compounds. Others were of uncertain status.

Davy suspected he could isolate elements from potash, but he was not looking for potassium as such. There was no concept {potassium} that would allow him to be seeking a substance that would be so conceptualised. Indeed, he is reported to have suspected that potash might contain phosphorus, or sulfur 'united to' nitrogen (Siegfried, 1963, p. 249). So, Davy was successful in isolating an element from potash, but it was a new element he had not anticipated. Davy also isolated sodium, another new metallic element. However, categorising potassium and sodium as metals, something taken for granted by chemists today, was not as straightforward given the state of chemical knowledge of the time. As Seigfried reports:

> Davy systematically determined the more obvious properties of the new metals and established not only their similarity to each other, but also their unique differences from other substances. Of primary interest to Davy was the relation of these metals to the generally accepted system of chemical knowledge. Davy asked, "Should the bases of potash and soda be called metals?" In spite of their low density, he decided that they should, for all their other properties were analogous to that class of bodies. With the concurrence of a number of "eminent scientific persons" he suggested that the names potassium and sodium be assigned "as implying simply the metals produced from potash and soda". (Siegfried, 1963, p. 250)

There are a number of chemical concepts that are used in telling this story. One is the concept {potassium}. That concept, {potassium}, did not exist on earth prior to Davy's work. What was discovered, or in a sense created (by the novel experimental set-up), was *a new phenomenon*: some minute globules of something unfamiliar that burst through the crust and caught fire when an electric current was passed through some moist potash. The phenomenon was observed, and needed to be conceptualised. It was conceptualised as (probably) being an element – something no doubt facilitated by the conceptual framework brought to the observation. Davy was seeking to find elements within a substance he assumed, or at least suspected, was not itself elemental.

5.4.1 The Creation of Potassium

Siegfried writes "minute globules *of potassium* burst through the crust of potash and took fire" but that is a reconstruction in the light of what happened after the event – what has been referred to as a 'rhetoric of conclusions' (Schwab, 1958). That is not to say it is incorrect, simply that in terms of 'science in the making' (Shapin, 1992) what was seen by Humphry was not potassium (that is, not yet potassium), but some globules of something uncharacterised and as yet nameless. Only with the benefit of hindsight can we say this was potassium.

In this example, it is probably fair to say not only that (the concept) {potassium} did not exist on earth prior to Davy's work, but also that (the metal) potassium did not exist on earth prior to Davy's work. We might consider this an example of where natural science identifies something that is artificial in the sense of not existing naturally. Given the high reactivity of potassium, it seems very unlikely any of the pure metal had ever existed 'naturally' on earth (or indeed anywhere else in the universe), in the sense that gold, silver and copper may be found 'native'. Potassium, like all the heavy elements, is considered to be have been formed in supernovae – where the conditions are hardly suitable for pure metal to condense.

This latter point actually highlights a complication of one of the most important concepts in chemistry – {element}. Whilst the elements heavier that hydrogen and helium are considered to have been formed in stars through nuclear reactions – they did not at that point of initial formation exist as elements in the sense of pure substances. Under the conditions of their formation matter exists as plasma – the atomic nuclei were formed amid a 'mist' of electrons – but did not condense into the more familiar phases of matter under those conditions. So, elements found 'native' on earth – such as carbon, sulfur, nitrogen – were formed in stars, but not as elemental substances. The elements only formed later, elsewhere, in very different conditions. The elements sodium and potassium were also formed in stars, but (as far as we know) never became elemental substances on earth until Davy extracted them from their compounds.

So, in a sense Davy did not so much discover (the elemental substance) potassium as create it. This raises the questions of whether we should consider *the element* potassium actually existed before the first sample of the elemental *substance* potassium was created. (It is possible, of course, that some alien chemist elsewhere in the vast cosmos deserves credit for producing potassium before Davy, but the same issues arise – potassium probably does not form naturally anywhere in the universe, so someone (some being) first created it.) (What are now recognised to be) compounds of potassium certainly existed – but perhaps in a sense they were compounds of an element yet to exist. So in a sense these compounds preceded one of their constituent elements.

This interpretation is avoided if one considers potassium as an element to exist as soon as potassium nuclei are formed in the universe, regardless of the absence of potassium the substance. It seems even an apparently well-defined and understood concept such as {element} may appear to be a fuzzy concept (see Chapter 4) when interrogated closely.

5.4.2 The Construction and Development of the {Potassium} Concept

So, the birth of the chemical concept {potassium} was not located at the moment Davy observed the globules, as at that point what was conceptualised was a substance being formed, which was possibly an element, but – if so – one as yet unidentified. It might perhaps have been an element, and – if so – perhaps one already known, or perhaps one not previously reported. The conceptualisation was developed with the characterisation of potassium, as more of its properties were detailed. That characterisation was initially undertaken by Davy, but in the two centuries since our conceptualisation has developed further as new properties of potassium have been determined. For example, Davy (who died in 1829) was in no position to appreciate that potassium was radioactive, nor was he in a position to speculate on the isotopic composition of the element.

At the start of the twentieth century Campbell and Woods reported that potassium was radioactive (Campbell *et al.*, 1907). The discovery of radioactivity itself had modified the chemical concept {element} as it added a new property that elements, although it appeared only some elements, might have. The concept {potassium} was itself modified by the identification of potassium as one of the radioactive elements. After this discovery, the chemical concept {potassium} had shifted beyond the concept {potassium} that Davy had developed.

The chemical concept {element} had for many years implied that each element was associated with an atom of specific type, which when atomic structure became known implied a particular configuration of subatomic components. The discovery by J. J. Thomson that some of the then-known elements may actually be "non-elementary" has been described as "entirely accidental" (Harkins and Liggett, 1923, p. 75), something noticed as a side effect of an investigation into the positive-ray spectrum of neon. This led to the suggestion in 1915 that an element with a non-integral atomic weight, that is, "an element whose exact element (so-called atomic) weight differs by more than 0.1 per cent from a whole number" (p. 75), might be a mixture of several isotopes.

The 1922 Nobel prize for chemistry was awarded to Francis Aston for demonstrating isotopic composition of a range of elements using the mass spectrometer (Anon., 1923). It is quite difficult for modern chemists to put themselves back into the mindset of the time before isotopes were known. So, for example, one of the best-known teaching examples used in secondary school is chlorine with its atomic mass of c.35.5 – which from a contemporary perspective already offers very strong evidence that what is being measured is a weighted mean of several different isotopes of the element. But at the time Aston was working it was a very real question "whether its accepted atomic weight represented the true mass of its individual atoms [*i.e.*, non-integral] or was merely a statistical mean" (Aston, 1920, p. 218). Aston answered this question, reporting "strong lines [in the mass spectrum of chlorine] corresponding to masses 35, 36, 37, and 38, all of which were whole numbers to the accuracy of experiment. *There was no indication whatever of a line corresponding to a mass 35.46"* (1920, p. 218, italics in original). The emphasis given to that latter statement reinforces how this result related to a genuine scientific question of the time, rather than something that would now be taken for granted by chemists and advanced students. The conceptual relationship between atomic structure and atomic masses, that to a first approximation the (isotopic) atomic mass is the sum of the proton and neutron numbers (which is part of the target knowledge expected of secondary school chemistry students today) was less clear at the time.

Regarding potassium, Aston showed in 1921 that the less problematic atomic mass of c.40 for potassium, neatly approximating a whole number, appeared to be due to a mixture of (at least) two isotopes – the spectrum suggested these were K^{39} and K^{41}. Further work suggested that neither of these appeared to be the source of the radioactivity of potassium. It was found that potassium also contained the isotope K^{40} at "about 1 part in 8300" (Harkins and Liggett, 1923, p. 178) and it was reported that the K^{40} isotope "is responsible for the entire known activity of potassium" (Smythe and Hemmendinger, 1937, p. 182).

The paper reporting this definite finding is a wonderful example of 'science in the making' (Shapin, 1992) or 'science in action' (Latour, 1987), in that it offers a warts-and-all account. One problem was that potassium ions produced, in order to measure their mass:charge ratio in the apparatus, damaged the target: "K^+ ions have a very destructive effect on any target upon which they impinge, so that not only the potassium but also the material of the target is sputtered on all neighboring objects" (p. 179). This problem was addressed by angling the target, and by largely encasing it in a box. Despite this, it was found that when collecting a single sample using the apparatus "the collected sample is always mixed with tungsten sputtered from the target" that there seemed to be "no easy way of eliminating" (p. 180). The ion beam damaged other parts of the apparatus, requiring some parts to be replaced after each run.

Organising a suitable Geiger–Müller counter to detect the beta radiation emitted by the potassium led to a "few unsuccessful experiments" before suggestions from a colleague (a Mr Harper) were followed (p. 179). Among various other technical challenges described, the authors report that "unfortunately the resolving power went bad several times during the collection of the K^{39} sample, perhaps because a different batch of emitter and new heater insulation were used, so that it is contaminated with an unknown amount of K^{40}" (p. 181).

These points are useful in reminding us of the way that science is empirical, and so chemistry develops through an iterative interplay between theory and observation. As has been reported in more contemporary examples (Latour, 1987; Knorr Cetina, 1999) the actual process of scientific discovery involves a good deal of judgement about when what is observed should be taken as an indication of what is being looked for (or of something else that is novel and noteworthy), and when it is rather an artefact of the limitations of apparatus and technique.

One simple version of the nature of science treats Karl Popper's (1934/1959, 1989) falsificationist account as providing a straightforward criterion for scientific progress. So, it is often suggested that no number of confirmations prove a theory beyond question (which is fair enough as experimental tests always underdetermine any theory), yet a single refutation is enough to reject a hypothesis from consideration. Of course, actual science bears little resemblance to such a simplified model. Such 'refutations' only exist in hindsight, in just the same way Davy only discovered potassium in hindsight.

That is, some years later the text books may refer to some critical experiment, but it only became seen as a critical experiment in the light of a wider programme of research and the scholarly critique and debate that accompanies and follows publication of the research results. In practice, getting no results, or getting the 'wrong' result (according to the hypothesis motivating a study), are often not seen as an indication that the hypothesised effect is absent, or different in form to what is anticipated, but rather that the effect looked for requires refinement of the observational technique: in particular of the experimental manipulation of the conditions needed to observe what is sought.

This process is then some way from the naive model of scientific enquiry often drawn from Popper (1934/1959, 1989) where the failure to observe a hypothesised effect should be seen as a refutation. In real science, it actually takes a good deal of professional judgement to decide when an apparent, *prima facie*, refutation should be considered as an actual refutation, rather than an indication that some more refinements are needed to an experimental set-up. Scientific investigations are not usually one-off experiments, but series of iterations of procedures where (during the provisional stages) theoretical knowledge is brought to bear on understanding of techniques, largely to evolve the technique rather than develop the theory. By comparison, much practical work in school and even undergraduate chemistry is a one-shot activity, before moving on to a quite different practical the next session, giving little feel for the actual nature of scientific research.

5.4.3 The Shifting {Potassium} Concept

So, although there was a chemical concept {potassium} established by Davy's work, the conceptualisation that Davy had of potassium changed, developing over time from his initial isolation of the new substance through his characterisation of its properties. So, the discovery of potassium, as Davy came to conceptualise it, extended over this programme of work. Moreover, the concept {potassium} developed by Davy might be considered deficient by comparison with the concept {potassium} as understood by a modern-day research chemist. As one example, atomic theory was not fully established in

Davy's time, and certainly if the modern concept {potassium} encompasses that potassium commonly exists as a mixture of largely K^{39} and K^{41}, but also with a small proportion of the radioactive isotope K^{40}, then Davy's concept {potassium} was not equivalent to what might be considered the canonical chemical concept {potassium} today.

That is not the same as saying that Davy's potassium was materially different from what would be considered potassium today. Empirically we cannot know whether Davy's samples had the isotopic composition of samples formed today. Theoretically, it seems obvious that the isotopic composition of potassium has not substantially changed over a few centuries (*i.e.*, there is no known feasible mechanism that could lead to such change), but there is no way to test Davy's samples today. From a chemical perspective, there is no reason to doubt that the potassium Davy produced was a mixture of K^{39} and K^{41} with a much smaller level K^{40}. The point being made here is not that the material samples judged to be potassium today are any different from those examined in Davy's laboratory, but rather that just because the same term, potassium, is used, that does not mean they are conceptualised the same way. The {potassium} concept of contemporary chemistry evolved, with modifications, from the concept {potassium} deriving from Davy's own investigations.

5.4.4 Potassium *sans* Isotopes

From the perspective of modern chemistry, then, we can *infer* that Davy's potassium had much the same isotopic composition as any sample made by similar processes (from similar starting materials) today. However, it could be argued that there is a sense in which Davy's samples of potassium cannot be said to have an isotopic composition because isotopic composition is not something directly observable, and so given, in nature, but rather is the outcome of particular manipulations and conditions that allow certain measurements to be made (Shapin and Schaffer, 2011; Chang, 2012). These manipulations and conditions were not available to Davy.

This is the kind of argument that sometimes leads to some natural scientists having little time for sociology or philosophy of science. However, such claims can be based on detailed analysis of historical and contemporary cases of scientific discovery, which show that as procedures evolve and become standardised, objective scientific results emerge that can only be fully understood within the wider framework of the system (of scientists, laboratories, apparatus, techniques, standards, theories, grants, journals, *etc.*) that produces them. All scientific results are in a sense constructions rather than simply discoveries.

When one reads some of the case studies that have been undertaken by historians, sociologists, and anthropologists exploring how science actually progresses (Latour and Woolgar, 1986; Latour, 1987; Collins, 1992; Knorr Cetina, 1999; Mol, 2002; Shapin and Schaffer, 2011) such a perspective can be very convincing. This should not lead to the abandonment of a belief in science as an objective activity, and the adoption of a view discussed by some commentators (Habermas, 1988) that science has no more claims to objective truth than other cultural activities – but it does remind us that all scientific results are only meaningful in the context of a particular theoretical background, including theories of instrumentation, technical compromises, experimental errors, analytical approximations, *etcetera*.

This example of the chemical concept {potassium} changing as new scientific investigations become possible can be reinforced with many other examples. One here will suffice – that of crystal structure. Just as Davy had no way of ionising potassium atoms and measuring subsequent deflections in magnetic fields, he also lacked any means to subject samples of potassium to beams of X-rays and to detect diffraction patterns produced due to the radiation interacting with the regularly spaced atomic centres in the metallic structure. However, in the twentieth century, X-ray diffraction was used to attempt to establish the crystalline structure of the lattice. Posnjak reported:

> While not definitely established, it is probable that potassium crystallizes in the cubic system. A careful determination of the density of the alkali metals has been made...giving for potassium at 20° the value of 0.862. Introducing this value in the usual equation...we find that $n = 2.02$. This is within experimental error a whole number, as it should be. Accepting, conversely, that two atoms are associated with the unit cell of potassium, the density of potassium calculated by the above formula assumes the value 0.851, which is in good agreement with the direct determination. The only possible cubic arrangement for two like atoms in the unit cell is the body-centered one... (Posnjak, 1927, p. 357)

So, if we do not quibble about the difference between the integer 2 and the measured value 2.02, or that between 0.851 and 0.862, the evidence suggested that potassium had a body-centred cubic lattice (a structure inherently less compact than the face-centred cubic or hexagonal close-packed structures found in many metal crystals). So, again, something new was learnt about potassium, so that the chemical concept {potassium} was again modified. More associations were created embedding {potassium} more firmly in a net of related concepts, and increasing the 'content' of the concept (as this term was used in Chapter 2). The concept {potassium} not only refers to a low-density metal extracted from potash, but one with two common isotopes and a less common radioactive isotope, and a body-centred cubic crystal structure (at least at 20 °C). This is different from Davy's concept of potassium that necessarily lacked these associations.

In relation to these particular aspects, it seems the chemical concept {potassium} today is different from the {potassium} concept immediately after Davy's work, but without being inconsistent with it. In this case it also seems likely that any substance that would be judged to be a sample of potassium applying one of these two versions of the concept would also be

The Origin of a Chemical Concept: The Ongoing Discovery of Potassium

considered potassium when applying the other. However, given that concepts are used to make discriminations, the contemporary canonical {potassium} concept offers more conceptual content to support such discriminations (what is, and what is not, potassium) than the {potassium} concept that Davy was able to develop.

The contemporary concept is not in disagreement with Davy's concept (where they can both be applied), but rather is much more specific. Potassium will still have the same density or melting temperature as it ever did, but it can now also be discriminated through such means as X-ray data, mass spectrometry, *etcetera*. A more specific concept that does not distinguish phenomena differently from a less specific version suggests something about the entities the concepts can be applied to. This is indicated schematically in Figure 5.2. It seems the concept {potassium} describes an ontologically discrete cluster of entities within the diversity of entities observed in the universe.

Figure 5.2 is intended to imply that although Davy's concept {potassium} was more limited (and so in principle less exclusive) than the current chemical concept {potassium}, it was sufficient to support the same discriminations. That is, the characterisation of potassium in Davy's time through such properties as melting temperature was sufficient to distinguish potassium from anything that is not potassium (which we might denote {potassium}). This would not necessarily be expected in all cases of increasing

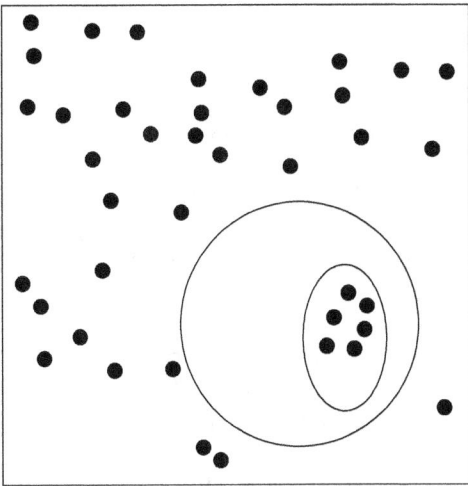

Figure 5.2 The developing conceptualisation of potassium. The outer box nominally represents the universe, which contains a range of things (a small sample of which are shown as dots). The circle represents Davy's concept of potassium – that supported a discrimination between samples of potassium (things with certain specified characteristics) and other things (*i.e.*, {potassium} – anything 'not potassium'). Since Davy's time, the concept has developed to include additional characteristics that further specify (*i.e.*, the oval) what counts as a sample of potassium.

conceptual content, but reflects the ontological basis of the classification of elements as being discrete in nature. So, by comparison, advances in anthropology have led to shifts in the way hominid species are conceptualised such that the same fossil bone specimens might have been be assigned to different species at different points in the development of the concepts {Homo sapiens}, {Homo habilis}, {Homo erectus}, *etcetera* (Lewin, 1989).

5.4.5 Developing the {Metal} Concept

Another chemical concept that was mentioned in the account of Davy's discovery of potassium was that of {metal}. Davy recognised that properties of potassium and sodium matched those of other metals, suggesting that these new substances were metals, so that the concept {metal} applied to them. Yet potassium (and for that matter sodium) were not typical of metals known at that time – besides density, referred to in Siegfried's account (quoted above), relatively low melting points, and high reactivity with air and water are obvious examples of how these new metals were unlike iron or copper or most other metals then known. (Of course, mercury was known by the alchemists, and it has an even lower melting temperature than potassium.)

So, Davy faced the genuine question "should the bases of potash and soda be called metals?" (Siegfried, 1963, p. 250). The choice of the word 'should' offers an interesting ambiguity here. If we consider the concept {metal} to be an accurate description of a kind of natural entity (a natural kind, as discussed in Chapter 3), then the question becomes:

- "should potassium and sodium be called metals [because they are metals]?": that is, the answer is 'yes' if we decide that potassium and sodium belong in the natural category of metal.

However, if we accept that concepts are mental ways of organising experience, then 'should' does not reference some inherent ontology existing in nature (*e.g.*, there are metals) and the question facing Davy becomes:

- "should potassium and sodium be called metals [because they are best understood as metals]?": that is, the answer is 'yes' if we decide that the most useful way of conceptualising potassium and sodium is within the category we conceptualise as metals.

Siegfried writes that "In spite of their low density, [Davy] decided that they should [be called metals], for all their other properties were analogous to that class of bodies" (Siegfried, 1963, p. 250). The implication here is that 'metals' was a class of bodies with certain common characteristics. Analogy is a common technique used by scientists, that involves looking for structural similarities in different systems (Hesse, 1959). If the properties of these new substances were mapped onto those typical of those substances already admitted into the category (*i.e.*, as characterised by the existing {metals} concept) then the potential analogues (potassium, sodium) failed to map onto the target in terms of one important property, *i.e.* density, but mapped well onto the target in terms of other properties considered relevant, so potassium and sodium 'should' – on balance – be considered metals by analogy with the other substances considered as metals.

Even if sodium and potassium had been newly discovered metals with properties that fitted well with the range of characteristic properties of previously known metals, the identification of sodium and potassium as metals would have modified the concept {metals} a little by expanding the known examples of metals – the concept {metal} would have new associations with other concepts ({potassium}, {sodium}): for example, 'potassium is an example of a metal' (see Chapter 2). So, the content of the canonical {metal} concept would have been enlarged. However, the identification of these relatively low-density substances as metals modified the chemical concept {metals} somewhat more substantially by admitting two more elements that were less prototypically metals.

This chapter has looked in some detail at aspects of the origin of one chemical concept, {potassium}, which might be considered to refer to a natural kind, something that is presented to us in nature as an inherent category: as a grouping that reflects some particular essence (LaPorte, 2004). Even in such an 'elementary' case it appears the origin of the modern chemical concept is not straightforward. The notion of 'natural kinds' is explored in more detail in Chapter 7. Before that, the next chapter discusses the example of another familiar concept, {acid}, where the concept's development cannot be seen largely as simply adding new associations to the canonical concept.

References

Anon, (1923), The Nobel Chemistry Prizes for 1921 and 1922, *Sci. Progr. Twentieth Century (1919–1933)*, **17**(68), 635–636.

Aston F. W., (1920), Mass-spectra and the atomic weights of the elements, *Sci. Prog. Twentieth Century (1919–1933)*, **15**(58), 212–222.

Campbell N. R., Wood A. and Moulin M., (1907), La radioactivité des métaux alcalins, *Radium*, **4**(5), 199–202.

Chang H., (2012), *Is Water H_2O? Evidence, Realism and Pluralism*. Dordrecht: Springer.

Collins H., (1992), *Changing Order: Replication and Induction in Scientific Practice*. University of Chicago Press.

Fensham P. J., (2004), *Defining an Identity: The Evolution of Science Education as a Field of Research*. Dordrecht: Kluwer Academic Publishers.

Gilbert J. K., (1995), Studies and Fields: Directions of Research in Science Education, *Stud. Sci. Educ.*, **25**, 173–197.

Glasersfeld E. V., (1989), Cognition, construction of knowledge, and teaching, *Synthese*, **80**(1), 121–140.

Gobet F., Lane P. C., Croker S., Cheng P. C., Jones G., Oliver I. and Pine J. M., (2001), Chunking mechanisms in human learning, *Trends Cognit. Sci.*, **5**(6), 236–243.

Habermas J., (1988), *On the Logic of the Social Sciences*. Cambridge: Polity Press.

Harkins W. D. and Liggett T. H., (1923), The discovery and separation of the isotopes of chlorine, and the whole number rule, *J. Phys. Chem.*, **28**(1), 74–82.

Hesse M. B., (1959), On Defining Analogy, *Proc. Aristotelian Soc.*, **60**, 79–100.

Knorr Cetina K., (1999), *Epistemic Cultures: How the Sciences Make Knowledge*. Cambridge, Massachusetts: Harvard University Press.

Laland K. N. and Hoppitt W., (2003), Do animals have culture?, *Evol. Anthropol.: Issues, News Rev.*, **12**(3), 150–159.

LaPorte J., (2004), *Natural Kinds and Conceptual Change*. Cambridge: Cambridge University Press.

Latour B., (1987), *Science in Action*. Cambridge, Massachusetts: Harvard University Press.

Latour B. and Woolgar S., (1986), *Laboratory Life: The Construction of Scientific Facts*, 2nd edn, Princeton, New Jersey: Princeton University Press.

Lewin R., (1989), *Bones of Contention: Controversies in the Search for Human Origins*, London: Penguin.

Matthews M. R., (2000), Appraising constructivism in science and mathematics education, in Phillips D. C. (ed.), *Constructivism in Education: Opinions and Second Opinions on Controversial Issues*, Chicago, Illinois: National Society for the Study of Education, pp. 161–192.

Miller G. A., (1968). The magical number seven, plus or minus two: some limits on our capacity for processing information. *The Psychology of Communication: Seven Essays*. Harmondsworth: Penguin, pp. 21–50.

Mol A., (2002), *The Body Multiple: Ontology in Medical Practice*. Durham, NC: Duke University Press.

Piaget J., (1970/1972), *The Principles of Genetic Epistemology* (W. Mays, Trans.). London: Routledge & Kegan Paul.

Popper K. R., (1934/1959), *The Logic of Scientific Discovery*. London: Hutchinson.

Popper K. R., (1989), *Conjectures and Refutations: The Growth of Scientific Knowledge*, 5th edn, London: Routledge.

Schwab J. J., (1958), The Teaching of Science as Inquiry, *Bull. At. Sci.*, **14**(9), 374–379.

Shapin S., (1992), Why the public ought to understand science-in-the-making, *Public Underst. Sci.*, **1**(1), 27–30.

Shapin S. and Schaffer S., (2011), *Leviathan and the Air-Pump: Hobbes, Boyle, and the Experimental Life*. Princeton, New Jersey: Princeton University Press.

Siegfried R., (1963), The Discovery of Potassium and Sodium, and the Problem of the Chemical Elements, *Isis*, **54**(2), 247–258.

Smythe W. R. and Hemmendinger A., (1937), The Radioactive Isotope of Potassium, *Phys. Rev.*, **51**(3), 178–182.

Taber K. S., (2009), *Progressing Science Education: Constructing the Scientific Research Programme into the Contingent Nature of Learning Science.* Dordrecht: Springer.

Taber K. S., (2014), *Student Thinking and Learning in Science: Perspectives on the Nature and Development of Learners' Ideas.* New York: Routledge.

The Royal Society, (2011), *Neuroscience: Implications for Education and Lifelong Learning.* London: The Royal Society.

CHAPTER 6

Conceptualising Acids: Reimagining a Class of Substances

The idea of acids has been used since ancient times. The primary characteristic of an acid was its sour or sharp taste, and vinegar (*i.e.*, the material produced when wine soured) was the most common example. It was also recognised that when acids were mixed with some other substances (such as lime), a spontaneous bubbling, efflorescence, was produced. By the time of Robert Boyle, it was generally considered that acids were substances that had a sour taste and produced efflorescence with alkalis (Ruthenberg and Chang, 2017).

6.1 Formation of a Concept of Acids

So, at some point, the idea that vinegar had a particular kind of taste, and produced a particular effect when poured on lime, was developed into a category of a type or class of substances, 'acids'. The {acids} concept could be used to make discriminations based on the two characteristics of taste and reaction with alkalis (relying on another chemical concept, {alkalis}, that also allowed discriminations of membership/non-membership of a group of substances).

Acids were also found to change the colour of certain vegetable dyes from blue to red. The admission of this association into the concept {acid} paved the way for certain dyes to be used as indicators – and so provided a link with a new concept {indicator} as a class of substance allowing acids to be identified. The content of the {acid} concept (see Chapter 2) was increased.

Conceptualising Acids: Reimagining a Class of Substances

It seems that if there could be a clear answer to the question of when acids were discovered, then it is unlikely to be accessible to us with our limited historical records. At the point where it was recognised that there was a class of substances, including, but not limited to vinegar, which had sufficient in common (they tasted 'sharp' like vinegar) to be worth considering as similar in some pertinent way, people were using an acid {concept}.

A sour taste and reaction with alkalis is certainly still part of a modern {acid} concept, but clearly the concept of {acid} used by chemists today goes beyond taste, reaction with alkalis, and even the potential for detection with indicators. Indeed, in terms of laboratory work, the original defining characteristic of acids (by taste) is seldom applied on health and safety grounds, and perhaps would no longer be a very good basis for discriminating the set of substances currently considered acids.

6.2 Theoretical Elements of the {acid} Concept

Concepts may be based on perceptual features alone. For example, although the notion of a sphere can be defined and characterised in various ways (a solid body, minimal volume, no apices or edges, all radii equal, constant curvature of the surface, obeying particular formulae for surface area and volume...), it is possible for someone to form a concept {sphere} based on noticing perceptual similarity between a range of objects that appear spherical. This would be enough to make discriminations of those objects seen to be similar in this way, as opposed to other objects (*i.e.*, what we might *describe* verbally as distinguishing spherical objects from other, namely, non-spherical, objects – although having the ability to describe the process in language is not essential for this).

Science has been characterised as going beyond empirical observations to relate what is observed to some level of theoretical interpretation that can support explanation and prediction (see Chapter 5). Even before modern scientific work, the acid concept had acquired a theoretical component. Acids were considered to contain particles that had sharp points, thus explaining their sharp (*i.e.*, sour) taste. Chang (2012a, p. 691) characterises the definition of an acid given in Dr. Johnson's dictionary of the English language (1755) as "an interesting mix of the everyday concept of acidity and a mechanistic metaphysical notion", comprising of the point particles, the 'piercing' taste, and its action on the colour of syrup of violets (a mixture made by heating – but preferably not boiling, as that spoiled the taste – the flowers of violets with sugar in water).

Although the notion of a sharp substance containing sharp particles can be considered theoretical (there was no empirical observations of such pointy particles), it seems to fall short of being scientific. Indeed, it almost seems a magical conception, but it does seem intended as an actual mechanical explanation (there are particles, they are pointy, and they are sharp enough to penetrate human tissue). It is interesting that 'sharp taste implies sharp particles' reflects a common alternative conception that many learners develop that the macroscopic properties of materials can be explained by positing that the material is comprised of particles with those same properties (*e.g.*, Taber, 2001) – so glass particles would be transparent for example (Ben-Zvi *et al.*, 1986). More scientific notions of what made something an acid looked to relate the class of substances considered acids with properties that could be demonstrated in the laboratory.

In discussing how two (then) relatively recent notions of the acid (those labelled as Brønsted–Lowry and Lewis, as discussed below) were related, Kolthoff put it thus:

> When the properties of substances become known and it appears that several substances have properties in common, they are usually classified as a group. A definition based upon the composition or the common properties of the substances is then given to characterize such a group. In the interpretation of the properties of such a group of substances it is usually necessary to develop another definition which is based not upon facts but upon conceptions or theories. (Kolthoff, 1944, p. 51)

Kolthoff immediately went on to comment: "Most definitions are usually more or less limited in scope and sometimes ambiguous. This is especially true of definitions which have been given to characterize the categories of acids and bases" (p. 51).

6.3 Lavoisier's Theoretical Account of Acids

Lavoisier's chemistry (or, perhaps, the Lavoisiers' chemistry, as Madame Lavoisier, Marie-Anne Pierrette Paulze, was his coworker as well as his wife, and it is not clear how much credit she deserves for 'his' ideas) is best remembered for the conceptualisation of combustion as linked to oxygen rather than phlogiston – even if the abandonment of phlogiston was accompanied by a role for the equally mercurial caloric (Chang, 2012b). However, it has been argued that a theory of acids was central to the Lavoisier chemical theory. As with his ideas about combustion, oxygen was a focus. Oxygen was seen as the 'acid generator' or generator of sharpness (taken as synonymous with sourness). Today, Lavoisier is much better remembered for his ideas about the role of oxygen in combustion (largely coherent with contemporary chemical thinking) than for those about the role of oxygen in acidity.

For Lavoisier, acidity depended upon the presence of oxygen, or an oxygen principle, that could be present to different degrees. That is, some compounds or radicals would have higher levels of oxygen than others, and higher levels were associated with acidity. So, sulfuric acid was more acidic than sulfurous acid because it had a higher level of oxygen (in today's terms, H_2SO_4, *cf.* H_2SO_3). According to this theory, the presence of oxygen was not enough to make something acid; but oxygen was a prerequisite for acidity. That is, the presence of oxygen was a necessary but not sufficient condition for acidity.

Conceptualising Acids: Reimagining a Class of Substances

Now, in retrospect, this may seem a rather dubious theory as one of the acids well known in Lavoisier's time, and among the best known to school children today, was what was then called muriatic acid, or what is known today as hydrochloric acid. Muriatic acid clearly fell within the category created by the {acids} concept in terms of its chemical properties, but did so despite an absence of oxygen.

However, that would be an ahistorical judgement, as although no one had shown muriatic acid did include oxygen during Lavoisier's lifetime (or since, of course), neither had it then been demonstrated definitively to be an exception. Lavoisier was aware of a good many acids known to include oxygen, and there were some examples where the inclusion of oxygen was yet to be demonstrated, but was to be expected from the theory. So, what seems a refuting example with the gift of modern chemical hindsight, was at the time covered by the principle that absence of evidence cannot be taken as evidence of absence.

So, we might suggest that this {acid} concept included both a principle that acids contained oxygen and (in apparent contradiction) included the example of HCl as one of the acids: yet that is mischievous as the notion of 'HCl' is a modern updating. It is better, then, to suggest that this {acid} concept included both a principle that acids contained oxygen and included the example of muriatic acid (now recognised as HCl) as one of the acids. Later, Davy (who's work in relation to the discovery of potassium, was discussed earlier in Chapter 5) demonstrated that muriatic acid did not contain oxygen, and suggested that some substances considered acids only actually became acids when dissolved in water. Kolthoff (1944) noted how in 1815 Davy "rejected the view that iodine pentoxide is an acid and maintained that it becomes an acid only by combination with water" (p. 52). So, in this sense, HCl is not an acid, even if we consider it an *acidic* gas that when dissolved gives hydrochloric acid, $HCl_{(aq)}$, in aqueous solution.

Crosland (1973) has noted "the main acids known, or at least the main acids analysed, in Lavoisier's time were the oxyacids. In modern terms, therefore, one might restate a part of Lavoisier's theory by the apparent tautology that 'all oxyacids contain oxygen'" (p. 307). However, Crosland also notes that "[a]ny proper theory of acidity had to account not only for the resemblances between acids but also their differences. Lavoisier achieved this by means of a dualistic theory, considering acids as compounds of oxygen (responsible for their acidity) and a radical (responsible for the peculiar characteristics of the acid)" (p. 317). In some ways, this dualist conception reflects the way acids are often presented in schools today (although with a focus on hydrogen rather than oxygen).

Clearly Lavoisier's concept of acids went beyond the everyday concept: it was a theoretical concept that was meant to explain a good deal of empirical evidence, and could be used as the basis for further laboratory investigations. So, this {acid} concept should be considered a scientific concept. Lavoisier's ideas were not universally accepted by his contemporaries, but were widely influential, and indeed are often considered the basis of the 'revolution' that led to modern chemistry (Siegfried, 1988). So here there is a chemical concept of {acids} that had been 'discovered', albeit somewhat at odds with canonical chemistry today.

There are parallels with this case and that of the 'discovery' of potassium discussed in the previous chapter – both involve the conceptualisation of an ontological category of substance. However, where potassium was considered a single substance, the {acids} concept was always seen as referring to a class or group of substances.

6.4 Acids and Hydrogen

More familiar to modern school children is a quite different theoretical notion of acids, that due to Arrhenius, that associates acidity not with oxygen but with hydrogen. The Arrhenius model suggests acids give rise to hydrogen ions in aqueous solution. Here, then, is a distinct chemical concept {acids}, quite different from that of Lavoisier.

Kolthoff (1944) describes how following the work of Davy, as discussed above, and Dulong (perhaps best known today for the Dulong–Petit law relating to specific heats), Liebig (immortalised through the Leibig condenser, based on a design that he helped develop) offered a definition of acids based on hydrogen rather than oxygen. This reflected the empirical evidence then available, and was then developed theoretically by Arrhenius:

> [it was] Liebig, who in 1838 defined acids as "compounds containing hydrogen, in which the hydrogen can be replaced by metals." This definition, in a somewhat extended form, became generally accepted. It is based upon *experimental facts* referring to *composition* (acids contain hydrogen) and *reactivity* (hydrogen replaced by metals; acid and base react with formation of salt and water).
>
> In the qualitative and quantitative *interpretation* of acidity and basicity it appeared that the classical definition was too limited. In order to account for the properties of acids and bases it was necessary to formulate certain *conceptions*. This was done by Arrhenius, whose definition of acids and bases is based on the *theory* that acids when dissolved in water dissociate into hydrogen (hydronium) ions and anions, and that bases when dissolved in water dissociate into hydroxyl ions and cations. (Kolthoff, 1944, p. 52, italics in original)

The differences between the Lavoisier {acid} concept and the Arrhenius {acid} concept raise the question of whether it makes sense to talk about the 'discovery' of these concepts. Acids may be discovered, but can concepts? If we think that acids are substances that dissolve in water to produce hydrogen ions it may make sense to ask how *this* ('fact') was discovered to be the case: yet by the same token, we would probably not today ask how Lavoisier *discovered* that all acids included oxygen (as we know of exceptions, and do not see this as a valid notion). It seems a perversion of the usual use of language to ask how someone discovered something if we do not accept the discovery.

We might be tempted to put the word discovery in ironic speech marks: how did Fleischmann and Pons 'discover' cold fusion? How did Prosper-René Blondlot 'discover' N-rays? How did Nikolai Fedyakin 'discover' polywater? How did Lavoisier 'discover' the oxygen principle that was a key feature of his {acid} concept?

6.5 Current and Historical Scientific Concepts

If we judge Lavoisier's theory (and so his {acid} concept} flawed, but consider Arrhenius's theory (and so his {acid} concept) as canonical, or, at least, more canonical, we might have fewer reservations about asking how Arrhenius came to *his* discovery. To do this we have to occupy a historical moment. There are plenty of ideas in science that have at one time been considered canonical, but have later been discredited and de-canonised; plenty of generalisations and law-like principles that have been later found to be approximations or have limited ranges of application (*e.g.*, the central dogma of molecular biology, Crick, 1970) and plenty of ideas that may have seemed heretical (*e.g.*, jumping genes, McClintock, 1950) that have later entered the canon.

If we are to adhere to a strict scientific attitude, then we need to adopt a humility about the current scientific canon. All scientific ideas, no matter how widely accepted, no matter how well supported by our current interpretations of broad bodies of evidence, need to be seen as somewhat provisional; as potentially revisable; as, in principle, open to reconsideration in the light of new evidence or of new ways of conceptualising the existing evidence. To think otherwise, would be scientistic arrogance. So, if we feel that one can only actually discover what actually exists, then we should perhaps avoid referring to the *discovery* of theoretical entities.

6.5.1 Inventing New Concepts

It might make more sense to suggest that the processes involved in developing chemical ideas are not so much matters of discovery than of invention. So, Lavoisier invented a way of thinking about combustion and acidity, which entails one {acid} concept, and Arrhenius invented a different way of thinking about acids, which entails a different {acid} concept. The Arrhenius {acid} concept is no less the result of an act of imagination, even if it was informed by many additional years of chemical research and looks more 'right' to the modern chemist's eye.

The finding that a number of substances have the sour, or sharp, taste associated with vinegar can perhaps be considered a discovery (*i.e.*, an empirical discovery). The conceptualisation of this pattern in terms of a category of substances, including vinegar, but also encompassing other substances found to be like vinegar in this sense – acids collectively – as sour substances, was not a discovery, so much as an invention. It was the imagining of a possibility. As far as that goes, it might be argued that a concept that is little more than forming a category of sour substances is not so different from forming concepts of blue objects or noisy sounds (*i.e.*, perceptual, and atheoretical). A concept based on such observable qualities familiar in everyday life has been labelled a quotidian concept (Chang, 2012a).

Both Lavoisier and Arrhenius went beyond the quotidian concept, to offer scientific concepts, indeed chemical concepts, as their models offered accounts of *why* one substance would be an acid, but not another. Their concepts were therefore less like a notion of blue objects than an explanation of why blue objects have this perceptual similarity. These concepts were inventions in the sense that someone, drawing upon known chemical data, imagined what might make some substances acids, and posed a hypothetical construct that could be tested by further laboratory work. Their theories were different, and moreover mutually inconsistent. Arrhenius's concept {acid} includes hydrochloric acid, as does the quotidian concept, whereas Lavoisier's concept {acid}, would (it is now recognised) exclude hydrochloric acid.

This might suggest a judgement that the Arrhenius concept {acid} is scientific whereas the Lavoisier alternative is not; or that the Arrhenius version is 'more scientific' than the Lavoisier concept. Yet that takes advantage of hindsight, where surely what makes a concept scientific is its development though the iterative processes of science: perhaps something like observe, reflect, imagine, consider, test, repeat *ad nauseam*; or, at least, observe, reflect, imagine, consider, test, repeat-as-indicated.

It seems more appropriate to recognise the Lavoisier {acid} concept as a chemical concept, albeit one that is no longer canonical – no longer part of the accepted canon of the discipline. So, we might distinguish between concepts that are scientific and those that are not (the former being formed through the processes of science); and then also between those scientific concepts that are *historical* and those that are *current*. We should also make another discrimination, and not necessarily equate a current scientific concept with a canonical one. There is a time lag between the proposal of a scientific concept, and its admission into the canon, and prior to that point it will not be known whether the concept will come to be seen as canonical. So, many concepts that are formed (*i.e.*, invented) by scientists and that derive from the systematic, iterative processes of science (and so, in the usage here, are scientific concepts), never become canonical. In retrospect, we might judge these as historical, but whilst they are still current they remain candidates for being adopted by the community of scientists.

6.5.2 Concepts Reaching Canonical Status

That is, once a concept is, through the consensus of a scientific community, rejected as flawed or simply unhelpful, then we can see it as no longer current. That community might be the community of chemists if the concept clearly falls under the discipline of chemistry; or perhaps by the community of biochemists, or some other subgroup, if it is a more specialised concept.

So, there is an asymmetry here. All canonical chemical concepts are current (as being canonical means they are retained as part of the canon of the discipline) but not all current chemical concepts – those with currency within the dialogue maintained by recently published research papers and other forms of disciplinary discourse – are canonical. Yet, to be considered current, when not canonical, is to be a candidate for admission into the canon, as once a point in time has been reached where it is clear that this will not happen (N-rays, polywater), the currency has passed.

This treatment assumes that there are canonical concepts that are adopted by consensus within a disciplinary community. This seems a reasonable assumption, at least within research traditions (Kuhn, 1970), yet considering the nature of concepts as mental entities (as suggested in Chapter 2) raises questions about how robust such an assumption is. I will address that question later in the book (in Chapter 9), but trust it will not be too much of a spoiler to reveal here that despite the notion of canonical concepts presenting some problems, I retain is as a useful referent. The notion of canonical concepts conveys a sensible meaning, and does useful work when discussing the teaching and learning of chemical concepts.

Current concepts that are not canonical are not necessarily those that will never be canonical, but rather a mixture of those that will become widely adopted, and those that will fall into disuse without ever having achieved canonical status. Which of these candidates will become promoted to canonical status is (and must be) an open question. It might be helpful to have a term then to distinguish these levels of currency – such as canonical concepts and mooted concepts.

So, there are three relevant distinctions here:

(i) everyday concepts and scientific concepts;
(ii) historical (scientific) concepts and current (scientific) concepts;
(iii) canonical (current scientific) concepts and mooted (current scientific) concepts.

Historical, as well as current, scientific concepts either may, or may not, attain canonical status (although for understandable reasons it is mostly those historical concepts that were at one time canonical that are widely known today). Scientific concepts therefore fall on a grid (see Figure 6.1).

The apparent symmetry of Figure 6.1 is likely to be broken by the way we would use these descriptors. A scientific concept we would label mooted has never yet reached canonical status – but if it is a current concept this may change. For a historical mooted concept that never reached canonical status to now become canonical it would have to be rediscovered (or reinvented or reclaimed) by modern chemists and so shift to once again being current and mooted before it could become canonical. This raises the issue of a whether

scientific concepts	mooted	canonical
historical		
current		

Figure 6.1 A classification of scientific concepts.

a currently mooted concept could ever really be the same as, rather than just in some ways reminiscent of, a historical concept, given how intervening advances in scientific knowledge would inevitably colour a current conceptualisation (*e.g.*, consider how the {potassium} concept is inevitably different today from what it was to Davy, as discussed in Chapter 5).

6.5.3 Extending the Concept of Acid

If the Lavoisier {acid} concept is historical and not currently canonical, we might ask whether we would identify the Arrhenius {acid} concept with the canonical chemical {acid} concept. Readers of this book are perhaps less likely to think of Arrhenius when reflecting on the chemical concept of acid (perhaps associating him more with kinetics and his rate equation) and probably more likely to think of the Brønsted–Lowry and Lewis definitions of acids.

If we consider that chemistry as a science is progressive (Lakatos, 1970), then the Brønsted–Lowry and Lewis theories should presumably be considered advances on Arrhenius, in the way we might judge the Arrhenius {acid} concept, let us denote it $\{acid_{Arrhenius}\}$, as an improvement on $\{acid_{Lavoisier}\}$. If we think progress in science is identified with some kind of closer approach to the 'truth' (Miller, 1974) of how nature is (which is perhaps how many scientists themselves think), then the currency of the Brønsted–Lowry and Lewis definitions of acids would lead us to assume they are 'better' than the Arrhenius version in offering an account of the truth of how nature is. They should support an {acid} concept (or concepts, as discussed below), which allows (objectively) 'better' discriminations of acids from acids (non-acids). That is, we should make fewer errors, either false positives or false negatives, so it becomes less likely that chemists either identify as an acid a substance that is 'not really' an acid, or fail to recognise some known substance that is 'really' an acid as an acid. (However, as suggested above, this implies there is a correct way to conceptualise nature – a way things really are – that scientists seek to uncover, when it may sometimes be more sensible to consider scientists to be seeking the most useful or productive conceptualisations rather than absolutely correct ones.)

Conceptualising Acids: Reimagining a Class of Substances 97

So just as the {acid$_{Arrhenius}$} concept is an advance over the {acid$_{Lavoisier}$} concept as it supports more accurate discriminations between acids and acids (non-acids – *e.g.*, muriatic acid would be excluded from being considered an acid by a modern interpretation of Lavoisier's definition, but not by Arrhenius's definition), then we might expect that we could find similar examples to show that the theories of Brønsted–Lowry and/or Lewis discriminate between acids and acids more accurately than the Arrhenius theory, and so provide the basis for a better conception of 'acid': a better chemical {acid} concept. It is worth noting at this point that I will be suggesting below that *in this particular sense* it is not clear that chemistry has made progress in modifying the concept {acid}. We need to be wary here of potential circularity. To decide whether one definition or another offers a better basis for discriminating acids from acids, we need to have some independent way of making those discriminations that is authoritative.

6.5.4 There Is Progression from the Lavoisier Concept to the Arrhenius Concept

So, we might wish to compare the Lavoisier {acid} concept and the Arrhenius {acid} concept. To be fair to Lavoisier, we should consider associations accepted as part of the canonical acid concept at the time he was working (Kolthoff, 1944, p. 51):

> "Lewis (1746) [note the date – not the Lewis of the Lewis acid model discussed below] gave a more elaborate characterization of acids, bases, and salts which in many respects is similar to a more modern definition: Acids are substances which (upon proper dilution) taste sour, which effervesce with chalk and alkali carbonates, and which form salts with such substances. Acids turn syrup of violets red, while alkalies turn this syrup blue (7)."

We might note that three of these characteristics can be observed perceptually (taste, effervescence, colour change), and one relies on less direct means of chemical analysis (forming salts with alkalis). Muriatic acid met these criteria, so fell within the category defined by this concept. Lavoisier suggested this was because it had a high oxygen content, and Arrhenius because it gave rise to hydrated hydrogen ions when dissolved in water. Here, then, we have a substance that is unambiguously an acid, but that fits one theoretical account and not the other. The Arrhenius {acid} concept, {acid$_{Arrhenius}$}, can therefore be considered a theoretical advance over the Lavoisier {acid} concept, {acid$_{Lavoisier}$}, as the theoretical account better matches the empirical distinction, and so the shift between these ideas is progressive. (I should point out that I am illustrating this argument with one example, and anyone with a serious interest in the question of whether {acid$_{Arrhenius}$} can be considered an advance over {acid$_{Lavoisier}$} should not be convinced by examining only one example.)

6.5.5 Have We Progressed Beyond Arrhenius in Discriminating Acid from Acid (Not Acid)?

The next question therefore becomes whether the $\{acid_{Brønsted-Lowry}\}$ or $\{acid_{Lewis}\}$ concepts offer similar advances over $\{acid_{Arrhenius}\}$. To consider this, we have to appreciate why Lowry and Brønsted, and Lewis, offered revised accounts of acids. Kolthoff explains that the proposals of Lowry and Brønsted sought to extend the class of substances that would be considered acids compared with $\{acid_{Arrhenius}\}$, given that chemists had explored solution systems based on solvents other than water (such as liquid ammonia). Therefore, the Arrhenius definition that was part of the $\{acid_{Arrhenius}\}$ concept,

> appeared too limited when phenomena of acidity and basicity were studied in non-aqueous solvents. This led Lowry and Bronsted to the more general definition which is based upon the conception that an acid may be considered as a combination product of a proton with a base, while a base is defined as a substance which can combine with a proton to form an acid. This definition limits the group of acids to substances which contain a transferable proton (Kolthoff, 1944, p. 52).

This might be characterised as saying that Bronsted and Lowry suggested that there were other chemical systems involving substances dissolved in non-aqueous solutions that *behaved analogously* to how acids behaved when dissolved in water, as proton donors, and that *by analogy* these should also be called acids.

It seems a choice was made here that was not inevitable: it was decided that these analogous systems were similar enough in nature to familiar acid–base systems that they should be considered as belonging within that category. Alternatively, it would have been possible to have recognised the similarity, but not looked to extend the scope of the {acid} concept (and so the {base} concept), and given these analogous reagents new categories: acidalogues and baselogues (or something more catchy), which would have introduced new concepts ({acidalogues}, {baselogues}), and left the then canonical {acid} concept substantially unaltered.

The 'substantially' here recognises that if we consider concepts to be embedded in extensively connected networks (see Chapter 2), this would still modify the {acid} concept somewhat by changing its pattern of associations/connectivity. The discovery/invention of the analogous systems would offer new associations that expanded the conceptual content of {acid}, but without fundamentally changing it. Centrally, it would not alter the discriminations that would be made with the canonical {acid} concept. This point will be developed in more detail for the Lewis proposal, which I now turn to.

Kolthoff explains the basis of Lewis's model to be somewhat similar considerations to those used to justify the Brønsted–Lowry development,

> G. N. Lewis was the first to point out that many substances which do not contain a proton react in aprotic media with bases with the formation of

compounds which have some properties similar to those of a salt formed by interaction of a neutral molecule (Brønsted) acid and a base. Lewis, quite rightly [sic], considers the Brønsted definition too narrow and bases his definition upon the view that a base is a substance which can donate an electron-pair to form a conjugate bond and that an acid is a substance which can accept an electron-pair from a base to form a coordinate bond. Neutralization occurs when an acid and a base combine (Kolthoff, 1944, p. 52).

This could be characterised in the following terms:

(i) Substances that are recognised to be acids react with substances in another class (bases) to produce substances in a third class (salts).
(ii) In some chemical systems, some other substances, that do not fall within our existing category of acids, react with bases to produce substances that can be considered similar to salts ('to have some properties similar to those of a salt').
(iii) So, let us extend the class of salts to include these compounds that have some properties similar to those of a salt, to give a more inclusive class of salt-like substances.
(iv) Then let us consider that any substance which reacts with a base to give a salt-like product is an acid. Now, by this definition, those 'other' substances that did not fall within our existing category of acids, but that react with bases to produce substances that are considered salt-like, should now be considered acids.

In effect, the concept {salt} has been substituted by the superordinate concept {salt like entities} that includes new substances now considered to in effect be salts as they seem analogous to established salts; and then we have modified the concept {acids} by an analogy masquerading as an established identity (*i.e.*, if it reacts with base to give a salt, it is an acid).

6.5.6 An Alternative History – Forming New Concepts Rather Than Expanding the Range of Existing Ones

By this last comment I mean that in point (iv) above the statement "neutralisation occurs when an acid and a base combine" seems at first sight uncontroversial: as we have seen this was a canonical understanding long before Brønsted, Lowry and Lewis, presented their ideas. Yet the products of neutralisation reactions had been limited to salts, and so the preceding point, (iii), shows this is not quite as it may seem. Prior to the Lewis proposal the term neutralisation referred to salt-generating reactions between acids and bases, so the concept {salt} was essential to the definition; yet in point (iv) a different concept is being used.

That is, a criterion that had been 'produces a salt' was relaxed to something like 'produces a salt-like substance'. This relies on the argument that the products of a reaction between a Lewis acid and a base, an adduct or complex, is sufficiently similar to a salt not to quibble over the distinction.

The chemical community seems to have accepted that argument, but one might imagine that this could have been contested and an alternative conceptualisation come to be preferred.

For the moment, I will instead introduce a new term, saltoids (by analogy with metalloids) for compounds that have some properties similar to those of a salt. Now we have

(i) (As before) Substances that are recognised to be acids react with substances in another class (bases) to produce substances in a third class (salts).
(ii′) In some chemical systems, some other substances, that do not fall within our existing category of acids, react with bases to produce substances that can be considered similar to salts. We could call these products *saltoids*.
(iii′) So, let us form a superordinate category of saltals to include both the established class of salts and also the saltoids.
(iv′) Then, as a salt is produced by the reaction of an acid with a base, let us consider by analogy that any substance that reacts with a base to give a saltal (that is, either a salt or a saltoid) an acid: by which definition those 'other' substances that did not fall within our existing category of acids, but which react with bases to produce saltoids, should now be considered acids.

This makes it clearer that any new substances are in effect being introduced into the category of acid by analogy rather than in terms of characterisation through the established definition of neutralisation. Previously, our defining characteristic was based on the reaction:

$$\text{acid} + \text{base} \rightarrow \text{salt}$$

That is, something that reacts with a base to give a salt is an acid. This implied that in the scheme

$$\text{some reactant} + \text{base} \rightarrow \text{saltoid}$$

the reactant is not an acid, as a saltoid is not actually a salt. Whereas, now, however, the defining equation for neutralisation has been modified to:

$$\text{acid} + \text{base} \rightarrow \text{saltal}$$

So something that reacts with a base to give a saltal (*i.e.*, *either* salt *or* saltoid) is an acid;

In which case

$$\text{some reactant} + \text{base} \rightarrow \text{saltoid}$$

is (as a saltoid is a saltal), subsumed under the more general equation

$$\text{some reactant} + \text{base} \rightarrow \text{saltal}$$

which defines this 'some reactant' as an acid, given the modified definition.

Conceptualising Acids: Reimagining a Class of Substances

This also suggests that rather than simply change the definition of acid (again), it would have been possible instead to have defined a new category of substance for those reactants similar to acids such that, with acids, these related compounds formed a more encompassing group (see Figure 6.2), that is,

(iv″) Then as a salt is produced by the reaction of an acid with a base, let us consider any substance that reacts with a base to give a saltal (that is, either a salt or a saltoid) an acidal: by which definition those 'other' substances that did not fall within our existing category of acids, but that react with bases to produce saltoids should now be considered acidoids, and so acidals.

So, we have the general rule that

$$\text{acidals} + \text{bases} \rightarrow \text{saltals}$$

In addition, we have the more specific rules that

$$\text{acids (one class of acidals)} + \text{bases} \rightarrow \text{salts (one class of saltals)}$$

$$\text{acidoids (another class of acidals)} + \text{bases} \rightarrow \text{saltoids (another class of saltals)}$$

Let me be clear that I am not for a moment suggesting that this is what *should* have been done: that instead of expanding the class of acids (and so fundamentally modifying the chemical concept {acids}), Lewis *should* have created ('invented') four new concepts (those here labelled {saltoids}, {saltals}, {acidoids}, {acidals}). Rather, the purpose of this analysis is (a) to make explicit the change obscured by the rhetorical 'sleight of hand' of apparently retaining an existing definition for neutralisation, but actually modifying it by widening the class of product; (b) to highlight that the decision to expand the class of acids (and so substantially modify the chemical concept {acid}) was only one of the options that could have been taken at that point. Instead, the existing {acid} concept *could* have been augmented by new concepts

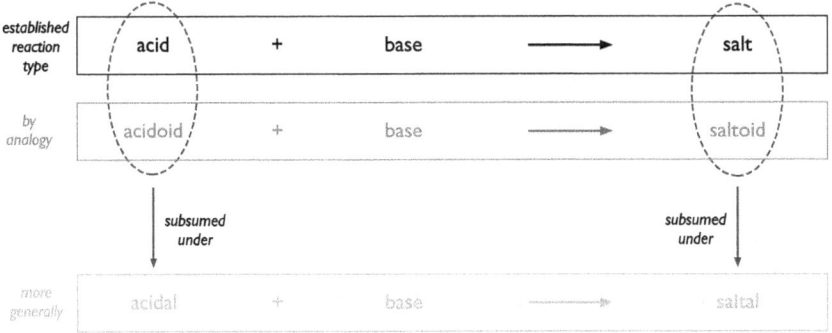

Figure 6.2 A counterfactual conceptualisation – new concepts could have been introduced, rather than making existing ones more inclusive.

rather than modified to admit in the class of acids substances that were not previously included.

6.5.7 Lewis Revised the {acid} Concept

By taking the more economical option Lewis arguably spared chemistry students having to learn additional concepts and their associated classes of substance. The serious question (and the point the preceding counterfactual history was intended to highlight) is how Lewis changed the chemical concept of {acid} through this proposal.

His suggestion could be understood to have the potential to

(a) add new content to the existing acid {concept} by providing additional definition/characterisation/associations to support discriminations of which substances are acids;

and/or

(b) change the discriminations made in determining which substances were and which were not acids by showing that substances that had previously not fitted the criteria for being considered acids should now be so designated;

and/or

(c) change the concept {acid} such that the accepted nature of acids changed under the modified concept.

Clearly, Lewis did (a) by introducing specific references to electrons pairs that had not been pertinent before. Clearly, he also did (b), as electron-pair acceptors that were not proton donors 'became' (*i.e.*, to be discriminated as) acids. The more interesting question concerns (c). The distinction between (b) and (c) might be paraphrased as

(b) we knew what we meant by acids, and we just got our discriminations wrong by missing some;

or

(c) we were in principle wrong about what should count as an acid, and we have corrected the concept.

So, the question is: did chemists here only change our knowledge of acids by finding out about more (in much the way that the discovery of potassium added to the list of known metals, without needing to fundamentally change our notion of a metal) or is this more like how the modern concept of interpenetrating, fuzzy and tenuous atoms is quite distinct from the solid, discrete and indivisible atoms of Democritus (Taber, 2003)?

This returns to the point raised earlier about the extent to which science makes progress by getting closer to some kind of truth, by developing an account of nature that is better in terms of being more accurate. If we do believe this is what is meant by scientific progress; and if we also accept that scientific knowledge is technically always provisional in nature, and is open to being revisited if new evidence or interpretations become available; then we can never be sure that science is making progress (our evaluations are themselves provisional): what appears like progress may later come to appear a cul-de-sac (but then, later still, might be restored as the best available account; but then, even later...).

Most scientists would feel such a strict view of the provisional nature of scientific knowledge is over-cautious, and that, in some areas at least, we can be pretty confident that our accounts of nature are better than previous accounts – Darwinian natural selection offered a better account than the acquired characteristics of Lamarkianism (even if the latter was perhaps not so completely wrong as once thought); the oxygen theory of combustion offers a better account than the phlogiston theory (even if the latter may have offered a useful way of looking at some phenomena that was discarded prematurely according to Chang (2012b)); the germ theory of disease was surely 'closer to the truth' than a theory that blamed a lack of balance in the four bodily humours.

Acknowledgement of progress seems reasonable in selected examples, but can we apply this to the developments in terms of acid theory. We might ask:

1. Does the {acid$_{Arrhenius}$} concept pick out acids in nature more accurately than the {acid$_{Lavoisier}$} concept?
2. Does the {acid$_{Brønsted-Lowry}$} concept pick out acids in nature more accurately than the {acid$_{Arrhenius}$} concept?
3. Does the {acid$_{Lewis}$} concept pick out acids in nature more accurately than the {acid$_{Brønsted-Lowry}$} concept?

Recalling the point about avoiding tautological arguments, we can only answer these questions if we have independent means of discriminating acids from substances that are not acids by which to make judgements. Otherwise we are free to pick the Lavoisier definition of an acid, in which case {acid$_{Lavoisier}$} is clearly superior to {acid$_{Arrhenius}$} and this aspect of chemistry has not been progressive at all. We might reject that particular evaluation as {acid$_{Lavoisier}$} does not allow us to discriminate all those substances that we can identify as acids because they taste sour, give efflorescence with alkalis, neutralise bases to give salts, and turn particular vegetable dyes red (*e.g.*, muriatic/hydrochloric acid).

Yet, we cannot use the same reasoning for the later comparisons (2, 3) – by moving beyond the previous notions of what counted as neutralisation, and going beyond the aqueous systems in which recognised indicators functioned, and by including substances that modern chemists are unlikely to want to/be prepared to/be allowed to taste, these traditional criteria become less pertinent. If we wanted a theory that tied in with the traditional class of acids, then Brønsted–Lowry and Lewis have done chemistry a disservice by developing {acid} concepts that lead us to discriminate as acids some substances that do not fit with the traditional understanding.

6.5.8 Are Acids a Natural Kind, or a Chemical Convenience?

We might then ask if the traditional criteria were inadequate for picking out substances that should be included in a natural kind 'acid', as they missed a lot of acids, and theoretical advances were needed to better be able to discriminate the substances that 'really' are acids? I imagine most readers who have persevered this far with this account might be dissatisfied with the view that there is an especially important natural kind called acids, and over centuries chemists have refined their {acid} concept to better catalogue this natural group of substances. At least, this does not seem a sensible interpretation to this author.

There is a vast number of substances found in nature, and an even greater number of substances that can be synthesised in the laboratory, and when these substances are characterised in terms of their physical, and – in particular – chemical properties, a whole range of similarities and differences can be discerned, many of which are matters of degree rather than clear dichotomous or ordinal distinctions.

Usefully (at least, in terms of potential applications), nature is wondrously messy and variable in this regard. So, we can set up all kinds of categories that are natural kinds in the sense that they have necessary, not contingent, characteristics (see Chapter 7 for a discussion of the idea of natural kinds), and of course we can label these categories as we wish, but it becomes meaningless to ask *what is the correct notion of an acid as acids occur in nature*. Acid is a conceptualisation, and according to the conceptualisation we adopt, we get different membership of the category. The question of which version is 'right' is not pertinent.

So, we can clearly see $\{acid_{Arrhenius}\}$ as an improvement over $\{acid_{Lavoisier}\}$ in terms of the traditional properties of the group of substances denoted acids; but this falls down when considering $\{acid_{Brønsted-Lowry}\}$ and $\{acid_{Lewis}\}$ as the protagonists have shifted from trying to offer a theory that better identified traditional acids to offering a new notion of *what makes up a useful chemical category*. Here, the argument is not that Arrhenius got it wrong (parallel to the widely accepted argument that Lavoisier got it wrong) but that given new chemical contexts that had been explored it might be useful to reinvent the category, and so change to a different {acid} concept.

So, rather than identifying the truth about 'acids in nature', chemists have made arguments more about what is useful, convenient, or intellectually satisfying. An analogy might be with the concept {voter} used to discriminate voters from voters (non-voters) in a modern state. So, in the UK for example, at one time 'voters' meant men over the age of 21 who had substantive

Conceptualising Acids: Reimagining a Class of Substances 105

property. At various times the level of wealth needed was changed, then women over 30 were franchised, and later women over 21, and now men or women over 18 years of age (or for elections in Scotland for the Scottish government, but not for constituencies in the UK parliament there, 16). The notion of 'voter' in the UK is a 'social kind' and not a natural kind (a term discussed in Chapter 7) as the qualifications can be changed arbitrarily. One can argue that it is morally right for women to have the same voting rights as men, but there is nothing 'natural' about a 19-year-old having a vote when her 15-year-old sibling does not. In some ways, chemists have over the last century treated the {acid} concept as open to the kind of changes that have occurred with suffrage: changes may have been made with very good reasons, but not because we have previously misidentified members of a kind ('voter' or 'acid') that is clearly presented to us in nature.

6.5.9 Is There a Canonical {acid} Concept in Chemistry Today?

So, according to the perspective developed by Brønsted and Lowry, acids and bases are understood in terms of reactions where one substance (the acid) donates a proton, that another substance (the base) accepts. (A pedantic reader might well spot a rather serious problem in that formulation, which will be considered in the next chapter.) This definition is not restricted to solution chemistry based on aqueous systems – so, for example, reactions occurring in solution in liquid ammonia can be included within this way of thinking. By contrast, the alternative Lewis conceptualisation defines bases as electron-pair donors and so acids as electron-pair acceptors.

The Brønsted–Lowry and Lewis definitions are often taught today in chemistry courses in secondary/high school and university, and it is probably reasonable to consider them canonical. These ideas are scientific in nature, in current usage, and widely accepted among chemists. This raises the question of their relationship with each other, as well as with other notions.

The core of the Brønsted–Lowry acid {concept} is based on what happens to protons, and the core of the Lewis {acid} concept on what happens to pairs of electrons, so they clearly were distinct concepts when first mooted, but we might always ask if it is possible for distinct concepts to become consumed under more encompassing concepts. Something of this sort happened with Maxwell's work, after which electricity, magnetism, and (what is now referred to as) electromagnetic radiation, could be seen as different aspects of 'electromagnetism'. So, we might consider whether the concepts {acid$_{Brønsted-Lowry}$} and {acid$_{Lewis}$} have simply become incorporated into a currently canonical chemical concept of {acid}.

More pointedly, it might be asked *if* there is indeed a modern canonical chemical concept {acid}, and if so, is it either (a) basically the Brønsted–Lowry {acid} concept or (b) basically the Lewis {acid} concept or (c) something else, and – if something else – does it encompass one or other of these specific acid concepts?

One option is that we are dealing here with two different facets of the same thing (*cf.* manifold concepts – see Chapter 4). This has precedence in scientific thinking, as when scientists attempted to formulate quantum mechanics. When Schrödinger's wave equation and Heisenberg's matrices were first proposed they were clearly very different conceptualisations of the focal topic, and a question that might have been asked is which approach actually 'worked' and was a useful representation of the phenomena of interest. Today these different approaches are considered to be complementary approaches, rather than one having triumphed and the other having been discarded. If this were parallel to the situation with the Brønsted–Lowry and Lewis {acid} concepts then we might conclude that Brønsted and Lowry, and Lewis, had found alternative and complementary approaches to defining acids that could both be included in the same {acid} concept.

Now, it was suggested in Chapter 2 that one of the core attributes of a concept, arguably its most fundamental function, is that it allows us to make discriminations, so on that basis it seems the chemist's {acid} concept needs to be able to distinguish (within chemical disciplinary discourse, if not everyday discourse) acid from acid (*i.e.*, anything that is not an acid). If the Brønsted–Lowry definition of an acid and the Lewis definition of an acid lead us to make the same discriminations (all these things are acids; these are all bases; everything else is neither an acid nor a base) then they could be part of the same {acid} concept, just as being sour and turning vegetable dyes red were once different facets of the same {acid} concept, without causing any difficulties.

Even if sourness, a sharp taste, was the first characteristic associated with acids, modern approaches to health and safety in school means that it is more likely today that colour change with indicators will be used as the primary indicator (sic) of acidity, and of course such indicators are now normally acquired from commercial suppliers such that their connection with vegetable extracts may not be obvious to students. Litmus paper is familiar to many school students, and the 'litmus test' has entered common parlance as an idiom for any critical test, but this does not mean that most school children, or members of the wider population, appreciate litmus derives from lichens rather than being just some vague product of the chemical industry. This does not negate the effect on vegetable dyes or the sourness association being part of an {acid} concept, although I would imagine there are acids recognised in modern chemistry that no one has ever tested for taste. This might include those so-called 'superacids' reported to be "up to billions of times stronger than sulfuric acid" (Olah, Prakash and Sommer, 1979, p. 4414).

Chang (2012a) argues "there are two accepted definitions of "acid" (due to Brønsted–Lowry and Lewis) and that various textbooks do not exactly agree about the relation between these two definitions" (p. 693). For Chang, "the point is not just that there have been historical fluctuations in the concept of acidity. The messiness still persists" (p. 693). The argument here might be understood, in terms of the focus of this volume, that in modern chemistry there are two, alternative, canonical {acid} concepts, relating to the

Brønsted–Lowry definition or model and the Lewis definition or model, and as these definitions do not clearly lead to the same set of discriminations they cannot be seen to be different facets of the same canonical concept, {acid}.

When Kolthoff (1944) offered his account, an intention was to suggest harmony between the {acid$_{Brønsted-Lowry}$} and {acid$_{Lewis}$} concepts. Yet Kolthoff acknowledged that "a strict application of Lewis' definition to acids containing a proton (Bronsted acids) would make it questionable whether such substances should be called acids" (p. 54), and so recommended that "acids which satisfy the Lewis definition are called Lewis acids or proto-acids. ... Substances like the proton, boron trichloride, *etc.*, are proto-acids" (p. 54). This allowed him to claim that when this "distinction is made between a proto-acid and an acid, there is no conflict between the Lewis and the Bronsted classifications of acids and bases" (p. 55). Despite needing this qualification, Kolthoff concluded that "there is no pedagogical difficulty involved in presenting both concepts as a harmonious pair" (p. 56), which should be reassuring for all teachers reading this presentation.

Chang puts it well:

There are a few operational tests of quotidian acidity (*e.g.*, sourness, corrosiveness, and color turning of litmus and other reagents) and a few paradigmatic substances that pass these tests clearly. And then there are various other acids that pass some of these tests but not others. The ability to neutralize alkalis, or bases, is another important quotidian operational test of acidity; however, it would be a mistake to try to elevate this test above all else as a fail-safe criterion, as that only works out if we turn it into a tautology by defining a base as whatever neutralizes an acid (Chang, 2012b, p. 700).

Chang's last point raises the question of how we might define neutralisation, apart from being a reaction between an acid and a base (introducing scope for another tautology). Chemists do not mean producing a neutral product when they refer to neutralisation, counter to what many students – not unreasonably – infer (Schmidt, 1991). If to neutralise an acid means something like to exhaust its reactivity as an acid, then bases would include not only hydroxides, carbonates, and also some oxides (*i.e.*, basic oxides and amphoteric oxides) but some metals (not, by definition, the noble metals, and, indeed, largely the base metals). None of this is likely to cause problems for chemists reading this book – we all know that the term 'acid' can be used in different ways, and that what counts as an acid depends upon which definition (and so which {acid} concept) we adopt at a particular time.

I suspect that most readers of this book have a manifold {acid} concept. This would recognise the everyday use in terms of a highly corrosive liquid, such that caustic soda might commonly be considered an acid, whilst DNA is clearly (in the vernacular sense) not; and distinguishes this from 'the' chemical meaning that leads to a different (if somewhat overlapping) set of discriminations. The reader will also know of the Brønsted–Lowry and Lewis definitions of acid, and appreciate they have somewhat different ranges of convenience, *i.e.*, contexts in which they are usefully applied.

Some readers may have given thought to the extent to which the Brønsted–Lowry and Lewis definitions define different sets of acids and wondered whether this means that they are components of distinct {acid} concepts. Perhaps we might imagine a Venn diagram where Brønsted–Lowry and Lewis acids are subsets of a wider acid set. Chemistry teachers (who have been successful chemistry students) have been able to conceptually navigate around such complications. Yet, I hope this account demonstrates that there are potential problems here for novices if they are being asked to learn about 'the' acid concept.

6.6 When Are Chemical Discoveries Made?

The examples explored in some depth in this and the previous chapter raise issues about the simplistic notion that chemists make discoveries about how nature is, and their canonical concepts reflect this. In the case of potassium, it seems reasonable to say that Davy discovered the element potassium, and clearly in becoming convinced this was a previously unknown chemical substance he formed (invented) a concept {potassium} which was a way of thinking about what we today think of as potassium.

Potassium has (it seems safe to assume) not changed since it was first isolated, although the purity of samples available has surely changed, as has the range of means available to characterise this substance potassium. Potassium today (the type of stuff the label refers to) is no different from potassium at any time in the past or future, but the canonical concept {potassium} today would seem different from the chemical concept {potassium} as formed by Davy, and that had become canonical before various developments (instrumental and theoretical) that added to our knowledge of potassium (such as the examples discussed in Chapter 5). The canonical concept changed because it acquired new associations, and was able to be applied in new ways, and indeed could be used to make discriminations between potassium and ¬potassium (*i.e.*, anything other than potassium) in ways that could not have been conceived of when the original {potassium} concept was mooted. This need not be problematic, but rather simply reminds us to be careful not to be ahistorical and confuse {potassium$_{\text{Davy-at-first-isolation}}$} with {potassium$_{\text{canonical today}}$} by treating 'the' {potassium} concept as though it has been a unitary and fixed conceptualisation over time. 'The' acid {concept} seems more problematic.

Potassium seems to clearly be a specific type of stuff that exists in nature. The potential for different isotopic compositions (and therefore for different radioactive properties of different samples), certainly complicates this. Yet, for most chemical purposes we can conceptualise potassium clearly enough to readily distinguish potassium from ¬potassium (*i.e.*, anything that is not potassium). It is less clear that acids exist as a class of substances in nature

Conceptualising Acids: Reimagining a Class of Substances 109

in quite the same way. It was suggested in Chapter 5 that the concept {potassium} describes an ontologically discrete cluster of entities selected from among the things found in the universe. I would suggest here that the concept {acid} describes a much less ontologically discrete cluster, and indeed there are choices to be made in identifying the cluster (see Figure 6.3). There seem to be various ways of conceptualising what exactly counts as an acid: alternative {acid} concepts, that may be applied in chemistry. Whereas 'potassium is potassium', acids are not just acids in the same sense. The {acid} concepts commonly used today discriminates different groups of substances than did the concepts {acid$_{Lavoisier}$} or {acid$_{Arrhenius}$}.

There are substances recognised as acids today that would not have been acids at earlier periods in the history of chemistry, and that is not simply because the substances were not known, or had not been characterised. We could see this as chemists having got better at discriminating acids from acids (*i.e.*, non-acids), that is at correcting previous 'false negative' cases – but it seems more reasonable to see the changes of conceptualisation not as a matter of correction, but simply as *changes of mind*. In the case of {acid$_{Lavoisier}$} we might say chemists came to realise the concept needed adjusting as it did not match empirical evidence, but at other times chemists did not seek to *correct* the acid concept in use as much as substitute a new one that was considered to be more convenient or useful. The claims of {acid$_{Brønsted–Lowry}$} and {acid$_{Lewis}$} to be progressive over {acid$_{Arrhenius}$} seem to be concerned with greater utility and not greater verisimilitude.

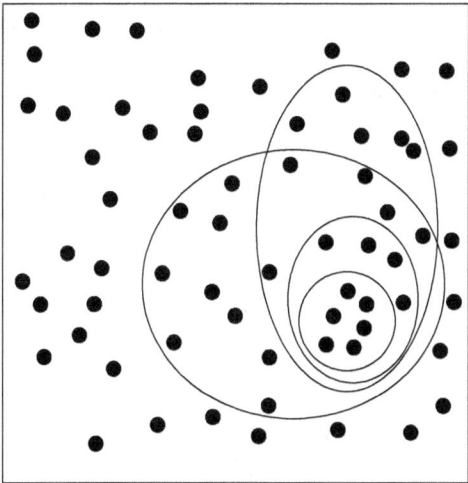

Figure 6.3 New {acid} concepts have expanded the ranges (as represented by the ovals) of those substances identified as acids, from those acids historically recognised using quotidian criteria (the circle), such that discriminations between acid and acid will depend upon which {acid} concept is used. (Compare with Figure 5.2 where a concept develops without changing the discriminations made.)

Given the importance and currency of the notion of acids in chemistry, it might seem obvious that there must be a current canonical {acid} concept. But, if so, it does not seem so clear what exactly it is: whether it is basically (sic) {acid$_{Brønsted-Lowry}$}, or {acid$_{Lewis}$}, or something else? Given that {acid$_{Brønsted-Lowry}$} and {acid$_{Lewis}$} do not support precisely the same set of discriminations between acid and acid it does not seem that we can simply equate {acid$_{canonical}$} = {acid$_{Brønsted-Lowry}$} = {acid$_{Lewis}$}. Either there is a canonical concept, with auxiliary related concepts sometimes used in its place; or a canonical concept that subsumes other more specialised {acid} concepts also currently in use; or a complex canonical concept that shifts its defining characteristics depending upon chemical context; or – perhaps – there are actually different notions of what the {acid} concept actually is as used by different members of the community of chemists with different practices and research foci.

I do not think it is entirely clear which is the case, but my suspicion is that even if most professional chemists would consider there is one prime {acid} concept in use in chemistry today, they would not necessarily agree on entirely what that concept is and so exactly what counts as (and what does not count as) an acid. If that suspicion is correct, then arguably there is no chemical concept {acid} that is genuinely canonical. That might make teaching learners the canonical concept somewhat tricky.

References

Ben-Zvi R., Bat-Sheva E. and Silberstein J., (1986), Is an atom of copper malleable? *J. Chem. Educ.*, **63**(1), 64–66.

Chang H., (2012a), Acidity: The Persistence of the Everyday in the Scientific, *Philos. Sci.*, **79**(5), 690–700.

Chang H., (2012b), *Is Water H$_2$O? Evidence, Realism and Pluralism*, Dordrecht: Springer.

Crick F., (1970), Central dogma of molecular biology, *Nature*, **227**(5258), 561–563.

Crosland M., (1973), Lavoisier's Theory of Acidity, *Isis*, **64**(3), 306–325.

Kolthoff I. M., (1944), The Lewis and the Brönsted–Lowry Definitions of Acids and Bases, *J. Phys. Chem.*, **48**(1), 51–57.

Kuhn T. S., (1970), *The Structure of Scientific Revolutions*, 2nd ed, Chicago: University of Chicago.

Lakatos I., (1970), Falsification and the methodology of scientific research programmes, in Lakatos I. and Musgrove A. (ed.), *Criticism and the Growth of Knowledge*, Cambridge: Cambridge University Press, pp. 91–196.

McClintock B., (1950), The origin and behavior of mutable loci in maize, *Proc. Natl. Acad. Sci.*, **36**(6), 344–355.

Miller D., (1974), Popper's Qualitative Theory of Verisimilitude, *Br. J. Philos. Sci.*, **25**(2), 166–177.

Olah G. A., Prakash G. K. S. and Sommer J., (1979), Superacids, *Science*, **206**(4414), 13–20.

Ruthenberg K. and Chang H., (2017), Acidity: Modes of characterization and quantification, *Stud. Hist. Philos. Sci. Part A*, **65–66**, 121–131.

Schmidt H.-J., (1991), A label as a hidden persuader: chemists' neutralization concept, *Int. J. Sci. Educ.*, **13**(4), 459–471.

Siegfried R., (1988), The chemical revolution in the history of chemistry, *Osiris*, **4**, 34–50.

Taber K. S., (2001), Building the structural concepts of chemistry: some considerations from educational research, *Chem. Educ. Res. Pract.*, **2**(2), 123–158.

Taber K. S., (2003), The atom in the chemistry curriculum: fundamental concept, teaching model or epistemological obstacle? *Found. Chem.*, **5**(1), 43–84.

CHAPTER 7

Concepts and Ontology: What Kind of Things Exist in the World of Chemistry?

7.1 Chemical Concepts and the Kinds of Things We Perceive in the World

In Chapter 3, four important classes of chemical concept were introduced. Three of these were concepts of the kinds of objects (entities) and events (processes) and their qualities (properties) that interest chemists. (The fourth category, referred to as meta-concepts, is discussed in Chapter 8.) Observation is of course itself a kind of process, and observing or measuring properties of entities often involves setting up the conditions for a particular process to occur: dissolving to test solubility; melting to find melting temperature, and so forth. Processes of classifying particular examples of entities and events within chemical typologies usually involves determining properties. These different types of concepts are therefore often intimately connected when actually doing chemistry.

It was suggested in Chapter 2 that concepts can be understood as mental attributes that allow a person to make discriminations. Concepts are therefore related to ontologies – ways of thinking about the kinds of things that exist in the world. All concepts can be understood as *abstractions*, that is, as abstract representations of classes of things (whether those things are considered real, or just imagined) or events, *etcetera*, and as *categories* that support potential discriminations. That is, if we hold ('have') a concept then we are able to meaningfully consider the question of whether something is, or is not (or perhaps is somewhat), within the category of instances of that concept.

7.2 Concepts and Typologies

An important type of concept found in chemistry concerns typologies – ways of classifying instances (of objects or processes) into categories. So, elements and compounds are classes of substances; oxides (already a class) may be divided into subclasses such as acidic/amphoteric/neutral/basic (which of course depends on their properties). In terms of concepts, chemists are operating with the concept {oxides} that discriminates some substances from others; and applying a new concept relating to properties ({nature of oxide}) to the class of substance discriminated by the {oxide} concept to form new subsidiary concepts: for example, {amphoteric oxide}.

There are many such examples in chemistry. Elements may be divided a number of ways into type by position on the periodic table (group or period or block); bonds may be classified as ionic, covalent, metallic and so forth; organic compounds (itself one type compared to inorganic compounds) may be classed into groups such as hydrocarbons, sugars, amines, alcohols; and the hydrocarbons may be classed as alkanes, alkenes, alkynes; as linear or cyclic, as saturated or unsaturated, as aliphatic or aromatic, and so forth. All of these cases, and many others, involve the establishment of new subsidiary concepts: the concepts {period 3 elements} and {noble gases} subsumed under, but distinguished within, the entities discriminated by the concept {elements}; the concepts {monosaccharides} and {disaccharides} subsumed under, but distinguished within, the entities discriminated by the concept {sugars}; and so forth.

These typologies then can operate at different levels. For example, the concept {substance}, which is fundamental to chemistry, allows us to distinguish substances from non-substances. This is also an area where chemical terminology is at odds with vernacular language. In everyday discourse, a piece of wood would be considered to have substance where a chemist might protest it is better described as being material as wood is not *a* substance.

Oxygen and ammonia would be considered substances where, *inter alia*, air, distillation, wood, and activation energy are not considered substances. The concept {substance} applies to oxygen, but not air. This example immediately invites a further discrimination: activation energy and distillation are not substances in a sense that may seem more profound or obvious than the way in which air and wood are not substances.

7.2.1 Nesting of Concepts

This is because air and wood are materials, and the {substance} concept is primarily used in discriminations made in relation to materials. So, although it is true that distillation and activation energy are not substances (and so it is perfectly valid logically to seek to apply the concept {substance} and exclude them from that category 'substance' – indeed to see them falling under the concept {substance} that is the conceptual category for anything that is not a substance), we would not usually expect to be (consciously) applying the {substance} concept here, as it is most useful in making discriminations between different materials – and distillation and activation energy are not material.

As an analogy, we might consider the discriminations that neither (i) whales nor (ii) daffodils are fish. Children (and some adults) may have a concept {fish} that largely selects for animals that live in water, as whales do – and so may include the so-called 'shell-fish' which would not be included under the scientific concept of 'fish', but rather under other concepts such as {molluscs}, {crustaceans}, or {echinoderms}. It may make good sense to have a concept {animals that live in water} – but denoting this concept {fish} adopts a label that scientists use for a different {fish} concept.

So whales are then sometimes informally considered fish although the canonical, scientific, concept {fish} does not include whales, as they do not breathe through gills, and they give birth to live young, and so forth. It is also true that daffodils are not fish either (*e.g.*, they also do not have gills or produce roe), but it seems more salient to exclude them simply because they are not animals. Fish are a type of animal, so something that is not an animal cannot be a fish, regardless of more specific criteria for distinguishing fish from fish (*i.e.*, non-fish).

Concepts, therefore, can often be considered to be nested within other concepts, as in set theory where one set may contain other sets (*e.g.*, members of the set of positive integers are all members of the more inclusive set of natural numbers). The concept {fish} can be considered to be nested in the concept {animals} – so logically there is an asymmetry here:

- all fish are animals;
- some, but not all, animals are fish;
- something that is not a fish, may or may not be an animal;
- something that is not an animal, is not a fish.

Something that is an animal could, but need not, be a fish: so, a seahorse is an animal that is a fish, and a whale is an animal that is not a fish. Something that is not an animal is definitely not a fish, so a daffodil, not being an animal, cannot be a fish.

Applying the same logic to the relationship between the concept {material} and the concept {substance}:

- all substances are materials;
- some, but not all, materials, are substances;
- something that is not a substance, may or may not be a material;
- something that is not a material, is not a substance.

Something that is a material could, but need not, be a substance: so wood is a material but is not a substance. Something that is not material, cannot be a substance, so as activation energy is not material, it is not a substance.

These relationships will seem obvious to chemistry teachers, and should be to more advanced students, but understanding logical relationships has to develop in children (Luria, 1976), and it has been shown that even adults very commonly make errors in tasks requiring this kind of logic (Evans, 2008). Teachers do not always take time to make such (obvious) relationships between concepts explicit to learners. Often, in teaching, we may assume we do not need to emphasise logical relationships of these kinds, as they seem to so obviously follow from our definitions. Yet, such inferences may not be so readily explicit to some learners – especially when they are attempting to make sense of unfamiliar information that may be fully occupying their working memories (Danili and Reid, 2004).

7.3 Do Chemical Concepts Represent Natural Kinds?

It was suggested in Chapter 3 that the term natural kind refers to a class of object that is found to exist in nature. (See also the discussion about whether acids are a natural kind in Chapter 6.) That is, nature seems to present to us different examples that appear to be the same type of entity. There is assumed to be something about a natural kind that makes its members essentially, necessarily, members of that kind, rather than contingently so.

7.3.1 The Importance of Natural Kinds in Science

Science commonly adopts a metaphysical assumption that nature presents us with natural kinds – with sets of entities that are not just contingently found together, but that are intrinsically similar because they share some common essence. In labelling this a metaphysical assumption, I am suggesting that this is something we assume when we set out to do chemistry or another natural science – something that makes science possible – rather than something we find out by doing science. This does not mean that scientists know in advance what particular natural kinds they will discover and characterise – but science, as we understand it, could not proceed without the assumption that nature offers us phenomena that can often be understood as examples of discrete natural kinds rather than as an array of unique objects and events with continuously variable properties.

Chemistry is one of what William Whewell (1857/1976) called the 'inductive' sciences. Induction is the process by which we can justify making *general* statements based on limited *specific* evidence. The problem that induction cannot be justified purely in logical terms has long been recognised, and, following the work of Karl Popper (1934/1959), was indeed widely acknowledged as being intractable. The problem may be posed in terms of how many specific instances do we need to examine in order to make a valid general statement that also applies to all the instances of that type not yet examined? For example, if you picked up a decaying leaf in your garden, one of perhaps thousands, and found a woodlouse underneath it, you would probably not immediately assume that there would be a woodlouse beneath every decaying leaf in the garden. But what if you turned over a second leaf to find a woodlouse, and then a third, and a fourth, and a fifth – finding woodlice each time. How many confirmatory instances are required to ensure that a conjecture such as 'there is a woodlouse beneath each decaying leaf', always applies?

This is in important issue, as science relies extensively on such generalisations. Millikan argues that,

> A science begins only when, at minimum, a number of generalizations can be made over instances of a single kind, for example, over instances of silver, or instances of humans, or instances of massive bodies, or instances of, say, moments in the American economy. For those disciplines systematic enough to be clearly labeled as sciences, the kinds studied typically belong also to some higher category being, say, kinds of chemical, kinds of animal, or kinds of national economy, and so forth, each higher category supporting generalizations of the same or similar types. For example, for the most part samples of each element and compound that the chemist studies are uniform with respect to melting and boiling point, specific heat, quantitatively expressed dispositions to combine chemically, tensile strength, color, odor, electrical conductivity and so forth.... The result is that in the case of many sciences, observations need to be made of only one or a very few exemplars of each kind studied in order to determine that certain properties are characteristic of the kind generally. If I have determined the boiling point of diethyl ether on one pure sample, then I have determined the boiling point of diethyl ether. If the experiment needs replication, this is not because some other sample of diethyl ether might have a different boiling point but because I may have made a mistake in measurement (Millikan, 1999, pp. 48–49)

This begs the question of why we can be so sure that another sample of ether would have the same boiling temperature.

We might imagine a natural scientist, a logician, and a sceptical philosopher, visiting the local pond. The scientist might proclaim, "see that frog there, if we were to dissect the poor creature, we would find it has a heart". The logician might suggest that the scientist cannot be certain of this as she is basing her claim on an inductive process that is logically insecure. Certainly, every frog that has ever been examined sufficiently to determine its internal structure has been found to have a heart, but given that many frogs, indeed the vast majority, have never been specifically examined in this regard, it is not possible to know for certain that such a generalisation is

Concepts and Ontology: What Kind of Things Exist in the World of Chemistry? 117

valid. (The sceptic, is unable to arbitrate as he simply refuses to acknowledge that he knows there is a frog present, or indeed that he can be sure he is out walking with colleagues who are discussing one, rather than perhaps simply dreaming about the whole episode.)

The use of induction here, assumed by the logician to be the basis of the scientist's claim, might take the form:

1. a great many frogs have been examined, at different times and places, in sufficient detail to know if they have hearts;
2. each and every one of these frogs, without exception, has been found to have a heart;
3. therefore, we can assume all frogs have hearts;
4. this, in front of us, is a frog,
5. therefore, it has a heart.

Steps 3–5 represent the familiar syllogism that allows a valid conclusion to be deduced from valid premises. So, for example, if we accept (i) that all humans are mortal; and that (ii) Socrates was a human; then we logically must accept that (iii) Socrates was mortal – as this is necessarily true given the premises. However, deduction of this kind is only useful given sound premises, as we can see by:

(i) noble gases do not form chemical compounds;
(ii) argon is a noble gas;
(iii) therefore, argon does not form chemical compounds.

The *logic* here is certainly sound enough, but the conclusion depends upon the premises, and since chemists have abandoned the view that noble gases never form chemical compounds, the deduction is clearly invalid.

The conclusion that (5 above) some chanced upon frog has a heart cannot be deduced from the premise (3 above) that all frogs have hearts *if* that premise is itself not sound, and statement 3 relies upon induction (from statements 1 and 2) that is itself logically incomplete. Despite this technical problem with the logic, I imagine most readers are still siding with the scientist's claim. So, can we be confident this particular frog has a heart, without ourselves being heartless enough to cut it open to see?

7.3.2 Induction, Bias, and Prejudice

It may intuitively seem induction is perfectly valid, as science seems to rely on it a good deal, but, clearly, we would be less confident in induction in many other contexts – for example, might we yet find a decaying leaf that has no woodlouse beneath it? We might be critical of ready applications of induction in other contexts. Imagine a young child who met their first specimen of a particular national or social or ethnic or other group, and then met a second different individual of the same group (and perhaps even then a third), and as a result concluded that "all [placeholder 1] are [placeholder 2]". Would responsible parents feel the need to question such an inductive inference?

If placeholder 1 was 'teacher' and placeholder 2 was 'wonderful', this might not seem overly problematic, even if possibly an over-generalisation. Yet placeholder 1 might be 'blonde' or 'Hispanic person' or 'gypsy' or 'homosexual' or 'deaf child' or 'Australian' or 'person with mental illness' or 'adopted child', *etcetera*, and placeholder 2 could be 'dumb' or 'cruel' or 'smelly' or 'dishonest' or 'insane', *etcetera*. Humans are quite quick to make such generalisations – and then reinforce them through confirmation bias (Nickerson, 1998). Any responsible parent would surely feel the need to challenge such a notion. People are different in all kinds of ways, and we cannot taint (or indeed eulogise) a whole group of people because of a few unfortunate examples (or indeed paragons). That road leads to prejudice, discrimination, and, sometimes ultimately in extreme cases, even genocide.

A question was posed above: 'How many confirmatory instances are required to confirm a hypothesis like this as generally true?' Human psychology is such that in practice people often behave as if the number is very small indeed – a person who has only ever met two lawyers (as far as they are aware) and found both of them 'slippery' could very easily form a generalisation about a whole professional group that substantially influences their future behaviour regarding seeking and considering legal advice. Science is, of course, meant to engage the kind of critical thinking that avoids such tendencies.

Yet, if it is clearly invalid to adopt such stereotypes, then how can we assume all frogs have hearts – or for that matter expect all samples of concentrated sulfuric acid to act as dehydrating agents? How can Millikan claim, "If I have determined the boiling point of diethyl ether on one pure sample, then I have determined the boiling point of diethyl ether"?

Millikan is humble enough to consider there could be a flaw in his work, but if his characterisation was procedurally valid then, assuming careful and accurate working, the answer to the question 'How many confirmatory instances are required to confirm a hypothesis like the boiling point of diethyl ether is generally true?' is none. Having measured the boiling point of ether once, Millikan does not consider this as the basis for a hypothesis about other samples of the same substance, but rather (because they will be samples of *the same substance*, the same kind) as the basis for a fact.

A key question then seems to be: under what circumstances can it be assumed that a property measured for some instance or specimen (something conceptualised as a member of some type or class) also applies to all other instances of the same type? If a chemist measured to be of height 1.8 m determines the boiling temperature of some diethyl ether, why do we feel secure assuming we know the boiling temperature of any pure sample of that substance – but not the height of any chemist who may carry out a replication?

7.3.3 Conceptualising Kinds Assumes Some Essential Properties

The response is that we are not actually here using a process of induction that assumes if we look at sufficient examples we can make a generalisation across the class: rather we are using theoretical considerations. We assume that 'frog' is a kind or type such that as part of its essence (related to what makes it a frog) it necessarily has a heart. Boiling temperature is, on theoretical grounds, considered an essential property of a particular substance in a way that height is not considered an essential property of chemists. Particular substances, or types of living thing, or types of mineral, have been considered to be 'natural kinds', types found to exist in nature with certain essential qualities.

Frogs are a type of animal that is part of a larger group that share the same kind of circulatory system based on blood vessels and a heart to pump the blood around. We do not have to test every frog, because this type of anatomy and physiology is necessary (essential) to being a frog: it is part of the essence, the very nature, of frogness. So, the hypothetical scientist was arguing that the frog is a natural kind: a particular type of thing that exists in nature and that has certain necessary aspects, such that as long as we know we have a frog, we know these aspects will be found. So, the logical chain is something like:

1. frogs are recognised as making up a natural kind;
2. one of the necessary features of frogs is the presence of a heart;
3. therefore, we can assume all frogs have hearts;
4. this, in front of us, is a frog;
5. therefore, it has a heart.

This is not generalisation by traditional induction, but generalisation by deduction from theoretical considerations. The theoretical considerations concern whether frogs are a kind, and whether we know what essential characteristics this kind has – as well as whether we can empirically recognise a frog. Behind this, however, is the metaphysical assumption that natural kinds exist, and do indeed have particular essential, and so necessary, characteristics.

7.4 Natural Kinds in Chemistry

As chemistry is a natural science, it may seem obvious that any typologies used in the subject will represent natural kinds. We might even suggest that this is the defining characteristics of natural sciences, that they concern natural kinds, as opposed to the social sciences that concern those much more ephemeral kinds that depend upon human convention – class, ethnic identify, gender, institutions, cultural conventions, *etcetera*. Science tells us what counts as an element; but what counts as a school or a teacher or a curriculum subject is open to social consensus, or will alternatively depend upon the local cultural context, or could even be relative to a particular observer. Science is objective: it seeks to discover facts about the natural world that are independent of the perspective of the particular observer.

Underpinning the notion of natural kinds, then, is a metaphysical assumption about the nature of reality. It is assumed that if we examine the different things that exist in the natural world, we can place those things into a finite set of categories that reflect essential differences. Essential here means in terms of some inherent essence, rather than through some arbitrary or purely contingent process. It is certainly possible to set up a category along the lines of 'the first twenty things I see after waking in the morning' (or 'objects mentioned in the lyrics of Beatles songs', *etc.*), but what ends up in the category would be contingent: it might include an alarm clock, if I use one, but not otherwise; it could (but would not necessarily) include a dressing table, a coffee cup, a toothbrush, a magazine, a carpet, a ceiling lamp, and so forth. These things are not considered to have anything essentially in common, and another person, or even the same person on another day, might well have a different set of objects included. The things I'm likely to see on awaking are contingent on the things that have been brought into my house, and none of them are necessarily there (*i.e.*, I could have chosen not to have them in the house, or indeed I could choose to sleep in a different room, or at a local hotel).

We might note that, in terms of what was suggested in Chapter 2, this imaginary example involves the formation or application of a concept. The concept {the first twenty things I see after waking in the morning} can be acquired, understood, and applied. However, it is not a concept of much value in chemistry – it has no basis for becoming a canonical concept used as a matter of general practice in a field of activity. Similarly, *inter alia*, a helter-skelter, a guitar, the sun, an octopus, a silver hammer, your hand, the taxman, a clean fire engine, a blackbird, the USSR, the Amsterdam Hilton, Henry the Horse, and a barber showing photographs – as well a lot of acorns and holes (one of which was being fixed) – can be seen to be associated through a concept, but hardly represent a natural kind.

There might seem to be more coherence if the criterion used were blue objects, or objects with mass more than 20 kg, but still these would not usually be considered to reflect natural kinds. Such criteria may seem less arbitrary, but are still somewhat contingent. Many blue objects (some front doors) could easily have been green. Many bookcases have a mass a little less than 20 kg, whilst other similar ones have a mass a little more than 20 kg. Science, however, does seek to identify *natural* kinds, kinds of things that share some essential nature. Two obvious candidates might be the chemical elements (like potassium) and other natural substances (perhaps acids) and biological species and other taxa (such as the frog).

Naturalists identify individual living specimens to be members of specific (sic) species of living things. The species level is nested within other more encompassing categories such as family and order. The classification of living things might seem a strong example of natural kinds, and one that is familiar even to young children. A lion is not a tiger, nor for that matter is it an elephant or an elephant shrew. It seems obvious to us that different lions have something in common that is not arbitrary. A lion, a lioness, and a lion cub can be clearly discriminated by inspection, and in some ways the lioness may appear more similar to a cougar than her mate with his mane – but it is considered that lions have something essential in common that make lions comprise a natural kind, and cougars a related, but distinct, natural kind.

7.4.1 Species as Natural Kinds? A Warning from Biology

People, not just scientists, tend to naturally (sic, automatically) notice kinds in nature: for example, kinds of mineral, kinds of meteorological conditions (*e.g.*, types of clouds), and perhaps most obviously, kinds of living thing (Keil, 1992). When children, we all readily notice and learn that the world contains different kind of living things. There are birds and horses and dogs and fish and so forth. We come to recognise levels of classification without difficulty: this animal is a dog, and also a Labrador; this creature is both a sparrow and a bird. Later, when we study science in school we find that such distinctions are made formally by scientists, although not always in ways that entirely fit with informal everyday use (Medin and Scott, 1999): so mushrooms should not be considered plants, for example.

More advanced study might lead us to realise that the recognition of species and other higher-level taxa is not so straightforward. When I was at school, it was considered that the dinosaurs, the 'terrible lizards', as a group became extinct around 66 million years ago at the time of the formation of the Cretaceous–Paleogene (aka KT, Cretaceous–Tertiary) boundary, but since then 'lizard' has become a questionable category of natural kind, whilst many biologists now claim that birds are technically extant (rather than extinct) dinosaurs.

Having learnt that the main orders of vertebrates were fish, amphibians, reptiles, birds, and mammals, it appears that not only might birds be considered reptiles, but that reptiles are by some biological criteria not actually an essential kind (that is a kind with a particular essence). Even fish are not exempt. Leaving aside the tendency of the term fish to sometimes be used in a vernacular sense of sea creature (to include whales and 'shell-fish' for example), it seems that by some criteria fish do not share a particular essence as a group, as some fish are more closely related to members of other groups than they are to some other fish (LaPorte, 2004). Guinea pigs are no longer seen as members of the mammalian group of rodents. In addition, these are just some examples from the vertebrates, among the most familiar groups of animals to most people.

Of course, such changes of mind could be explained simply as biologists learning more about the natural world and so getting better at identifying and characterising species and other taxa. A helpful perspective is to treat the notion of biological species as a scientific hypothesis (Knapp, 2017), in that when a scientist proposes a species this is a hypothesis about a certain regularity in the natural world: a hypothesis that is then the basis for further investigation. So, we might perhaps think that there are indeed distinct, discrete natural kinds of living thing in the world, but it is just not always obvious what their essential qualities are, and so which groupings are genuine natural kinds.

Certainly, before Darwin's work it was widely assumed that species were natural kinds that had a unique origin, so that all members of a species were blood relatives of one another, but had no such relationships with members of other species (LaPorte, 2004). This might be considered one of the 'hard-core' assumptions that had guided naturalists. Hard-core assumptions are so taken for granted in a research programme that they are assumed as tenets or axioms, and not challenged within that programme (Lakatos, 1970). They are in effect treated as metaphysical for the purposes of that research tradition: they cannot be empirically tested as all concepts and results within the programme are interpreted in the light of what follows from those assumptions. (So, the existence of natural kinds may be considered a hard-core assumption of science more generally, and 'changes in mind' about specific kinds may be expected in a science as it progresses, as long as the core idea itself is not abandoned.)

Modern scientific thinking, post-Darwin, suggests that there are no absolute distinctions between species. Darwin himself thought he had done biology a service in offering the perspective on the biota suggested by his theory of natural selection. Descent of different groups from common ancestors, should (Darwin thought) have brought an end the interminable wrangling about whether particular groups were 'really' different species or actually varieties of the same species. For Darwin, understanding the origin of species suggested there could be no absolute distinct essence of any particular biological grouping that meant there would always be an absolute distinction between specimens of one species and another. Attempting to find such an essence would be a "vain search for the undiscovered and undiscoverable essence", as instead scientists actually set out judgements of species "for convenience" (Darwin, 1859/2006, p. 757).

Darwin, in effect, abandoned an existing research programme and initiated an entirely new one where the concept of species had to be understood completely differently. As is well documented, scientific shifts that are that revolutionary bring about changes in the concepts used in an area of science that many other scientists struggle to accept and cope with (Kuhn, 1970). This is something it will be useful to bear in mind when considering the major changes in student thinking sometimes needed for learning canonical chemical ideas (see Chapter 15). So, arguments about whether a particular specimen is, or is not, an example of a certain species, or arguments about whether certain distinctions reflect different species, or rather varieties within a single species, may be questions of which classificatory system is most useful, or, as Darwin suggested, 'convenient', rather than which is – in

Concepts and Ontology: What Kind of Things Exist in the World of Chemistry? 123

some absolute sense – right. This seems to have resonances with the discussion of {acid} concepts in Chapter 6.

7.4.2 The Elements as Natural Kinds

Species are, of course, the business of biology, and so understandably messier, ontologically speaking, than chemistry. A good chemical example, element, is surely more clear-cut. The distinction that chemists make between hydrogen, and helium, and lithium, and so forth, are objective in that any competent chemists, given a suitably equipped laboratory, could agree over (a) whether a sample was of one of the elements, and if so, (b) which element it was. Even if the bench chemistry did not enable entirely clear distinctions in some cases, mass spectroscopy would offer a clear means of analysis.

There is much less scope here for matters of opinion over whether a pure sample of a substance is a specimen of phosphorus rather than a specimen of sulfur than there is in determining whether a fossil bone is a trace of, say, one of the very last specimens of Homo heidelbergensis or one of the very first specimens of Homo sapiens. In chemistry, we do not need to impose dubious interpretations on evidence to avoid dealing with samples of 'element 20.5' or 'element almost 90'. Rather, nature offers regularity that seems unambiguous.

The reason that it is possible to be more unequivocal in this case is that elements can be well characterised such that a whole range of characteristic properties can be defined (*e.g.*, see Chapter 5). One has to be careful here not to ignore the bootstrapping logic sometimes involved in practice in applying such ideas. As has been well described by Chang (2004), often in science we make pronouncements about properties by applying the use of instruments that, to some extent at least, depend upon us being able to assume what we mean to test. So, for example, to determine melting points we need a trustworthy way of measuring temperatures.

Chang describes this in some considerable detail in his discussion of the development of thermometry. A less adequate summary of his account might be that to measure temperature we need instruments able to distinguish one temperature from another. Therefore, we need to identify thermometric properties that are valid and can be applied reliably. It may seem obvious (that is, today, with the role of hindsight informing our scientific education) that such properties exist: the length of a thin column of mercury, the volume or pressure of a contained sample of gas, the conductivity of an alloy, the potential difference between pairs of junctions of dissimilar metals, the colour spectrum of emitted radiation – whatever.

Of course, we can only be confident that these properties are valid and reliable indicators of temperature once that proposition has been successfully tested through measuring the property carefully across a range of temperatures. But, to do that, we need a suitable thermometer to determine accurate temperatures at which these thermometric properties are measured – one that is based on a valid and reliable thermometric property and has been carefully calibrated across the range of temperatures where we wish to use it. So, it seems, to develop a valid and reliable thermometer scale we need to use an existing valid and reliable thermometer based on a valid and reliable thermometer scale.

In practice, we have moved slowly through various iterations of increasing quantificational sophistication using a process that had built in error-detection checks – something that indeed could be considered a description that applies widely in science. For example, a school student can make a simple thermometer from a boiling tube of water bunged with a glass tube of narrow bore. The boiling tube can be immersed in large beakers of water that can be felt (directly perceived) to be cold, tepid, warn, and hot, and the change in volume of the water in the boiling tube can be detected. This demonstrates a potential thermometric property, even if it is well known that our sensory system does not support making precise judgements in such matters (perception being biased to detect changes, thus the well-known demonstration that after placing one hand in very cold water, and the other in hot water, and then moving both into the same vessel of tepid water, the sensation is that the hands are in water at different temperatures). The next step might be to introduce some reasonably reproducible standard of cold – perhaps a beaker of ice-water – and mark up a reference reading on the thin column of water in the capillary tube.

Similar considerations apply in many scientific contexts. So, in chemistry we often quote properties that are characteristic of pure samples of specific substances. Yet to identify such substances and characterise the samples as pure, we appeal to these very properties that we claim such pure samples have. Reference tables will tell us the density, conductivity, specific heat capacity, melting temperature, *etcetera*, of a pure sample of some substance under certain conditions – but it only became possible to establish such characteristics as matters of reference once samples of sufficient purity were available to be tested. That, of course, depended upon a means of characterising pure samples so they could be distinguished from impure samples.

That is, once we have a way of being sure what a pure sample of some substance is like, which itself depends on being sure we have a sufficiently pure sample, we can measure its properties and so set out what that pure substance is like – which then enables us to be confident of the purity of samples we might wish to characterise. For example, to be confident that the density of potassium at 20 °C is 0.862 $g\,cm^{-1}$ one would have to be confident in measuring mass, volume, and temperature, as well as confident one had fairly pure samples of potassium (without relying on measuring its density). Clearly science here has to work in an iterative way, and refine its techniques, and, in parallel, refine its conceptualisations in the process.

7.4.3 Are Natural Kinds Just the Operation of a Mental Bias?

Underpinning the notion of natural kinds is the metaphysical assumption that if we examine the different things that exist in the natural world, we can place those things into a finite set of categories that reflect essential differences. I have characterised that as an assumption brought to, prior to, science rather than a deduction deriving from science. This might seem unreasonable: as scientists *notice* that natural things fall into obvious groups. Elements can clearly be shown to exist as specific kinds. Would it not be better to consider the notion that natural objects fall into natural kinds to be an empirically discovered fact?

This may seem the case. It is clear from our own experiences that we are able to infer group membership (*e.g.* giraffe) based on perceptual cues alone. It is a capacity of human cognition to make sense of experience in terms of groups and categories based on perceived similarities and differences. We might therefore argue that we are able to recognise natural kinds because (i) there are such kinds, and therefore (ii) it is an understandable adaption that humans have evolved to be able to recognise these kinds.

There is certainly good reason for humans to have evolved the apparatus to form abstract distinctions (that is, concepts) in relation to what might be termed 'practical kinds'. Some things in our environment are good to eat, and some are edible but offer little nutritional value, and others are toxic. Some places in our environment provide shelter and protection, and others are hazardous. Some individual animals we come across might be potential mates, and others might be predators. And so forth. Being able to recognise these 'practical kinds' is certainly very useful in making important distinctions. (Even if a male black widow spider may discover too late that 'potential mate' and 'potential predator' are not always mutually exclusive categories.)

An individual that was unable to notice or remember that these berries are very similar in shape and colour and taste to those I ate a while back just before I was very sick would be at a disadvantage. So might our hypothetical logician if he based actions on the realisation that it was logically unsafe to generalise from the fact that 'these berries are very similar in shape and colour and taste to those I ate a while back just before I was very sick' to an induction that 'berries with these properties are likely to make me sick'. The logician is right – these may be superficially similar berries, or perhaps it was a coincidence that the previous engagement with berries such as these coincided with sickness. However, even if it is a *logically* unsafe generalisation to damn this new sample of berries before testing them, it may be *practically* unsafe to wish to test the association by consuming the berries. During human evolution, erring on the side of underdetermined induction may have been a wise general strategy.

So, there are two points here of particular relevance to our assumption that nature can be authoritatively carved up ('at its joints') into natural kinds:

Although humans seem to have an innate tendency to recognise kinds in the natural world, and that ability would seem to have been important in our evolution and survival:

(a) We may be biased to over-generalise – to use this mental propensity both when it is appropriate, and on other occasions. This seems likely

if, in general, positing groups that had no intrinsic basis ('false positives') had less serious consequences for our survival than not spotting groups that were very pertinent ('false negatives').
(b) Kinds identified to be of practical importance may not map onto the natural kinds assumed in science.

As an example of the latter point, the natural kind 'tree' seems to be commonly picked out in human societies. 'Tree' seems to be a culturally useful grouping, so people readily learn a {tree} concept. The {tree} concept is used in the discourse community people are brought up into, and they readily perceive trees from trees (*i.e.*, non-trees): that is, they make appropriate distinctions according to the {tree} concept. Yet science does not consider 'trees' as a biologically significant group – it does not seem to be a natural kind of any significance to natural science. This does not mean that trees themselves are not important – but that when different types of plants are characterised according to what are considered scientifically significant features, essential features, trees are found to occur across different types. This does not mean that 'tree' might not be a useful practical kind (for example, in reforestation) and indeed in responding to global warming the {tree} concept may prove to be of great practical value.

7.5 The Use of the Definite Article in Relation to Chemical Kinds

One of the apparent liberties chemists take with language is to refer to 'the …'. This habit is applied widely. Consider the following examples:

The methane molecule is tetrahedral.
The copper is deposited on the electrode.
The bromide ion acts as a leaving group.
The ethene molecule is unsaturated.

The definite article, 'the', is then made good use of in chemical discourse. There are many examples in this book, such as:

"The 1922 Nobel prize for chemistry was awarded to Francis Aston for demonstrating isotopic composition of a range of elements using ***the*** mass spectrometer" (Chapter 5).

The definite article is also widely used in everyday discourse as well. It is widely used in two ways, which may be exemplified:

- The supreme leader has denounced chemistry as a source of magical wonder.
- The student dropped the conical flask.

In the first statement the definite article seems to refer to a definite entity (as the definite article should), which members of the language community would recognise. There is only one supreme leader, so when 'the' supreme leader is referred to, the referent is clear enough. It was not the deputy supreme leader who issued an edict, or the supreme leader's husband, or a local mayor – but THE supreme leader. The second example seems rather different. The world contains more than one student, indeed a great many, so the use of 'the' seems inappropriate unless there is some wider context to pick out 'the' particular student being discussed. So, what might the context be? I will suggest two options, confident that fluent readers of English will consider one more congruent than the other.

(1) The undergraduate laboratory was chaotic today. A lot of the students seemed rather distracted. The student dropped the conical flask.
(2) A professor, a student, and a university porter, went together into a bar. The professor took a manuscript with him, the student held a conical flask, and the porter carried a large bunch of keys. The student dropped the conical flask.

Thematically the first option seems more coherent. We can imagine that a student in a teaching laboratory would likely be holding a flask, and might drop it – especially if there was something of a chaotic atmosphere. There is no reason why a student should not go for a drink with a professor and a porter, but that seems a less common scenario. We would wonder why an item of laboratory glassware was taken along, as that seems abnormal and a little irresponsible.

Despite this, in terms of the use of language, the second story works better. 'The student dropped the canonical flask' refers to the particular student (and flask). There may be many such students and flasks in a busy teaching laboratory, so it breaks a rule of normal language use to refer to 'the' student without having picked one out. We need to correct this by preceding 'the student' with some way of foregrounding one as a subject, or instead using an indefinite article ('a'):

(1′) The undergraduate laboratory was chaotic today. A lot of the students seemed rather distracted. *One student was especially clumsy when handling a conical flask and a beaker whilst filtering a solution.* The student dropped the conical flask.
(1″) The undergraduate laboratory was chaotic today. A lot of the students seemed rather distracted. *A student* dropped *a conical flask*.

Yet we often seem to break these usual rules of when to use an indefinite article when we are talking about and teaching chemistry. This usually does not matter when one professional chemist talks to another, but when a chemistry teacher talks to a student this adds an additional burden in making sense of teaching. In some situations, the use of the definite article is fine because of the context. If a teacher is clearly talking about a chemical demonstration where a piece of copper is heated, then talking of "the copper being heated" should be clear enough, as it obviously means 'this specimen of copper here that I am showing you'.

What seems potentially more problematic is the use of the definitive article when the referent is not a specific individual specimen. Chemistry teachers will say things like "the ammonia molecule is pyramidal" when no ammonia molecule is either specified directly or can be inferred to be the case in point from the context. This probably does not seem problematic for the simple reason that it does not matter which ammonia molecule is being referred to: they are all pyramidal. So, statements such as the ammonia molecular is pyramidal; the chlorine atom readily accepts an electron; the K shell is nearest the nucleus; and the iodide ion is a good leaving group; *etcetera*, will be true regardless.

These statements 'work' in a way that some apparently parallel statements from outside of chemistry would not: the house has a blue door, the man walks with a limp, the baby sneezed all night, the bicycle has squeaky brakes, *etcetera*. Some houses have blue doors – many do not (as pointed out above, colour is not an essential property of doors). So, we should not say 'the house has a blue door' unless we have made it clear which house we are referring to. Yet, we do not need to say which particular water molecule is polar, as they all are (*i.e.*, it may be considered an essential quality of a water molecule). So, the question here is why a teacher would say 'the ammonia molecule is pyramidal' when they are not actually referring to a particular specimen, and the point they are making is actually that (all) ammonia molecule*s are* pyramidal.

We can think of some parallel uses in everyday life that would 'work': *the* car is a means of transport, *the* whale is a magnificent beast, *the* tax collector is a target for scorn, *the* football crowd is a tribal phenomenon, ... But then why might someone say that *the whale* is a magnificent beast, if presumably they mean that *whales are* magnificent *beasts*?

7.5.1 Talk About Ideal Prototypes

It seems that this links back to how people operate as if there are essential kinds such as natural kinds (as discussed above) or practical kinds that are treated as if natural kinds. 'The whale' does not, here, refer to the only whale, or to the particular whale previously referred to, or to the particular whale that can be inferred by the context of an utterance – as when passengers on a cruise ship are gazing at a passing lone whale. Indeed, it seems quite likely that 'the whale is a magnificent beast' is a viable utterance as the passengers peer at a school of passing whales! Only a pedantic person is likely to respond, 'Oh, which one?' As suggested above, we have built-in a tendency to recognise types, and view them as natural kinds.

This tendency is not limited to natural entities but also found with artificial types. When people form abstractions (*i.e.*, concepts) that allow

Concepts and Ontology: What Kind of Things Exist in the World of Chemistry? 129

them to discriminate different specimens as examples of the same kind (not just natural kinds, but any practical kinds) they often seem to identify the specimens with a kind of ideal placeholder that represent those individual things included by the concept. So 'car' is a class of object found in the environment and treated as though it is a type with an essence. There is something, or some things, which make an object a car, and so that all cars necessarily have (whatever this might be – not four wheels, not a petrol engine, not a rigid cover above the passengers… but something, or some cluster of things). The utterance that *the* car is a means of transport applies to this ideal placeholder, and so by implication to any individual object that shares the essential qualities necessary to be conceptualised (discriminated) as a car.

So 'the whale' that is '*a*' magnificent beast is not only (i) not a particular whale, but (ii) in a sense not actually a whale at all, but rather the idea of a whale that can stand for all whales – the prototypical object with whale essence. If this analysis is correct, then people are using something like Plato's idea of forms. For Plato, the real things that exist in this world are imperfect reflections of ideal forms. The form of a chair exists in another realm, and real chairs are chairs because they are imperfect but material copies of this form, and share in something of the essence of this prototype chair.

This may seem an odd idea in relation to chairs, and quite difficult to appreciate when Plato applies it to notions such as truth and beauty, but it actually seems a helpful idea in some cases. Consider the sphere – a geometric form that is perfect. Objects we come across that we may label as 'spherical' share in having the essence of the sphere, but are not actually spheres. A tennis ball, a raindrop, a planet – these may approximate to spheres, without actually strictly meeting the criteria for a true sphere. (This point is developed further in Chapter 9.) We can use the idea of a sphere and label things as spherical, and make ourselves be understood without reserving the terms for objects that match the perfect sphere form – that itself is never found in this world.

7.5.2 Which Methane Molecule Is Tetrahedral?

So, consider the claim 'the methane molecule is tetrahedral'. This could sometimes be used in reference to a specific molecule, but actually in chemistry we are seldom interested in single molecules as one molecule is usually of little significance by itself. We generally assume that students appreciate that – even when so much time in chemistry classes seems on the face of it to be focused on what is happening to specific molecules and ions and electrons that get represented in various ways. Teachers should perhaps check this assumption from time to time – to make sure their students share the convention.

During the period when I was writing this book, there was an incident where two people, a man (a former Russian intelligence agent) and his daughter, were poisoned in the English town of Salisbury when they came into contact with a banned nerve agent, and were considered to be critically ill for some time before recovering. A police officer called to the scene was contaminated and also had to be treated in intensive care. Some months later, in a nearby town, a couple fell ill after contact with the same nerve agent, and the woman died. It was speculated that the couple had come into contact with a small amount of residue missed by the initial decontamination measures in Salisbury – as a small dose of such agents can prove fatal.

During an interview on a BBC Radio 4 news programme (July 5th, 2018), Hamish de Bretton-Gordon, who brands himself as one of the world's leading chemical weapons experts, warned listeners that there may be risks to the public due to residue from the original incident in the area. Whilst that may have been the case, his suggestion that "we are only talking about molecules here...There might be a couple of molecules left in the Salisbury area..." seemed to suggest that even someone presented to the public as a chemistry expert might completely fail to appreciate the submicroscopic scale of individual molecules in relation to the macroscopic scale of a human being. It was later reported that the nerve agent had been discarded in a perfume bottle, and the woman who died had sprayed some on her wrist – no doubt considerably more than 'a couple of molecules'.

So maybe 'the methane' molecule stands for any of the methane molecules in this flask I am holding, or any of the methane molecules formed in this reaction I am demonstrating, or any of the methane molecules formed in this reaction whenever and wherever it occurs, or indeed – when we claim the molecule is tetrahedral – any methane molecule you might ever be interested in. Methane molecules are a natural kind so the {prototypical, ideal} methane molecule can stand in for any methane molecule, as what essentially makes something a methane molecule means it will be in its essentials just the same as 'the' methane molecule discussed.

By contrast, 'the' methane molecule will not be moving at 342 m s^{-1} – except when we are indeed referring to a specific methane molecule at a particular moment – as this is not part of the essence of methane molecules, and so not a necessary, essential property for something to be considered a methane molecule. But the methane molecule will have a mass 17 amu, which – excepting any isotopic variants – is an essential feature.

7.5.3 No Methane Molecule Is Tetrahedral

If tetrahedral means having the shape of the (sic) geometric solid figure, the tetrahedron, then the methane molecule is not actually tetrahedral. A tetrahedron is a solid body with four triangular faces, meeting at edges, and with four apices. A methane molecule, however, is a fuzzy ball of fields with no surface, so no faces, and no edges, and no apices. It is clearly not a tetrahedron. So, if *the methane molecular is tetrahedral,* then this must mean something else.

Presumably chemists recognise enough of the essence of being a tetrahedron in the geometry of a methane molecule to feel they justify the claim that 'it is' tetrahedral. Even the more careful wording that the methane molecule has tetrahedral geometry needs to be carefully interpreted given what tetrahedral geometry actually is. A student faced with the claim that the methane molecule is tetrahedral has to unpack quite a lot about the way we use language if they are to understand how we consider such a molecule tetrahedral.

In understanding that the methane molecule is tetrahedral the student needs to understand that 'the' is not being used in the way the definite article is usually used, and that stating the molecule is tetrahedral is not strictly meant to be taken to mean the molecule has the shape of a tetrahedron. It is more a claim that the shape of 'the' molecule (whatever shape might mean for a fuzzy ball of infinitely extending fields) has something of the essence of a tetrahedron.

There is a difference of degree, though, between many of the chemical cases and many of those that tend to be used in everyday discourse. For example, 'the car' that is a form of transport is clearly more general and vague that any actual object that is discriminated by the concept {car}. Objects judged to be cars may share the essential qualities of 'the car' but they also show a great deal of variety in other senses (number of doors, number of wheels, colour, *etc.*). However, in chemistry there are fewer contingent properties to complicate how some particular methane molecules we may be concerned with are distinct from an ideal of 'the methane molecule'

7.6 Constructing Typologies in Chemistry

Besides quanticles such as particular molecules and ions, specified elements and compounds are perhaps the strongest candidates for being treated as natural kinds. Groups such as second period elements, the halogens, transition elements, or main block elements; or such as salts, alkenes, acids, oxidising agents, or Grignard reagents, are also important in chemistry. The natural kind status of such groups may be less clear cut than with elements and compounds.

It could be argued that alkenes are a natural kind as they necessarily (essentially) are hydrocarbons that include a C–C double bond. Yet, this is more of a theoretical definition than something which is directly observed in nature. As we saw in Chapter 6, when considering the {acid} concepts, such groupings include some features of – if not arbitrariness, then certainly (as Darwin suggested for species) – convenience.

Even considering the halogens as a natural kind is less about what is directly given to observation in nature, and is more theoretical. We might seek to define the halogens in terms of a feature of electronic structure, but that clearly draws upon theoretical considerations – a student does not see the electronic structure when observing samples of the elements. We might prefer to define them in terms of their position in the periodic table – but nature did not present us with the periodic table: it was constructed by chemists. Different forms of periodic tables that have been mooted remind us that, again, some judgements about what is most useful to chemists have been involved here. We could seek to define the halogens as a type in terms of some similarities in physical and chemical properties – but it would be challenging to produce a simple account of how to recognise a halogen that clearly discriminated the halogens and excluded any non-halogen. {Halogen} is largely a theoretical concept. This means it is not readily available to novice learners. The same is true of, *inter alia*, {reducing agents}, {aldehydes}, {transuranic elements}, {redox reactions}, {chelating ligands} and {Friedel–Crafts reactions}.

A new student starting out on the study of chemistry will readily form concepts of beaker, test-tube, and conical flask, based on perceptual similarities. The similarities between fluorine and iodine; or between methane and octane, or between the oxidation of hydrogen and the oxidation of iron, are not directly available through the senses: they only become available within a complex conceptual framework of ideas.

Most of the types used in chemistry do not offer an immediate perceptual basis for forming concepts: but rather have only emerged from the iterative processes of science. Despite this, chemists and chemistry teachers may treat many of these types as natural types, and so, for example, refer to unspecified specimens of the class with the definite article when discussing what are considered essential qualities (so, *the* iodide ion is a good leaving group). That is, over an extended period of interplay between increasingly more sophisticated theory and increasingly more elaborate 'observations', certain patterns have been abstracted that inform concepts of classes and groups that are useful in chemistry. The motivation for prioritising such types of objects or events as of chemical significance may be far from obvious to the uninitiated, and so needs to be carefully developed in learners.

References

Chang H., (2004), *Inventing Temperature: Measurement and Scientific Progress.* Oxford: Oxford University Press.

Danili E. and Reid N., (2004), Some strategies to improve performance in school chemistry, based on two cognitive factors, *Res. Sci. Technol. Educ.*, **22**(2), 203–226.

Darwin C., (1859/2006), The Origin of Species, in Wilson E. O. (ed.), *From So Simple a Beginning: The Four Great Books of Charles Darwin*, New York: W. W. Norton.

Evans J. S. B. T., (2008), Dual-Processing Accounts of Reasoning, Judgment, and Social Cognition, *Annu. Rev. Psychol.*, **59**(1), 255–278.

Keil F. C., (1992), *Concepts, Kinds and Cognitive Development.* Cambridge, Massachusetts: MIT Press.

Knapp S., (2017), Carl Linnaeus: Naming Nature, *The Forum*: BBC World Service.

Kuhn T. S., (1970), *The Structure of Scientific Revolutions*, 2nd edn, Chicago: University of Chicago.

Lakatos I., (1970), Falsification and the methodology of scientific research programmes, in Lakatos I. and Musgrove A. (ed.), *Criticism and the Growth of Knowledge*, Cambridge: Cambridge University Press, pp. 91–196.

LaPorte J., (2004), *Natural Kinds and Conceptual Change*, Cambridge: Cambridge University Press.

Luria A. R., (1976), *Cognitive Development: Its Cultural and Social Foundations*, Cambridge, Massachusetts: Harvard University Press.

Medin D. L. and Scott A. (ed.), (1999), *Folkbiology*, Cambridge, Massachusetts: The MIT Press.

Millikan R. G., (1999), Historical Kinds and the "Special Sciences", *Philos. Stud.*, **95**(1), 45–65.

Nickerson R. S., (1998), Confirmation bias: A ubiquitous phenomenon in many guises, *Review of General Psychology*, **2**(2), 175–220.

Popper K. R., (1934/1959), *The Logic of Scientific Discovery*, London: Hutchinson.

Whewell W., (1857/1976), *History of the Inductive Science*, 3rd edn, Hildesheim: Georg Olms Verlag.

CHAPTER 8

Chemical Meta-concepts: Imagining the Relationships Between Chemical Concepts

The previous chapter discussed concepts used in chemistry to discriminate examples of particular types of things (conical flasks, samples of potassium, *etc.*), including things that cannot be directly perceived (ammonia molecules, double bonds). However, concepts do not only refer to things that we imagine to exist in the material world, as we also form concepts relating to our abstractions themselves (concepts related not to things and events, but to other concepts). That is, we describe the world using notions such as laws and principles. These are not things considered to exist in the environment like conical flasks, but abstractions such as perceived *relationships between* different concepts (that themselves reflect types of thing that are considered to exist). Similarly, when we form a model, this is a deliberate abstraction made through reflection on our concepts (*i.e.*, abstractions of kinds of things and events, the subjects of concepts discussed in the previous chapter). In Chapter 3, the term meta-concepts is applied to these types of abstractions.

So, in chemistry we use concepts of properties (*e.g.*, {temperature}, {pressure}, {volume}). Scientists then posit relationships between the properties, and these are often given labels (Boyle's law, Avogadro's principle, *etc.*) This involves forming concepts of these abstractions ({Boyle's law}, {Avogadro's principle}, *etc.*)

This is perhaps obvious to readers, but, in this book, I am seeking to make explicit some of the mental work done in conceptualising chemistry that can make learning chemistry so demanding, and so challenging for some learners. The concept {gas} is already an abstraction as it is a category that can cover a great many actual samples of material. A property such as temperature is also an abstraction – the {temperature} concept has associations with direct perceptions, but also with various kinds of measurements (and various theoretical ideas). Forming a concept of, say, Boyle's law, is more complex still as it is associated with not only the instances of stuff covered by the {gas} concept, but relies on relating other abstractions (such as the requirement of temperature not changing). So, such a concept inherently involves a range of other concepts, some of which are already quite theoretical as they do not directly relate to immediate, unmediated, sensory experience. This chapter considers several classes of such meta-concepts: laws and principles; models; and theories.

8.1 Concepts and Laws (and Law-like Principles)

There are many ideas in science that are labelled as laws. So, there is a notion of a canonical concept of {scientific law} – a concept that is (assumed to be) shared by practising scientists. Supposedly, laws are universal and have no exceptions, so that when scientists moot a law they are claiming to have discovered a regularity that is necessary rather than contingent. A paradigm case might be the notion of universal gravitation. This suggests that all massive particles (*i.e.*, all particles with mass) in the universe attract all others, and, moreover, the gravitational force between any two masses is proportional to the product of the two masses and inversely proportional to the square of the separations. As a law, this is supposed to hold for any two masses, anywhere, at any time – including at times and in places where the law may never be tested (and perhaps could never be tested). Of course, given that we cannot confirm such generalisations empirically, laws are (like any scientific claims) conjectural in nature – and so potentially open to revision in the light of new evidence or new ways of conceptualising the available pertinent data. So, the {scientific law} concept should allow us to discriminate laws, such as the law of universal gravitation, or the law of conservation of energy, and other mooted patterns, which might be less law-like in nature, such as the (now abandoned) claim that the noble gases do not form chemical concepts.

At one time, it was widely considered that the 'inert' gases could not as a matter of principle react to form compounds and that atoms of these elements did not form chemical bonds. That might have seemed quite law-like, as it was often assumed to be an absolute. We now know a range of 'exceptions' so that although the noble gases are found 'native' and do not tend to react readily, it is certainly not a law (or law-like principle) that they do not form chemical compounds.

Yet, in practice, the distinction between scientific laws and other kinds of principles, patterns and relationships (*i.e.*, applying the concept {scientific laws} to make such discriminations) may not be as clear-cut as this. Or, to put things another way, the way scientists have discriminated scientific laws from scientific laws (*i.e.*, everything else) has not always matched a definition of laws as precise, universal patterns. In choosing to adopt the label 'laws' scientists have sometimes made discriminations that now seem inappropriate (according to the {scientific law} concept presented above, with its conception that such laws are universal) or they have been applying somewhat different {scientific law} concepts that do not match the version used here.

There is a range of ideas in science that are usually assumed to be law-like, such as Coulomb's law, or the rule that the maximum occupancy of an electronic orbital is two electrons. However, the term 'law' is sometimes also applied to relationships that are considered to be *less precise* than a true law (to what is approximately always so) or that are considered to have a bounded range of application (and so not be universal). Of course, universality may be more relative than the term may immediately suggest. A law may be defined in terms of all nuclei with integer spin, and may apply 'universally' to such nuclei, but has a more limited range of application than a law that applies to all nuclei.

There are also scientific principles that seem law-like but are not usually referred to as laws. There is a literature exploring the nature of scientific laws (*e.g.*, Christie, 1994), such as when the term should really be used, and so forth, but here the focus is on scientific laws (including nominal ones we might doubt are strictly laws) and other law-like principles, met during the study of chemistry.

Some examples of the kinds of entities being discussed here include:

- Avogadro's law;
- Conservation of mass and charge;
- Chargaff's rules;
- Raoult's law.

The argument made here is that laws are relationships between different conceptual entities, and that in conceiving that relationship, a higher-level concept (or meta-concept) is formed of the law.

8.1.1 Avogadro's Law

Avogadro's law posits that at any particular temperature and pressure, equal volumes of all gases have the same number of molecules. This can be understood as equivalent to suggesting that one mole of any gas, at specified conditions, will occupy the same volume. So, for example, one mole of a gas, any gas, at 'standard' temperature and pressure has a volume of 24 dm^3. As a law, this is expected to apply to a sample of any gas.

There are a number of concepts used in chemistry invoked here, {volume}, {gas}, {temperature}, {pressure}, {moles}... Avogadro's law can then be seen as a relationship between different concepts. In terms of what a concept was set out to be in Chapter 2, there is an {Avogadro's law} concept. I am phrasing the account here as if there is an identifiable canonical version of a scientific concept, such as the {Avogadro's law} concept, which is common to members of the community of chemists. In Chapter 9, I suggest that such an

Chemical Meta-concepts: Imagining the Relationships Between Chemical Concepts 137

assumption about scientific concepts may not be strictly true; but that it is a widely adopted assumption that works well enough most of the time. The argument made here is not undermined if such canonical concepts are simply useful fictions, as it applies to chemists' actual scientific concepts of various laws, *etcetera*, even if these may not be uniform across the community.

In Chapter 2 it was suggested that concepts were mental entities making up aspects of knowledge; that were based on abstractions of experience; that are used as thinking tools; and (as they are mental entities) are only observable to others indirectly when applied in some way. It was suggested that concepts are nodes in a conceptual map or network – so they have relationships with other concepts. An understanding of Avogadro's law seems to meet these criteria. The law is certainly an abstraction, and the {Avogadro's law} concept can be used as a mental tool in chemistry, that is embedded in a network of other concepts such as {molecule}, {gas}, {volume}, and so forth. It may be less immediately obvious that the concept can be understood as the basis for discriminations. Where the concepts {molecule} and {gas} can clearly be applied in such a way – 'that is a gas, but that is not (*e.g.*, it is a solution)'; 'that is a molecule; but that is not (*e.g.*, it is an ion)' – or perhaps in practice 'that is a representation of a molecule; but that is not' – it may seem that Avogadro's law only connects such conceptual categories.

However, effective learning of Avogadro's law does actually involve developing a specific {Avogadro's law} concept. There are two ways that the {Avogadro's law} concept supports discriminations. Someone who has a canonical {Avogadro's law} concept should recognise that:

1. at any particular temperature and pressure, equal volumes of all gases have the same number of molecules;

and

2. equal volumes of all gases, at any particular temperature and pressure, have the same number of molecules;

are both statements of the law (as the law reflects a supposedly universal relationship, not a verbal formulation) but that, for example, various other superficially similar types of statements are not representations of this law. For example,

3. at a particular temperature, the current flowing through a metallic conductor is proportional to the potential difference across it;

and

4. at any particular temperature and pressure, equal volumes of all gases have the same density;

and

> 5. at any particular temperature and pressure, equal volumes of all gases have the same number of electrons.

Our first counter-example (3) is instead an example of a completely different law (*i.e.*, Ohm's law). So, a discrimination using the broader concept {scientific law} might include this statement, but a discrimination using the concept {Avogadro's law} should exclude it. A student's success in recognising that might be seen as evidence of some degree of concept acquisition (but of course it only directly shows the ability to link particular law names with their statements, and could even simply be based on recalling little more than 'Avogadro's law is about gases' or 'Ohm's law is about resistance'). It would of course be possible for students to discriminate the valid statements of the law (1, 2) from the other two counter-examples (4, 5) by memorising the formulation with limited understanding, which could likely allow someone to select a valid response from several options simply by recognising the form of words, as one might learn some lines from different poems.

Learning one formulation (*e.g.*, 1) should allow the second equivalent formulation (*e.g.*, 2, simply reordering some of the sentence clauses) as equivalent simply based on familiarity with the structure of the English language, even if there was limited understanding of the terms themselves. However, without a deeper understanding of the meaning of the law the two examples above that look clearly wrong to a chemist (4, 5) could be easily accepted as statements of the law (pressure, volume, gas, temperature... same *something*, so was it same density? Was it number of somethings, number of electrons, perhaps?)

Recognising that

> 6. Avogadro's law posits that at any particular temperature and pressure, equal volumes of all gases have the same number of molecules;

is equivalent to suggesting that

> 7. one mole of any gas, at specified conditions, will occupy a particular volume;

would require a 'deeper' conceptual understanding. Deducing that 'one mole of any gas, at specified conditions, will occupy a particular volume' from 'at any particular temperature and pressure, equal volumes of all gases have the same number of molecules' requires both some active processing of the statement (*i.e.*, some understanding, beyond simply recalling it, or recognising it) and relating 'the number of molecules' in the original statement to the {mole} concept.

So, one sense of having acquired the concept relates to recognising statements of the law, or to statements that might be considered equivalent to it. However, acquiring the {Avogadro's law} concept also allows the concept to be applied in other ways. For example, a student could be told that 1.00 g of hydrogen gas occupied 10.0 dm^3 of space under some particular conditions of temperature and volume, and the student could be asked which of the following masses of other gases would occupy the same space under the same conditions: 6.00 g of methane; 8.50 g of ammonia; 8.00 g of oxygen; 20.0 g of carbon dioxide. This requires more than knowing the law: one aspect of answering such a question requires using the law to make discriminations of which examples are consistent with the law and which are not. As suggested in Chapter 2, discriminating examples and non-examples of some class (here, examples consistent with Avogadro's law), relies on having developed a concept, here the {Avogadro's law} concept. Similar considerations apply for other laws.

8.1.2 Conservation of Mass

The traditional notions of conservation of mass and conservation of energy came to be seen, after the work of Einstein, as interconnected, for example that energy is always conserved but on the understanding that mass represents a form of energy. (So, the physical concept {energy} was modified to include 'mass-energy'.) This is an important development in some branches of science, such as in nuclear science where {mass defect} is an important concept. However, the mass changes that occur in chemical reactions (that is, the interconversions between mass and other forms of energy) are orders of magnitude smaller than the precision to which mass (or enthalpy) is usually measured, so – in effect – conservation of mass is still treated as a discrete law. This then is an example of a 'law' that is (in the way in which it is usually applied in chemistry) not absolute, but good enough for practical purposes. In practical terms, the conservation of mass is always obeyed in the chemistry lab. or industrial plant to a precision that is actually measured.

In a chemical reaction, the total mass of the products is the same as the total mass of the reactants. This is taken as a law-like relationship (although sometimes labelled as a 'principle') – admitting no exceptions. Any discrepancies are taken to indicate errors in the measurement process (*e.g.*, in school science, this may be not including the mass of a gas that has left a reaction vessel) rather than as challenges to the conservation principle.

A reaction can be defined by an equation, which when written in chemical formulae is usefully ambiguous (see Figure 8.1) – that is, it can as easily represent the reaction at the molecular scale, as much as in molar quantities (Taber, 2013). Conservation of mass can be understood in terms of the measured masses of the actual samples of reactants and products, or the molecular masses. Balanced formulae equations can be read as representing reacting masses in terms of molecular masses, and these numbers are scalable to the bench scale as molecular mass and molar mass are related by Avogadro's constant.

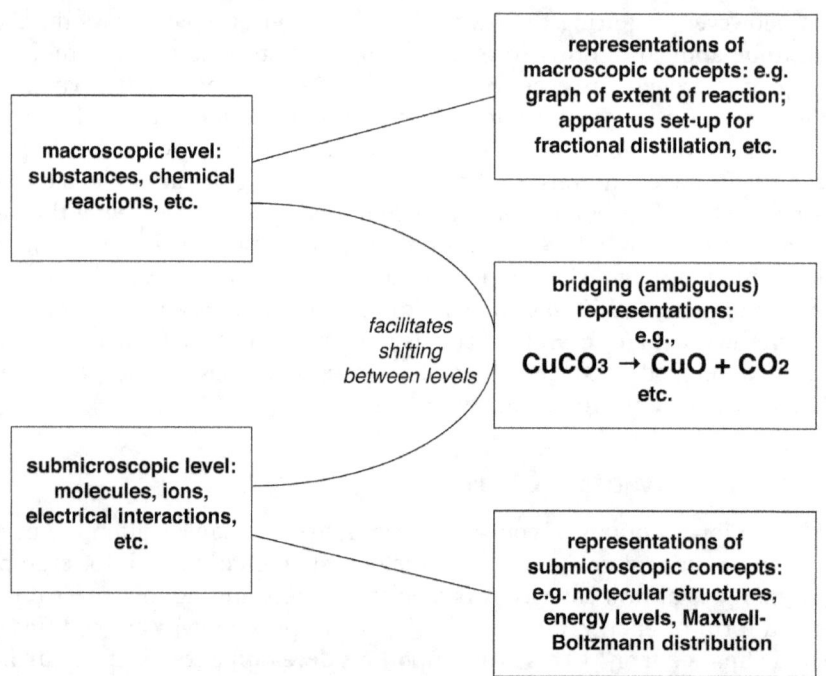

Figure 8.1 An example of a usefully ambiguous representation that bridges the macroscopic and submicroscopic conceptualisations.
Adapted from Taber (2013) with permission from the Royal Society of Chemistry.

Conservation of mass is a law in the strict sense that (within the parameters set out above) it is believed to necessarily apply. It is considered to apply universally, that is, throughout time and space, anywhere in the universe. It is a law in that sense. If at some future time it were to be decided by scientists that conservation of mass only sometimes applies that does not mean we got the law wrong, or that it is a 'law that sometimes applies' (an oxymoron), but rather that this is just a local regularity, and not a law of nature after all.

Reflecting the argument above, that learners develop a distinct {Avogadro's law} concept when they learn Avogadro's law, there is in chemistry a {conservation of mass} concept, and a related but distinct {law of conservation of mass} concept. That is, there is a {conservation of mass} concept (which of course links to the concepts {mass} and {conservation}) such that a person holding the {conservation of mass} concept can discriminate situations that are instances of the conservation of mass, and those that are not. This might be clearer by considering an example that might be examined:

copper carbonate (12.35 g) → copper oxide (7.95 g)

If a person had a canonical {conservation of mass} concept they would recognise that when discriminating between examples and non-examples of

Chemical Meta-concepts: Imagining the Relationships Between Chemical Concepts 141

conservation of mass, that this (as written) is an example of non-conservation. Someone who had a {conservation of mass} concept, but not yet a {law of conservation of mass} concept might be satisfied with that conclusion. As the chemical law of conservation of mass requires there are no instances of chemical processes that are not conservation of mass, a person with a canonical {law of conservation of mass} concept would *also* appreciate that as this example appears to be non-conserving, then it is excluded by a law of chemistry and – assuming that law – there is something wrong with this representation of a chemical change (see Figure 8.2).

To make it clearer that these are two distinct (if clearly related) concepts, we might posit a hypothetical (indeed unlikely, but not impossible) case where a person has a {law of conservation of mass} concept that is canonical, yet a {conservation of mass} concept that is not. Consider the following as a mooted representation of a (fictitious) chemical process:

copper bromide (22.34 g) → copper nitride (18.76 g)

This hypothetical person applies his or her {law of conservation of mass} concept, and decides this relationship meets the requirement of the law. How is this possible? Their concept tells them that all chemical reactions must be instances of conservation of mass, and they had used their (flawed) {conservation of mass} concept to decide this was an instance of conservation; so, then they judge this fits with the law using their {law of conservation of mass} concept. Perhaps their non-canonical {conservation of mass} concept discriminates between instances of 'conservation of mass' where the products and reactants have the same number of letters in their names and instance of conservation of mass (*i.e.*, non-conservation of mass) – where the

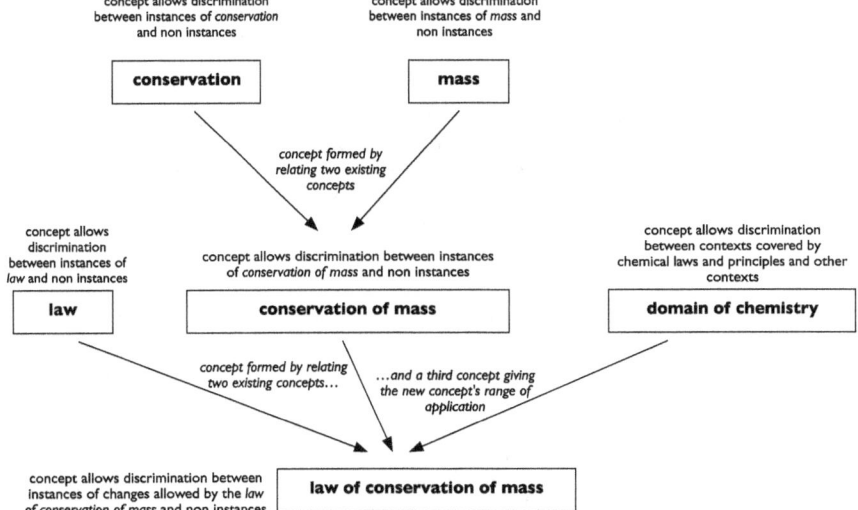

Figure 8.2 Forming a concept of the chemical law of conservation of mass.

number of letters in the name does not remain the same after the chemical change.

Granted, it seems a very unlikely example that someone would form a {conservation of mass} concept so flawed – although some of the alternative conceptions developed by students (Taber, 2014), see Chapter 14, may seem almost as odd to the expert chemist – but the point of the hypothetical example is to show there can be seen to be two distinct concepts at work here (see Table 8.1). As an analogy, we might imagine a child being told that they must not get out of bed to start opening presents on their birthday until after their alarm clock had sounded. Perhaps the child was awoken by a nearby burglar alarm very early in the morning, genuinely thought this was their alarm clock, and in good faith proceeded to start tearing the wrapping paper from the birthday presents. Technically the child has broken the rule, although they did not intend to do so.

We might suspect that someone accidentally breaking a rule does so because they do not understand that rule. However, in this case the flaw was not in misunderstanding the rule that had been laid down (stay in bed till the alarm sounds), but in applying the criterion essential to applying the rule (has the alarm sounded?). It is wrong to go a friend's house and remove their property without their knowledge and permission. Yet the absent-minded academic who leaves for home wearing the host's coat, thinking it is his own, does not break the rule because of any failure to understand the rule, and could likely give a very clear, if possibly tedious, explanation of the incident.

8.1.3 Chargaff's Rules

An example of a law-like regularity would be Chargaff's rules. These refer to the composition of DNA. If our canonical concept of {substance} in chemistry includes the principle that a substance has a fixed molecular (or equivalent)

Table 8.1 A hypothetical example to illustrate that the {law of conservation of mass} concept is distinct from the {conservation of mass} concept.

Discrimination made	Alternative {conservation of mass} concept	Canonical {conservation of mass} concept
Application of {conservation of mass} concept	Copper bromide: 13 letters Copper nitride: 13 letters 13 = 13 Mass is conserved	Copper bromide: 22.34 g Copper nitride: 18.76 g 22.34 g ≠ 18.76 g Mass is not conserved
Application of {law of concept of mass} concept	Law requires mass to be conserved Mass is conserved Therefore, law is obeyed	Law requires mass to be conserved Mass is not conserved Therefore, law is not obeyed

composition and structure, then DNA is not a substance and we should not refer to 'the' DNA molecule (perhaps even in the loose sense in which the definite article is commonly used to refer to ideal prototypes – see Chapter 7). Different DNA molecules have a good deal in common, but are not identical in terms of precise composition, molecular mass, or sequences of components. More correctly then, DNA is a term for a large class of technically different substances with strong similarities.

So, all DNA molecules have certain commonalities of structure – the sugar–phosphate 'backbone' is much the same in different DNA molecules, although it can vary in length. The two strands are always linked by pairs of bases, and these bases are always adenine, thymine, guanine and cytosine. These four substances are heterocyclic compounds, that is their molecular structures include rings, and these rings are not simply rings of carbon atoms (as in cyclohexane for example) but have more than one (elemental) type of atom included. Here, the rings include nitrogen as well as carbon. The molecular structures of thymine and cytosine include a single six-membered ring, including two nitrogen atoms. Substances whose molecules have this type of structure are called pyrimidines – uracil (found in RNA) is another pyrimidine. Adenine and guanine have larger molecular structures with a similar six-membered ring fused to a five-membered ring that is also heterocyclic with two nitrogen atoms. Substances with this type of molecular structure are known as purines.

The composition of DNA (or perhaps, technically, examples of the composition of some DNAs) was known in advance of its structural elucidation. Some samples of DNA from different organisms were analysed by Erwin Chargaff. He discovered that although the precise proportions of adenine, thymine, guanine, and cytosine would vary from one source to another, there were some regular patterns. In particular his analysis showed that the amounts of adenine and thymine extracted from a subject were always (approximately) matched, as were the amounts of guanine, and cytosine. That is, whilst for example the ratio adenine:cytosine might vary considerably from one organism to another, the ratio adenine:thymine was found to be close to 1 : 1, across a range of diverse species.

Chargaff's rules can be considered as one of the constraints that Crick and Watson had to consider in building a model of the chemical structure of DNA (an example of a model considered below). The rules also offered a useful clue: as adenine (a purine):thymine (a pyrimidine) :: 1 : 1 *and* guanine (a purine):cytosine (a pyrimidine) :: 1 : 1, then in general purine:pyrimidine :: 1 : 1 as well. As purines have much the same basic structure as each other but are larger molecules than the (structurally similar to each other) pyrimidines, then a structural unit of purine-pyrimidine would be similar size and overall structure whichever pairings were involved (A–C, A–T, G–T, G–C, adopting commonly used letter abbreviations), and the rules would be met if there were A–T and G–C units within the structure.

Now that this structural model is widely accepted, it can be used to explain the rules Chargaff inferred from his empirical studies. Given the accepted structure of DNA, then it logically follows that its composition obeys the rules:

number of purine units = number of pyrimidine units;
number of adenine units = number of thymine units;
number of guanine units = numbers of cytosine units.

This might now be considered as a law: given what we define as DNA, all DNA must have composition with these patterns, so (given current scientific understanding) there can be no exceptions. Chargaff himself, given the technical difficulties of carrying out the work, reported results that only closely approximated to the rules, but whereas the empirical work suggested approximate parities, theoretical considerations now require these equalities. What seems obvious now, was once just one possibly relevant idea of uncertain value – as was captured in one dramatised account of the discovery of DNA structure:

> EC: It intrigues me that you gentlemen think that you can establish a structure for DNA when you don't trouble yourselves about the differences between the bases. There are four of them, of course. I grant that's a lot to hold in the mind all at once.
> JW: Dr. Chargaff, I've read one of your papers. Don't your results show a consistent ratio between the bases in all the types of DNA you examined?
> EC: Yes.
> JW: Don't your results point to a pairing of pyrimidines and purines?
> EC: No. I shall say this slowly, because it involves some long words. My results show a rough equivalence between adenine and thymine, and between guanine and cytosine.

(Dialogue from the Film "Life Story" (Nicholson, 1987) – a dramatisation of the work to discover the structure of DNA. This section portrays Erwin Chargaff (EC) becoming frustrated with Crick and Watson (JW) when they ask him about his research, apparently without having taken the time to grasp the basic concepts involved.)

As with other examples discussed in this chapter here, it is clear that not only do Chargaff's rules link to a range of other chemical concepts – centrally {DNA}, {purine}, {adenine}, ..., but also {base}, {heterocyclic compounds}, *etcetera* – but that a learner can form a specific concept {Chargaff's rules}. Someone with a canonical {Chargaff's rules} concept would not only know and be able to apply the rules, but know they were the rules designated in this way. A holder of a {Chargaff's rules} concept would be able to discriminate between Chargaff's rules and Chargaff's rules (*i.e.*, anything that is not Chargaff's rules). As with the example of the law of conservation of mass, someone learning about Chargaff's rules who formed an alternative

Chemical Meta-concepts: Imagining the Relationships Between Chemical Concepts   145

conception (see Chapter 14) of the rules might apply their flawed {Chargaff's rules} concept to make inappropriate discriminations, *e.g.*,

> Chargaff's rules tell us that DNA always contains the same amount of the two purines, and the same amount of the two pyrimidines: the same amount of adenine as guanine, the same amount of cytosine as thiamine.

A student who had this (mis)conception might have generally canonical conceptions {DNA}, {purine}, {pyramidines}, {adenine}, {guanine}, {cytosine}, and {thiamine}, but not be able to discriminate the genuine rules from superficially similar ones, so the error here relates to the specific {Chargaff's rules} concept.

8.1.4   Raoult's Law and Its Deviations

Raoult's law concerns the vapour pressure of a mixture of liquids. It tells us that for an ideal mixture of liquids, the partial vapour pressure of each component will be given by taking the vapour pressure of a pure sample of that component (under the same conditions) and multiplying it by the mole fraction of that component in the mixture.

Despite needing a somewhat complex statement, the pattern reflected by the law is itself quite simple to understand. If one liquid, A, has a vapour pressure X under some set of conditions (*e.g.*, temperature, *etc.*), and another liquid, B, a vapour pressure of Y under the same conditions, then a mixture of the two would have a total vapour pressure that is a simple sum of the two partial pressures, which themselves are in simple proportions of that components' concentration in the mixture. So, a mixture of three quarters A and one quarter B will have a total pressure of $0.75X + 0.25Y$.

There is then a canonical concept {Raoult's law}, which might be expressed in the terms above, or some equivalent terms, such that it is possible to discriminate Raoult's law from Raoult's law (*i.e.*, anything that is not Raoult's law). The latter category would include, *inter alia*, lettuce, copper sulfate crystals, and Nobel prize medals. More significantly, someone with the canonical concept can discriminate statements of Raoult's law from statements of, say, Graham's law of diffusion, and, perhaps even more significantly, from superficially similar statements such as 'the partial vapour pressure of each component in a mixture of liquids will be given by taking the vapour pressure of a pure sample of that component (under the same conditions) and multiplying it by the molecular mass of that component', for example.

It was suggested above that the term 'law' should be reserved in science for generalisations that are considered to be universal, without exceptions, applying to all examples of the phenomenon it claims to refer to. So, the law of conservation of mass (see above) is considered to apply to all chemical reactions – rather than, say, all chemical reactions apart from those occurring on Thursdays, or those involving phosphorus, or those occurring in a strong magnetic field, or those carried out by unskilled school children and careless undergraduates. Because this is considered a scientific law, and as this is considered very robust scientific knowledge (*i.e.*, even those of us who make a point of emphasising the conjectural and provisional nature of scientific knowledge are pretty confident this notion is not going to need revising any time soon), when school children or undergraduates report apparent examples of non-conservation we do not tend to consider these as refutations, but more likely as shoddy laboratory work.

In this sense, Raoult's law would seem to be pretty secure as it is set up to be protected by tautology. Strictly, Raoult's law applies to ideal mixtures of liquids (those liquids discriminated by the {ideal mixtures of liquids} concept). An ideal mixture of liquids is one that shows certain properties, including obeying Raoult's law. So Raoult's law must apply to all ideal mixtures of liquids, as all ideal mixtures must obey Raoult's law. This may seem to make the 'law' rather pointless, along the lines of claiming that Keith Taber is the best teacher apart from all those teachers who are not Keith Taber (this reminds me of some wit who referred to a television programme getting the award for being the best situation comedy that year that was screened on Channel 4 and that had been set in a barber's shop).

There is a strong parallel here with the ideal gas law, which is also a scientific law, in that it is meant to admit no exceptions, but strictly refers to ideal gases. Real gases will approximate to the law, to differing degrees, under different conditions. A cynic might suggest that in terms of gases that actually exist, the ideal gas law is (rather than being a law that always applies) *a law that sometimes, almost, applies*. It can still be valuable to have knowledge of an approximate relationship that breaks down under some circumstances as long as it does so in a reproducible way so that it is possible to know when to expect the law to (almost) apply, and when it will not offer reliable predictions. However, it is questionable whether the term 'law' is justified for a regularity that is conjectured to always work exactly in some special non-existent conditions, and found to only sometimes approximately work in real cases. As there are no ideal gases, there are no instances where the ideal gas law strictly applies.

Theoretical considerations suggest that in principle the ideal gas law cannot be precisely accurate for real gases as it can be reconstructed on the assumptions that the particles in a gas take up a negligible volume compared with the volume of the gas itself and that the particles only interact through completely elastic collisions. Thus, the motivation for the development of the van der Waals' equation of state for a gas that includes terms to allow for the molecules having a volume, and the gas particles interacting with each other (neutral molecules in a gas, even if they have no net dipole, can form fluctuating transient dipoles and so there are van der Waals' forces between them).

In the case of Raoult's law, the theoretical interpretation considers how molecules in a liquid interact with each other. If, in a mixture of two liquids,

Chemical Meta-concepts: Imagining the Relationships Between Chemical Concepts 147

the molecules interacted in equivalent ways with both other same-substance and other-substance molecules, then the liquid could be ideal, and obey the law. Deviations from Raoult's law, the extent to which the vapour pressures for real mixtures deviate from the predictions of the law, are considered to be positive or negative depending upon whether the measured vapour pressure is greater or lesser than the value the law predicts. It might seem that it should be possible, in principle at least, to identify ideal mixtures of liquids, as the requirement relating to the strength of molecule interactions is not as unhelpful as the requirement for an ideal gas that it must have volumeless molecules that do not interact at all. Yet, that is only considering part of the system, the liquid, and ignores the vapour. Strictly, for Raoult's law to apply, the vapour must be an ideal gas.

It was suggested above that the canonical {Raoult's law} concept should allow the holder to discriminate between a statement of Raoult's law and some other statement, including one superficially similar. A canonical understanding of the chemical concept clearly includes other features discussed here – such as the notion of ideal mixtures of liquids, positive and negative deviations, and 'fugacity' (the corrected partial vapour pressure of a component allowing for its non-ideality as a gas). These can also be seen in terms of distinct concepts. Someone who held the concept {negative deviation from Raoult's law} should be able to discriminate instances of negative deviations from the law from ¬negative deviation from Raoult's law (that is, all other things that are not negative deviations from Raoult's law): *inter alia*, London buses, conical flasks, and – more pertinently, perhaps, non-deviations, and positive deviations, from the law. Again, learning a chemical concept, here the {Raoult's law} concept, is complex and nuanced.

8.1.5 Other Examples of Law-like Principles

Clearly there are many other examples of law-like regularities that have been identified and are applied in chemistry: the conservation of charge (and so by extension oxidation numbers – the balancing of oxidation and reduction steps in a reaction); the Dulong–Petit law; Le Chetalier's principle; Proust's law of definite composition; Dalton's law of multiple proportions; the Arrhenius rate equation; the Nernst equation . . . .

What each of these cases have in common is that they posit some abstract relationship that depends upon and links other concepts. Conceptualising these laws and principles requires the learner to appreciate a relationship, often between abstract or theoretical entities. The concept of the law is therefore embedded in a network of other concepts. Someone who had 'learnt' the concept could both apply the concept and recognise examples or incidences of the law, discriminating them from non-examples. This can clearly be a matter of degree, with different levels of understanding and sophistication.

It is also clear that, even if the relationship itself is understood as intended, conceptualisation would be flawed if the person held an alternative concept of the one of the other concepts subsumed under the law. So, someone who knew that the Dulong–Petit law suggested that the product of the specific heat capacity of an element and its molar mass is the same for all elements, would only be able to use the concept canonically if they had canonical understanding of the concepts related by the law. If someone held the alternative conception that the molar mass of an element was equivalent to the proton number of the atom of the element scaled up to be in grammes, or if they did not discriminate canonically the concept {product} from the concept {sum}, their {Dulong–Petit law} concept (even if phrased canonically) would be deficient accordingly. So, once we start to analyse what is involved in acquiring canonical chemical concepts, especially chemical meta-concepts, we can identify many ways in which such learning can go wrong.

8.2 Concepts and Models

Science develops many models of aspects of the natural world. Models are representations of (conjectured) aspects of reality and they are used as thinking tools (Gilbert and Osborne, 1980). They can support the development of explanations and can offer predictions for testing. Models are simpler than reality: deliberately excluding some features that are judged less relevant for particular purposes. As representations, they may take quite a different form from the reality being modelled. Accordingly, it is possible for several scientific models, representing the 'same' phenomenon, but with different foci, and taking different forms, to be quite distinct and yet still admissible as scientific models of that single phenomenon.

Younger students tend to be most familiar with material models, such as scale models (Treagust *et al.*, 2002) but, in science, some models are not realised in material terms: for example, they may be purely mathematical. Models are not in themselves concepts, but they draw upon scientific concepts, *and they are themselves conceptualised*. (That is, just as with laws, we can form a concept of the specific model itself.) Understanding models may require a student to appreciate the concepts underpinning them; and using a model appropriately requires a conceptualisation of the nature of model *qua* model. In parallel with what was suggested for laws and principles, learning about models involves forming concepts of specific models encountered, as well as developing concepts of the idea of models in general and the process of modelling.

One famous physical scientific model was built by Crick and Watson to represent the chemical structure of DNA, or, to be more precise, the structure of a salt of DNA (Watson and Crick, 1953). (The original model was deconstructed once no longer needed as a thinking tool – but there are replicas, so the following refers to the model as if it still exists.) The model is not meant to be identical to DNA, or to represent all aspects of DNA. The model is obviously very much scaled up from what is being modelled – a molecular structure. It is materially different as it was made from metal parts such as pieces of aluminium. The components were physically joined whereas what they represented was chemically bonded together.

The physical model was a way of giving material form to a structural hypothesis – it represented a particular arrangement in three-dimensional space, to test a conjecture about molecular structure that was constrained by geometry as well as empirical data relating to chemical composition (*e.g.*, see the discussion of Chargaff's rules above) and X-ray diffraction patterns. The complementary pairing of the two main strands (intended to reflect a symmetry group that was inferred based on X-ray data) was presented as offering a new hypothesis about how DNA is replicated *in vivo*.

Reference to 'the DNA molecule' is technically dubious – one of the key features of DNA is that there is not one (type of) DNA molecule in the ways there is one (type of) methane molecule. DNA comes in myriad varieties, which is why it can be considered to 'code' genetically. Yet, Crick and Watson did not set out to model every variety of DNA molecule, or even one particular DNA molecule, but just the general type of structure. This simplification was fine because they were looking for a proof of idea, not a model of any particular exemplar of DNA. Their model was very much shorter than actual DNA molecules (not in terms of absolute length, of course, but in terms of the number of repeats of the basic type of units conceptualised to comprise DNA) as that was sufficient to represent the features they wished to test.

So, Crick and Watson's model was not a concept, but clearly it was built as much from conceptual knowledge as from physical components. It made no sense without a {DNA} concept, nor many other concepts such as {ribose}, {purine}, {pyrimidine}, {hydrogen bond}, {helix} and so forth. For a student to be said to understand the model, a teacher would expect them to have reasonably canonical versions of at least a sub-set of these concepts. (Perhaps a biology teacher with a focus on the mechanism of DNA replication might be looking for a different sub-set than a chemistry teacher focusing on the molecular structure.)

That there is a concept of {DNA model} also seems clear. A student who says "this is DNA" when asked to identify a model may be assuming that it is obviously a model and that need not be stated. However, a hypothetical student who did not have a reasonably canonical concept of {scientific model} could not have a canonical concept of {DNA model} even if they were able to say "this is a model of DNA" as their labelling of the object as a model would not carry the necessary associations of the representative nature of scientific models discussed above – so perhaps for example they might deduce that real DNA is as rigid as the model they have identified, or that it comprises a double helix formed around along a long straight linear axis. That is, the concept {DNA model} is subordinate to the subsuming {model} concept and so will take on any deficient or alternative associations of a person's {model} concept.

Moreover, there are specific variant concepts of the {Crick and Watson DNA model}. The plurality here refers to discriminations one might make. In their paper in science Crick and Watson presented a model of the molecular structure of DNA. It was in effect a model of the same conjectured structure as the model they had physically built. But it was not the same model: building and shipping person-sized metal constructions with issues of scientific journals is not feasible. Indeed, in the physical model the arrangement and geometry of component parts are modelled in a fairly realistic sense (bits are joined together in pretty much the way Crick and Watson were imagining would be the case in the actually structure being modelled), but, in the paper, there is a schematic graphical representation of the double helix lacking details of the component parts, but supplemented by a verbal description offering more detail. Both of these very different – in terms of realisation – models were meant to present the mental model the scientists had formed.

So, one {Crick and Watson DNA model} concept refers to the representation presented in their research paper, and another {Crick and Watson DNA model} concept refers to the physical model they had built. Someone holding these two distinct {Crick and Watson DNA model$_{original\ physical\ model}$} and {Crick and Watson DNA model$_{published\ report}$} concepts and visiting the Cavendish Laboratory would have been easily able to discriminate with these concepts to divide the world into three sets of objects: the physical model (a set of 1 at the time); the published model represented on paper (in multiple printed copies then, now also available electronically); anything else in the universe.

Another {Crick & Watson DNA model} conceptualisation might refer to the original physical model, plus any other model constructed to be a replica of this. Someone visiting the Science Museum in London may use their {Crick and Watson DNA model$_{original\ physical\ model}$} concept to (mistakenly) discriminate that the model on display is the original, whilst another person aware of the deconstruction of the original model might discriminate that the model there is a Crick and Watson DNA model using their {Crick and Watson DNA model$_{models\ of\ the\ form\ of\ the\ original\ physical\ model}$} concept. It is also arguable that the concept {Crick and Watson's proposed scientific model of the structure of DNA} is what is really of most interest chemically, and that refers to the mental model that the physical and published models were intended to represent.

I imagine most readers will consider that they can readily discriminate between the scientific model of DNA that Crick and Watson developed, the physical model they used to test the idea (and any replicas of it), and the published account with its simplified diagrammatic representation; and, if so, they have (distinct) concepts of these distinct models. Perhaps this seems a trivial point, as chemistry experts are unlikely to conflate these different models or get confused between them. Learners, however, are relative novices, and have to be introduced to such distinctions – to develop a concept of DNA the material, a concept of 'the' general DNA molecule (and later perhaps concepts of specific variants), as well as of the scientific model of the structure and various representations (which can be seen as models realising aspects of the scientific model).

Mastering this clearly involves having a sophisticated {scientific model} concept (or else the physical model is likely to be seen *as* the scientific model) that we know is not available to many younger learners (Grosslight et al., 1991), and so has to be developed. Arguably, the scientific model of DNA is a conceptual model, and so only exists as a mental entity that is represented (and indeed itself modelled) in various material and symbolic forms – such that the scientific model can only be communicated indirectly through the various representations (an idea developed in Chapter 11).

8.3 Concepts and Theories?

Another key kind of idea met in chemistry, or indeed science more generally, is that of a theory. It has been found that school students often develop a concept {scientific theory} along the lines that a scientific theory is an idea or hunch a scientist has, but which has not yet been proven by testing (Taber et al., 2015). However canonically, the concept {scientific theory} refers to a well-established set of ideas positing relationships between different concepts that can be used as the basis of explanations and predictions.

Theories not only relate concepts to each other, but may include specific laws and/or models. So, for example, the Arrhenius rate equation takes the form of a law in that it suggests a rigid regularity. However, the rate law reflects an underpinning theory (involving modelling) of what occurs during chemical reactions that provides the conceptual basis for the actual relationship. As often in science it is too simplistic to see the Arrhenius rate equation as a deduction from empirical investigation or as *a priori* theorising – there is an iteration between theorising and practical investigations (*cf.* Figure 5.1) that leads to something that becomes canonical as part of the conceptual content of chemistry.

The link between concepts and theories relates to one of the ways concepts were described in Chapter 2. There it was suggested that concepts are usually linked to other concepts, and indeed we can think of concepts as nodes in conceptual networks. So, a particular concept such as {acid} (see Chapter 6), or indeed {oxidising agent}, {heterocycle}, {conical flask}, *etcetera*, is understood by its associations with other concepts, and these associations are part of the 'content' of the concept. So, part of the content of the canonical chemical concept {acid} is the relationship between substances discriminated by this concept as being acids, and those discriminated by the associated concept {base} as being bases; part of the concept is that acids are reactants in a type of reaction known as neutralisation – that can be discriminated by a {neutralisation} concept (that to be fully understood relies on understanding of another concept {chemical reactions}, and a related concept of {types of reactions} as 'neutralisation is a class of chemical reactions') – and so forth.

Within this perspective, theories can be thought of as particular networks of concepts and their associations that can be picked out from the boarder conceptual network. So, for example, the Brønsted–Lowry theory of acids (an example discussed in Chapter 6) can be characterised as a set of related ideas, which can be represented in a set of verbal statements, which define and associate a range of concepts. This particular theory includes as central concepts {proton donor} and {proton acceptor}, as well as including concepts such as {acid}, {base}, {neutralisation}, *etc.*

A theory is not a concept, then, but something more encompassing that depends upon concepts and imposes particular relationships on them. However, chemists do also form concepts of the theories themselves. That is, in developing a theory, a scientist works with particular established concepts, perhaps (deliberately or otherwise) offers a modified version of some of those concepts (see, for example, Chapter 6), and presents a particular conceptual framework that posits specified relationships between the concepts. Some of these links may already be known, but a new theory has some new conceptual content: new concept(s) and/or modified concept(s) and/or new associations between concepts and/or modified associations between concepts.

In this way, something new is created. What is new is not a physical object that did not exist before (as when scientists produced new synthetic products, new semiconductors, new superconductors, transuranic elements, *etc.*) but a mental entity. If this new entity is to be shared among the community of chemists then other chemists will need to conceptualise the new entity. The notions of sharing and communicating ideas deserves some closer attention, and will be discussed in later chapters (see Chapters 11 and 12), but there is clearly a sense that science proceeds in large part through the *sharing of new ideas* among scientists, and that chemistry teaching is largely about *communicating canonical ideas* to learners.

For a new theory to be successful it will need, at a minimum, to be recognised as a new chemical theory so that it can be the focus of discourse in the community. In practice, this process may be supported by naming the theory. A label may not be an essential criterion for having a theory, but pragmatically it may be difficult for it to be the subject of wide discourse without one. So, chemistry will have not only a new theory, but a new concept that likely has a recognised tag. There is a concept of {Brønsted–Lowry acid–base theory}, such that someone holding the concept can discriminate instances that are covered by the concept, and those that are not.

So, for example, someone who had the concept of {Brønsted–Lowry acid-based theory} might consider a statement such as "species with lone pairs of electrons that are able to donate these to form dative bonds can be considered as bases" as not being part of the concept {Brønsted–Lowry acid-based theory}. This statement ("species with lone pairs of electrons that are able to donate these to form dative bonds can be considered as bases") is not a statement of part of the theory, nor is it some equivalent statement. Given the nature of a theory, the judgement that *"species with lone pairs of electrons that are able to donate these to form dative bonds can be considered as bases" is not part of Brønsted–Lowry acid-based theory* need not be a rejection of the statement itself, just as *judging that the statement "one way a batsman can be dismissed is leg before wicket" is not part of the rules of association football* does

Chemical Meta-concepts: Imagining the Relationships Between Chemical Concepts 153

not imply denying the statement "one way a batsman can be dismissed is leg before wicket". Rather, it is a judgement that the statement is irrelevant in the current context and so has an indeterminate truth value in that context. This judgement is only possible when someone has acquired a {Brønsted–Lowry acid-based theory} concept.

Chapter 6 considered the issues of whether there is a canonical concept of acid in chemistry today, and, if so, whether it is some kind of hybrid which encompasses distinct models/theories/definitions of acids. Distinctions were drawn there between different acid concepts such as {$acid_{Brønsted-Lowry}$} and {$acid_{Lewis}$}. Considering concepts of theories, such as of the Brønsted–Lowry theory of acids, is clearly related to considering concepts of acids. However, these are not quite the same thing. An {acid} concept allows someone to discriminate whether something is an acid or not, and a specific {acid theory} concept allows someone to discriminate whether something is applying/part of/consistent with that particular theory (and of course the {theory} concept allows us to discriminate whether something really is a theory rather than something else, such as a hunch).

The concept {acid} (or perhaps, in the light of Chapter 3, *an* {acid} concept) would be used, for example, to decide whether trimethylborane should be considered an acid or not, whereas an acid theory concept, such as the {Brønsted–Lowry acid-based theory} concept, would be used to decide whether a statement such as 'trimethylborane is an acid because it can act as an electron-pair acceptor' is an application of a particular theory. Students of chemistry may be expected to make both kinds of discriminations, and so are expected to acquire both kinds of concepts.

8.3.1 Particle Theories

An especially important group of theories in chemistry are those that concern the structures of substances at a scale far removed from everyday experience, often referred to as 'submicroscopic' to imply that they relate to a scale well below what can be seen through the (optical) microscope. The concept {microscope} originally discriminated an instrument using optics to provide a visual image of the very small. The development of electron microscopes, involving somewhat similar principles based on using beams of electrons rather than visible light provided a new type of microscope, and so the {microscope} concept evolved to discriminate a wider range of entities (*i.e.*, as microscopes rather than microscopes – things that are not microscopes), which could be further discriminated using the subordinate concepts {optical microscope} and {electron microscope}. Detail can be seen in electron-microscope images (that is visual images obtained by processing the 'images' produced by the electrons) that cannot be seen using optical microscopes, so complicating the matter of just what scales are discriminated by the concept {submicroscopic}.

So-called 'scanning tunnelling microscopes' (STM) are instruments that produce images of surfaces that show individual atoms, so if they are considered to fall within the category discriminated by the concept {microscope} then (what are usually considered) submicroscopic features can be detected by (these) microscopes. If electron microscopes are broadly analogous to optical microscopes but using different frequencies of radiation, STM are quite different in principle, and work by a process that is more analogous to feeling one's way across a surface rather than looking at it.

There are some similarities to how the concept {microscope} developed and how the concept {potassium} developed, as discussed in Chapter 5. However, potassium as a substance is considered to be a natural kind (see Chapter 7), so although the {potassium} concept shifted as chemists discovered new things about the existing kind potassium, laboratory equipment is not found 'in nature' and is rather designed and constructed by people. The electron microscope was not discovered as a new form of microscope, but was a new technology informed by optical microscopy and theoretical considerations (*i.e.*, the relationship between the wavelength of radiation and the resolution possible in an image). At that point in time, then, there was a choice to be made whether to consider the electron microscope as a new form of microscope or a completely new type of technology, perhaps indicated by a different term (say an electron beam imaging device, EBID).

Similarly, the STM could have been given a different name – perhaps a NETJ scanner because it shows when electrons are 'near enough to jump' between the probe and the surface. It might be objected here that referring to electrons 'jumping' the gap is not very scientific language, as they do not actually jump, rather, they actually tunnel their way across. Perhaps we have become so accustomed to the 'tunnelling' metaphor being used here that it has become a 'dead metaphor' (Lakoff, 1987), and is now taken as a literal description. Words, of course, can mean whatever we want them to mean, and the issue in education is whether intended meanings can easily be communicated (see Chapter 12). Besides being another reminder of how concepts may shift, this actually relates to an issue of how to describe molecular scale entities in chemistry education.

8.3.2 Micro-/Macro-distinctions

A major issue that has been identified in chemistry education is how learners cope with studying a subject that is often discussed across two very different levels. These have been described as the macro-level (or macroscopic, or sometimes molar, level) referring to the scale at which chemicals are handled and observed in the laboratory, and the micro-level (or microscopic, or sometimes molecular, level) referring to the scale at which entities such as molecules, ions, electrons, and the like, are considered to exist.

Alex Johnstone (1982) pointed out that chemistry lessons often involved presentations of a mix of both the molar-level descriptions and the molecular-level accounts, and tended to also include specialised representations such as chemical formulae and reaction equations. Johnstone was

Chemical Meta-concepts: Imagining the Relationships Between Chemical Concepts 155

concerned that one of the barriers to effective chemistry learning was that there would be information overload given the limitations of learners' working memories (Johnstone *et al.*, 1994).

Examination of student explanations in chemistry suggested that students often seemed to be confused about the micro- and macro-levels, and how descriptions at the two levels were meant to link up (Ben-Zvi *et al.*, 1986; Taber, 2001). More recently the term micro-level has tended to be replaced in discussions of this issue by the term submicroscopic level – a level below that which can be observed with a microscope (see Figure 8.3). One motivation for this was the finding that novice students often underestimate the scale of molecules – and may assume that although molecules and ions are very small, they can be seen under the microscope. (Indeed, Chapter 7 discussed an example of the comments of a chemical weapons expert, someone whom might be presumed to be knowledgeable about chemistry, suggesting it may not just be novices who struggle to appreciate the extent to which single molecules are on a scale so far from familiar experience.) Interestingly, since Johnstone's original article on this matter, the prefix 'nano-' has come into wide use in contexts such as nanoscience and nanochemistry, and in hindsight it might have been better if Johnstone had chosen to compare

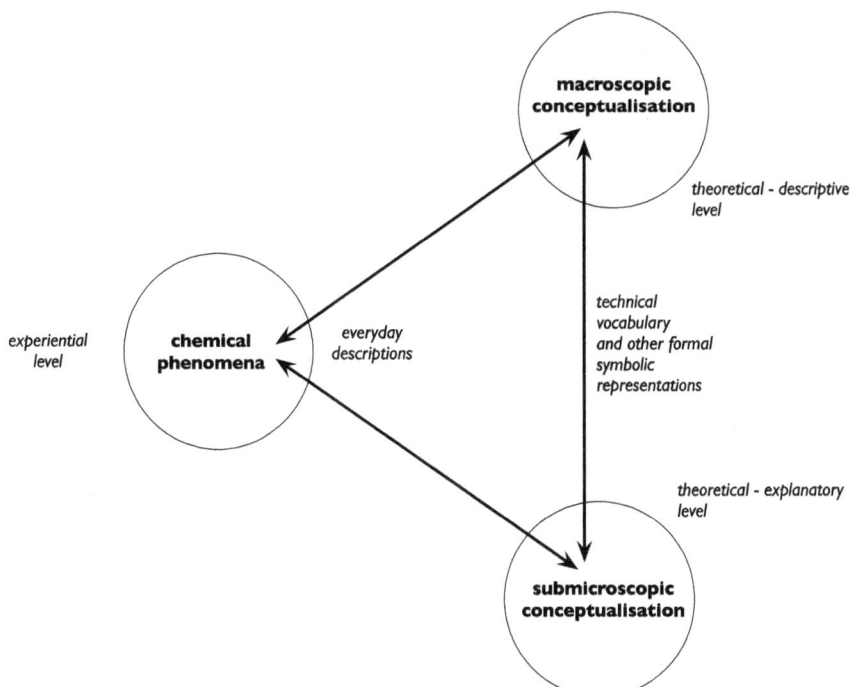

Figure 8.3 Three levels at which chemistry is conceptualised.
Reproduced from Taber (2013) with permission from the Royal Society of Chemistry.

the macro-level with the nano-level. But then, as the visionary William Blake noted, hindsight is a wonderful thing.

8.3.3 Relating the Two Levels

Particle theories are used in science to develop explanations of what can be observed macroscopically in terms of the structure and nature of substances on a much smaller scale. The particle theories posit the existence of particles composing materials at a submicroscopic scale, and suggest that the properties and interactions of these submicroscopic entities explain the properties observed at the scale of bench chemistry. In principle, the entities and their properties are conjectural – so theoretical – entities, even if – in practice – most chemists come to consider such entities as electrons, sodium ions, and methane molecules, and even sometimes 3p orbitals, to be as real as grains of salt, lumps of sulfur, or conical flasks.

The scope of these theories has become extensive – so they may be used to explain such phenomena as:

- The measured pressure of a fixed volume of gas increases when it is heated.
- A solution of sodium chloride conducts electricity.
- Copper salts are often blue or green.
- A mixture of the gases hydrogen and chlorine is stable in the dark, but reacts explosively in sunlight.
- Carbon, in the form of graphite, is an electrical conductor, but carbon, in the form of diamond, is an insulator.
- Octane has a higher boiling temperature than butane.
- Ethene readily undergoes addition reactions but benzene does not.
- Chromium forms compounds with various oxidation states.
- The rate of many chemical reactions approximately doubles with a 10 °C increase in temperature.
- If ammonium nitrate at room temperature is dissolved in water at room temperature, a solution is formed at a temperature lower than the ambient temperature.
- The solubility of sugar increase substantially with increasing water temperature, but the solubility of salt shows little change.
- Molecules of DNA can act as the source of genetic information in organisms.

This list of examples could be extended considerably.

8.3.4 What Do We Mean By an Observable?

Some of the examples in the list above clearly relate to what might be considered observations. Samples of copper salts can be directly inspected. That 'copper salts are often blue or green' could be a deduction from observing samples from a wide range of copper salts. Even here, some specialised conceptualisation is required. The concept {copper salts} relies upon disciplinary knowledge: copper salts are not given in nature for people to notice in the same way that trees or giraffes are picked out. The production of pure samples of copper salts that could be subject to observation relies upon identifying copper compounds in minerals, then separation and purification of samples of the compounds. So, inspecting a range of samples that are presented as 'copper salts' is only possible once someone is able to use the concept {copper salts} to discriminate instances of copper salts from copper salts (*i.e.*, all things that are not copper salts).

Despite electrical currents being ubiquitous in our technologically enhanced world, they are not directly observed (except, arguably, when we get a shock), but are only detected indirectly through their effects. The statement that 'a solution of sodium chloride conducts electricity' appears quite simple, but relies upon both discriminating using the concept {solution of sodium chloride} (that, in turn, requires identifying sodium chloride, and knowing what counts as a solution, so applying the concepts {sodium chloride} and {solution}), but also having an indicator of electrical current. We might think that someone could have a purely operational concept {conducting electricity}: we might imagine a hypothetical technician who discriminated conducting electricity from conducting electricity (*i.e.*, everything that was not conducting electricity) in terms of knowing that when she threw 'this' switch, and 'that' meter needle clearly shifted position, this was an instance of {conducting electricity}. That would reflect very limited understanding, and a more canonical concept of {conducting electricity} would relate to other concepts ({current} and {conductance}), and would also involve submicroscopic models of how the flow of current could be explained in terms of tiny charge carriers.

8.3.5 Can We See Salt Dissolve in Water?

Even the statement 'sodium chloride dissolves in water' that might seem to be purely based on observation, actually relies on conceptualisation (concepts are needed to decide what counts as sodium chloride, and as water, and what is meant by dissolve). It is everyday knowledge that salt, that is, table salt or common salt, dissolves in water, and not just some specialised notion only available to chemists or students of chemistry. We can all see that salt dissolves in water. Take a glass of water, add a tablespoon of salt, and watch and wait. If you are impatient then use hot water, and stir the mixture. The salt disappears.

Some chemists may be uneasy about the phrasing that the salt disappears, as 'disappears' is sometimes used to mean something ceases to exist (in the way dodos disappeared), but the salt ceases to be visible. So, what is seen is the salt ceasing to be visible – what is 'seen', is that the salt can no longer be seen! Laypeople sometimes describe this as the salt melting, which chemistry teachers inform students is not acceptable as melt has a technical meaning in science (even if the salt may appear to 'melt away' in the vernacular sense of something that seems to disappear from sight) that is different to 'dissolve'. A student who has learnt canonical concepts {melt} and {dissolve} can make the appropriate distinctions here.

A chemist who was pedantic might accept that the salt disappears, and point out that the water also disappears. At the start of the process what was visible was some water and some salt in the glass. The salt disappeared when it became a solute in solution, so we could argue that the water also disappeared when it became the solvent in a solution – what is visible in the glass now is not water, but a salt solution. The solution (a mixture) is a different material from the water (a single substance) so that has disappeared. There may be perceptual clues: bubbles of dissolved gas may be salted out (Taber, 1985), or there could be a slight visual effect approaching mild turbidity due to a concentration gradient giving different optical properties to solution regions of different density. However, it is quite possible that for many casual observers the solution would not appear any different from the water, besides the absence of salt grains at its lower surface.

So, deciding what could be observed depends upon conceptualisation. A young child would simply see the salt disappear. They might later learn that this process they are observing is called dissolving. Then later they may actually see salt dissolving in the water. The phenomenon has not changed, but learning has taken place, the brain has changed; so, the way sensory data has been processed has changed; and so, the way experience is presented to the conscious mind is different (see Chapter 13). The chemist; well, at least a pedantic one; may even learn to come to see the water disappearing as the solution forms.

At first sight the distinction between what we can actually observe directly, and the inferences we draw from actual observations, may seem clear. However, this has been challenged (Feyerabend, 1960/1999). Consideration of a range of examples might persuade us this is less a simple dichotomy, and more a matter of degree. If a chemist heats some white powder, and sees it turns yellow, then this seems a pretty clear example of direct observation. But what if the chemist was rightly conscious of the importance of safe working, and undertook the manipulation in a fume cupboard, observing the phenomenon through the glass screen. That would not seem to undermine our idea of direct observation – as we believe that the glass will not make any difference to what we see. Well, at least, assuming that suitable plane glass of the kind normally used in fume cupboards has been used, and not, say a decorative multicoloured glass screen more like the windows found in many churches. Assuming, also, that there is not bright sunlight passing through a window and reflecting off the glass door of the fume cupboard to obscure the chemist's view of the powder being heated. So, assuming some basic things we could reasonably expect about fume cupboards, in conjunction with favourable viewing conditions, and taking into account our knowledge of the effect of plane glass, we would likely not consider the glass screen as an impediment to something akin to direct observation.

Chemical Meta-concepts: Imagining the Relationships Between Chemical Concepts 159

Might we start to question an instance of direct observation if instead of looking at the phenomenon through plane glass, there was clear, colourless convex glass between the chemist and the powder being heated? This might distort the image, but should not change the colours observed. If the glass in question was in the form of spectacle lenses, without which the chemist could not readily focus on the powder, then even if – technically – the observations were mediated by an instrument, this instrument corrects for a defect of vision such that our chemist would feel that direct observation is not compromised by, but rather requires, the glasses.

If we are happy to consider the bespectacled chemist is still observing the phenomenon rather than some instrumental indication of it, then we would presumably feel much the same about an observation being made with a magnifying glass, which is basically the same technical fix as the spectacles. So, might we consider observation down a microscope as direct observation? Early microscopes were little more than magnifying glasses mounted in stands. Modern compound microscopes use more than one lens. A system of lenses (and some additional illumination, usually) reveals details not possible to the naked eye – just as the use of convex spectacles allow the longsighted chemist to focus on objects that are too close to see clearly when unaided.

If the chemist is looking down the microscope at crystal structures in a polished slice of mineral, then, it may become easier to distinguish the different grains present by using a Polaroid filter to selectively filter some of the light reaching the eye from the observed sample. This seems a little further from what we might normally think of as direct observation. Yet, this is surely analogous to someone putting on Polaroid sunglasses to help obtain clear vision when driving towards the setting sun, or donning Polaroid glasses to help when observing the living things at the bottom of a seaside rock pool on a sunny day when strong reflections from the surface prevent clear vision of what is beneath.

8.3.6 Has Anyone Seen the Cat?

A further step might be the use of an electron microscope, where the visual image observed has been produced by processing the data from sensors collecting reflections from an electron beam impacting on the sample. Here, conceptually, we have a more obvious discontinuity although the perceptual process (certainly if the image is of some salt crystal surface) may make this seem no different to looking down a powerful optical microscope. An analogy here might be using night-vision goggles that allow someone to see objects in conditions where it would be too dark to see them directly. I have a camera my late wife bought me that is designed for catching images of wildlife and that switches in low light conditions to detecting infrared. I have a picture of a local cat that triggered an image when the camera was left set up in the garden overnight. The cat looks different from how it would appear in daylight, but I still see a cat in the image (where if the camera had taken a normal image I would not have been able to detect the cat as the image would have appeared like the proverbial picture of a 'black cat in a coal cellar'). Someone using night-vision goggles considers that they *see* the fox, or the escaped convict, not that they see an image produced by electronic circuits.

If we accept that we can see the cat in the photograph, and the surface details of crystal grains in the electron microscope image, then can we actually see atoms in the STM image? There is no cat in or on my image, it is just a pattern of pixels that my brain determines to represent a cat. I never saw the cat directly (I was presumably asleep) so I have no direct evidence there really was a cat if I do not accept the photograph taken using infrared sensors. I believe there are cats in the world, and have seen uninvited cats in my garden in daylight, and think the camera imaged one of them at night. So it seems reasonable I am seeing a cat in the image, and therefore I might wonder if it is reasonable to doubt that I can also see atoms in an STM image.

8.3.7 Has Anyone Seen the Higgs Boson?

One could shift further from simple sensory experience. News media might give the impression that physicists have seen the Higgs boson in data collected at CERN. This might lead us to ask: did they see it with their eyes? Or through spectacles? Or using a microscope? Or with night-vision goggles? Of course, they actually used particle detectors.

Feyerabend suggests that if we look at cloud chamber photographs, we do not doubt that we have a 'direct' method of detecting elementary particles (Feyerabend, 1960/1999). Perhaps, but CERN were not using something like a very large cloud chamber where they could see the trails of condensation left in the 'wake' of a passing alpha particle, and that could be photographed for posterity. The detection of the Higgs involved very sophisticated detectors, complex theory about the particle cascades a Higgs particle interaction might cause, and very complex simulations to allow for all kinds of issues relating to how the performance of the detectors might vary (for example as they age) and how a signal that might be close to random noise could be identified (Knorr Cetina, 1999). No one was looking at a detector hoping to see the telltale pattern that would clearly be left by a Higgs, and only a Higgs. In one sense, to borrow a phrase, 'there's nothing to see'. Interpreting the data considered to provide evidence of the Higgs was less like using a sophisticated microscope, and more like taking a mixture of many highly complex organic substances, and – without any attempt to separate them – running a mass spectrum, and hoping to make sense of the pattern of peaks obtained.

8.3.8 Shifting from Observational Language

It seems, then, that there are challenges in determining what exactly counts as direct observation, if indeed such a distinction is even seen as meaningful. Nonetheless, there is a useful distinction (that can easily get lost in teaching, if we are not careful) between the description of an observation in

Chemical Meta-concepts: Imagining the Relationships Between Chemical Concepts

terms of what was perceived and how it is conceptualised. It was hinted above that this distinction is not so straightforward given the way human beings learn, because human brains modify themselves in response to experience, such that what is presented to the conscious mind as a perception is not raw sense data, but the outcome of some quite sophisticated processing involving the recognition of patterns and inference making (see Chapter 13).

In general, there is a need in chemistry to move beyond what can be directly observed to conceptualise observations in terms of chemical ideas (Taber, 2013). So, even if we acknowledge absolute distinctions may be problematic here, we can imagine statements that describe what is more directly observable in everyday language, and how we understand this as experts (see Table 8.2). So, we might distinguish the phenomenological description from a conceptualised description.

In these, and many other, situations, what can only be directly seen as a phenomenon by a novice will allow inferences to be drawn by an expert. The language of observation – the everyday descriptors such as powder, burning – are substituted by technical terms: names of substances, class of substance, processes such as types of chemical reaction. Sometimes with great familiarly these inferences may become automatic (this is discussed in Chapter 13), but even when they are the result of deliberate conceptualisation they come readily to the expert in a way that is not possible for a novice.

8.3.9 Applying the Submicroscopic Concepts

Commonly, chemists wish to offer explanation, or at least a description, of a chemical phenomenon in terms of submicroscopic models. However, this normally requires having a description in appropriate chemical terms. That is, in moving from the observation of phenomena to modelling the process in terms of particle models, it is normally necessary to first re-describe the observations by conceptualising them in chemical terms (see Figure 8.3).

Using particle models becomes second nature to most chemists. So, not only does a professional chemist come to see phenomena in chemical terms,

Table 8.2 Some examples of how observations are interpreted conceptually.

Phenomenological level	Conceptual level
The white powder turned blue when water was dripped onto it	Anhydrous copper sulfate absorbed water to give hydrated copper sulfate
The solution colour changed from blue to green	The end point of the neutralisation reaction was reached
The strip of metal burned giving out a bright white light	The metal reacted with oxygen from the air
The indicator lamp glowed	The solution tested conducted electricity, so the solute is an electrolyte

but likely often can immediately imagine what is going on (*i.e.*, what is theorised as occurring) at the submicroscopic scale. Familiarity with, and regular application of, particle models makes them readily available to build mental simulations – to imagine what is happening in terms of molecules, ions, electrons, orbitals, *etcetera*.

Consider a precipitation reaction, such as that between solutions of lead nitrate and potassium iodide. A novice would see two liquids mixed, and some solid settling out from the mixture. A chemist sees a precipitation reaction, and describes this in a specialised symbolic language, and visualises how ions of Pb^{2+} and I^- are sticking together, forming into clumps, and gradually settling as large crystal lattices beneath the solution of largely unaffected spectator ions (see Figure 8.4):

$$Pb(NO_3)_2(aq) + 2KI(aq) \rightarrow PbI_2(s) + 2KNO_3(aq)$$

Arguably, there is a sense in which the chemist 'sees' not only the reaction, but also, in the so-called mind's eye, the ions interacting in the solution.

Indeed, this expert ability to conceptualise and visualise chemical processes at two levels can lead to us using ambiguous language that is potentially unhelpful to novice learners. For example, in the preceding chapter it was stated that "according to the perspective developed by Brønsted and Lowry, acids and bases are understood in terms of reactions where one substance (the acid) donates a proton, that another substance (the base) accepts". However, a pedantic chemist would point out this is nonsensical – there would be no observable reaction if the acid substance donated one proton to the base substance rather than very many protons: the sentence is conflating molar substance and submicroscopic quanticle descriptions. The phrase is shorthand for something like: according to the perspective

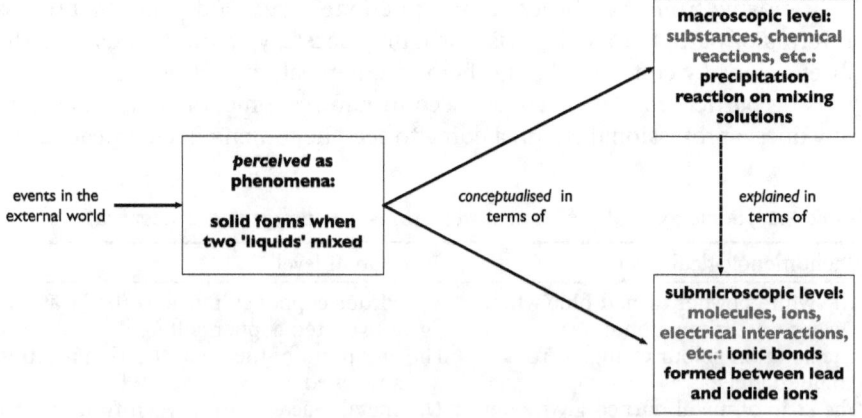

Figure 8.4 Conceptualising chemistry at two distinct levels.
Adapted from Taber (2013) with permission from the Royal Society of Chemistry.

developed by Brønsted and Lowry acids and bases are understood in terms of reactions where one substance (the acid) *comprised of molecules or ions that each* donate a proton that *molecules or ions comprising* another substance (the base) accept. Of course, that meaning was obvious to readers – but can we assume it would always be so obvious to novices such as students?

8.4 Conclusion

The argument offered in this chapter is intended to emphasise the extent to which conceptualisation – forming and developing concepts – underpins understanding of, and so learning about, chemistry. Chapter 7 considered concepts that could be understood as abstractions of sets of things, or events that indicate processes or properties, that could be grouped together – by forming concepts: {hydrogen}, {alkali metals}, {elements}, {reducing agents}, {temperature}, {combustion reactions} and so forth. We form concepts of theoretical entities ({ammonia molecules}, {p orbitals}, {neutrons}, *etc.*) as well as 'real', directly observable ones (*e.g.*, {conical flasks}).

This chapter has focused on the importance in chemistry of higher-level abstractions that relate to such concepts in various ways. The ideal gas equation involves {temperature}, {pressure}, {volume} and {ideal gas} concepts. Chargaff's rule relates to {DNA}, {purine}, {pyrimidine} concepts, as does the scientific model of the structure of 'the' DNA molecule. Lewis acid theory draws upon (and relates) a range of concepts such as {acids}, {base}, {electron pair}, {electron-pair donor}, *etcetera*.

The additional point stressed in this chapter is that in learning about these higher-level abstractions the student forms new concepts as well as relating existing ones. They are asked to form a {Chargaff's rule} concept, a {Lewis acid theory} concept, and so forth (which in principle allows them to discriminate examples of statements or applications of these higher-level abstractions from non-examples). The 'meta-concepts' not only associate other learnt chemical concepts, but – in order to be canonical – need to also link to nature of science concepts. So, for a student to develop a canonical {DNA structure model} concept or a {Lewis acid theory} concept, they not only need to form associations with other chemical concepts (*e.g.* the DNA structural model models a molecular structure) but associations with the higher-level concepts such as {scientific law}, {theory}, {model}, so that they make associations such as *the Crick and Watson DNA structure model is an example of a scientific model*. This type of association then becomes part of the content of their concept of a model, theory or law, and moreover for this aspect to be fully canonical, their {scientific law}/{theory}/{model} concept has to be sufficiently sophisticated. Knowing that *a picture of an atom shown as being like a tiny solar system is a representation of a scientific model* does not imply a canonical understanding for a student who thinks that a 'scientific model' simply means that the actual atomic structure has been scaled-up (and so the picture is intended as a realistic, if magnified, representation).

As with the book in general, my purpose here is to highlight some of the complexity of the chemistry that experts such as teachers tend to know well – so well that even though much of this may seem obvious when pointed out, we probably seldom stop to analyse the material we teach in these terms. Yet we are asking a good deal of students when teaching chemical meta-concepts, and hopefully reflecting on this complexity is valuable when planning classroom treatments to ensure that we remember to be explicit about those things that we may take for granted, and so take time to present material at a pace that reflects the complexity as it will appear to a learner. The remainder of the book shifts from a focus on chemical concepts themselves, with education in mind, to a more direct consideration of how chemical concepts are communicated and shared in science, and, in particular, in education.

References

Ben-Zvi R., Bat-Sheva E. and Silberstein J., (1986), Is an atom of copper malleable? *J. Chem. Educ.*, **63**(1), 64–66.

Christie M., (1994), Philosophers versus chemists concerning 'laws of nature', *Stud. Hist. Philos. Sci. Part A*, **25**(4), 613–629.

Feyerabend P., (1960/1999), The problem of the existence of theoretical entities, in Preston J. (ed.), *Knowledge, Science and Relativism* (vol. 3), Cambridge: Cambridge University Press, pp. 16–49.

Gilbert J. K. and Osborne R. J., (1980), The use of models in science and science teaching, *Eur. J. Sci. Educ.*, **2**(1), 3–13.

Grosslight L., Unger C., Jay E. and Smith C. L., (1991), Understanding models and their use in science: conceptions of middle and high school students and experts, *J. Res. Sci. Teach.*, **28**(9), 799–822.

Johnstone A. H., (1982), Macro- and microchemistry, *Sch. Sci. Rev.*, **64**(227), 377–379.

Johnstone A. H., Sleet R. J. and Vianna J. F., (1994), An information processing model of learning: Its application to an undergraduate laboratory course in chemistry, *Stud. High. Educ.*, **19**(1), 77–87.

Knorr Cetina K., (1999), *Epistemic Cultures: How the Sciences Make Knowledge*, Cambridge, Massachusetts: Harvard University Press.

Lakoff G., (1987), The Death of Dead Metaphor, *Metaphor Symbolic Activ.*, **2**(2), 143–147.

Nicholson W. (1987). *Life Story* (Film script).

Taber K. S., (1985), The presence of dissolved gas in water – demonstration by salting out, *Sch. Sci. Rev.*, **67**(238), 101–104.

Taber K. S., (2001), Building the structural concepts of chemistry: some considerations from educational research, *Chem. Educ. Res. Pract.*, **2**(2), 123–158.

Taber K. S., (2013), Revisiting the chemistry triplet: drawing upon the nature of chemical knowledge and the psychology of learning to inform chemistry education, *Chem. Educ. Res. Pract.*, **14**(2), 156–168.

Taber K. S., (2014), *Student Thinking and Learning in Science: Perspectives on the Nature and Development of Learners' Ideas*, New York: Routledge.

Taber K. S., Billingsley B., Riga F. and Newdick H., (2015), English secondary students' thinking about the status of scientific theories: consistent, comprehensive, coherent and extensively evidenced explanations of aspects of the natural world – or just 'an idea someone has', *Curriculum J*, **26**(3), 370–403.

Treagust D. F., Chittleborough G. and Mamiala T. L., (2002), Students' understanding of the role of scientific models in learning science, *Int. J. Sci. Educ.*, **24**(4), 357–368.

Watson J. D. and Crick F. H. C., (1953), Molecular structure of nucleic acids: a structure for deoxyribose nucleic acid, *Nature*, **171**(4356), 737–738.

TEACHING AND LEARNING CHEMISTRY CONCEPTS

CHAPTER 9

Accessing Chemical Concepts for Teaching and Learning

For chemical concepts to be taught and learnt, it is necessary for them to be identified and characterised. Teachers may have already mastered much of the material they are to teach, but most teachers of chemistry will feel that they cannot simply rely on their own past learning as the basis for teaching: that, sometimes at least, they will need to check on some of the conceptual content of the subject or topic they are to teach. Learners need to access canonical chemical concepts if they are to learn chemistry, and so teachers need to access canonical chemical concepts if they are to teach chemistry without simply assuming their own knowledge is sufficiently broad and infallible.

So, this chapter asks where can seekers of chemical knowledge expect to find the canonical concepts of chemistry. Three options are explored:

- scientific knowledge such as canonical chemical concepts can be found in texts, and in particular in the scientific literature;
- scientific knowledge such as canonical chemical concepts can be found in the shared knowledge of the scientific community;
- scientific knowledge such as canonical chemical concepts has an independent existence, and is accessible through rational thought.

9.1 The Problem of Locating Canonical Concepts

Chapters 5 and 6 examined the question of where chemical concepts came from, and discussed some historical episodes to inform a consideration of the origins and development of canonical chemical concepts. In particular, as examples, the development of the concepts {potassium} and {acid} were

---

Advances in Chemistry Education Series No. 3
The Nature of the Chemical Concept: Re-constructing Chemical Knowledge in Teaching and Learning
By Keith S. Taber
© Keith S. Taber 2022
Published by the Royal Society of Chemistry, www.rsc.org

outlined to show how the concepts shifted in response to the work of chemists. The question 'when was *the* [sic] concept {potassium} formed?' has no clear answer if it is assumed that 'the' concept being discussed is what might be considered the chemical concept {potassium} as used in chemistry today. The question 'when was *the* concept {acid} formed?' could reasonably invite the response 'which {acid} concept?' Clearly, similar issues arise with other chemical concepts.

The original {potassium} concept was generated and developed by Davy in response to his own observations in the laboratory, and later became further developed as other experimental work revealed new characteristics and properties of the substance that was the referent of the concept. Teachers do not, generally, however, generate the chemical concepts they teach about. Even those university teachers who have undertaken world-leading research that has opened-up whole new areas of study have usually had a role in positing or modifying only a very small proportion of the canonical chemical concepts presented in their teaching. In general, then, teachers are presenting and using in their teaching a wide range of concepts from within the chemical sciences that they have not themselves introduced or developed through the iterative scientific work of observing natural phenomena, suggesting hypotheses, and testing those ideas through laboratory experiments (*cf.* Figure 5.1). Chemistry is the iterative work of a community of scholars, and each contributor helps advance the communal enterprise.

It seems reasonable to ask where do teachers find the chemical knowledge, and in particular the canonical concepts, to be taught to their students? In part, this is clearly a matter of what they learnt in their own student days. Few teachers, however (unless I am unusually deficient in this regard), are able to teach purely by basing their classes on what they have previously learned as students. Teachers generally want to check things, and to learn about areas that have developed since their own studies or were not components of their own education but now appear in the curriculum they are to teach. Generally (and there is that qualification again) most people find that to be an effective teacher they need a deeper, and better integrated, knowledge of a subject than sufficed to succeed as a student (Kind, 2013). Students tend to be asked (for example in examinations) to respond to questions framed by subject experts, whereas teachers tend to be asked questions by people (*i.e.*, learners) with partial, and often flawed, understanding of the topic. Responding to the latter class of enquiries tends to require a more nuanced understanding of the focal topic, and much more flexibility in being able to frame and present the ideas. Thus, the common notion among teachers that someone only *really* understands a topic in depth when they come to teach it (Taber, 2009).

The question then is where might we find chemical concepts that we may wish to acquire. The answer perhaps seems obvious – the scientific literature. Chemical research is published in research journals, and reviewed in chemical review journals and then summarised and compiled into textbooks. The reason it was possible to present accounts of the chemical concepts {potassium} and {acid} discussed earlier in the book was because scientists report their work in research reports. The concepts we have and use, can be found represented in the primary journals of the subject. So, it seems new chemical concepts are reported in the research literature, taken up and applied in work reported in other studies, and subsequently those concepts appear in the books written for students learning chemistry.

At one level this is a perfectly reasonable account, which may suggest this chapter can end at this point having answered its question with a widely accepted and unremarkable answer. This is not suggesting the work involved would be simple – to research the concept {potassium}, for example (see Chapter 5), one would need to look at a range of studies undertaken over an extended period of time as the concept was developed. However, there is a more serious problem with this account. Indeed, to put the matter more definitively, journals and books do not contain canonical chemical concepts (Taber, 2013b).

9.2 Looking for Canonical Concepts in the Scientific Literature

There are two reasons why seeking canonical chemical concepts in the scientific literature may be problematic. One concerns the diversity of the literature itself. It does not offer a clear unambiguous set of knowledge claims. This means that in referring to the literature one needs to be selective – and so know how to discriminate mooted ideas that became and are still canonical ideas, from those that have been surpassed, refined, rejected or simply ignored as unproductive (*cf.* Figure 6.1). If you already know what the current canonical concept is, you can tell if you have found it rather than some other version of a concept, such as a historical concept no longer having currency: but that of course does not help you when the aim is to find out what the canonical chemical concepts now are. (So, the task resembles the iterative process by which science itself progresses, as discussed in Chapter 7, were we lack an independent arbiter of right answers.) In any case, the other more fundamental issue is that the literature does not actually contain any chemical concepts, or any other knowledge, but *only representations of* them, and this presents a more fundamental problem. First, however, there is the question of selecting the literature reporting the representations most relevant to teaching the current state of the science.

9.2.1 Lack of Coherence in the Literature

A little thought suggests that the scientific literature is an evolving body of texts comprising knowledge claims of diverse status: much that has been published in scientific journals is either conjectural (suggesting imaginative possibilities for further investigation, as one possible inference from findings) or has been supplanted by more recent work. Even if attention is limited to chemistry journals that are published by the most prestigious publishers, the range of material is too extensive and too diverse for individual chemists to be able to have an expert view, across the discipline, on which particular published accounts should, at this time, be considered canonical.

This links to the second suggestion (discussed in more detail later in the chapter), which is premised on science being a communal activity regulated through institutional processes (encompassing checks and balances) that leads to consensus. This is largely true. Yet, at any particular time, strong consensus (and, even then, not absolute consensus) tends to be limited to work that is no longer directly the subject of ongoing research interest. Areas of 'science-in-the-making' (Shapin, 1992) or 'science in action' (Latour, 1987) are by their nature more fluid, so the boundaries and affordances of concepts may best be considered 'under development'. A science teacher at school level may feel this is less of an issue than it would be for a university teacher as the school curriculum tends to focus on more established science. However, even when concerned with less advanced levels of chemistry education, there are not always unproblematic definitions available to characterise canonical concepts.

Thomas Kuhn (1970) considered the induction into science, the 'apprenticeship' of becoming a professional research scientist, and suggested that the ideas of science had to be learnt by working through standard examples, as learning formal definitions was not sufficient. We can see this, for example, with the concept {element}: one of the most fundamental in chemistry. This is a challenging idea to students first meeting it. Two standard types of definition met in school science tend to be:

- an element is a pure substance that cannot be split into anything simpler by chemical means; or
- an element is a substance that contains only one kind of atom.

An expert understands these definitions, and how to apply them, but arguably *because* they have already acquired the established concept (in relation to – and so in conjunction with – other closely linked concepts), not because of the inherent power of the definitions. Neither of these definitions are much value in representing the idea of a chemical element to a novice.

The first definition only makes sense to someone who already understands other chemical concepts, such as {substance}, what is meant as 'simpler' in this specifically chemical context, and what are considered *chemical* means. The second type of definition only makes sense once students have a grasp of the {atom} concept – and appreciate, for example, that *in this context* 'one type' implies the same proton number, but not necessarily the same neutron number: the same atomic number, but not the same atomic mass. In *other* contexts, atoms of the same proton number but different neutron numbers would be considered different types of atom. So, for example, in the process of ^{235}U enrichment, uranium (an element – so, by

Accessing Chemical Concepts for Teaching and Learning 173

one definition, containing one kind of atom) is separated into its isotopes (each containing different kinds of atoms).

Karl Popper (1945/2011) also noted how definitions had a limited role in science, seeing the definition of terms as useful in making communication more viable (as one word can substitute for phrases or longer descriptions), but as something that could only be applied as shorthand *after* the scientific thinking has been done. This suggests that a concept needs to be understood to some degree *before* it is useful to think about it in terms of a formal verbal definition.

9.2.2 Texts and the Nature of Knowledge

The common-sense notion that knowledge is found in books and in journals does not stand up to critical scrutiny. It is considered here that knowledge is a property of (and so requires) a knower (see Chapter 4), and so that only an agent with a sufficiently sophisticated mind can have knowledge. If concepts are mental entities (as suggested in Chapter 2), then they exist in minds. A person can know something, but a book or periodical article cannot. Therefore, a person can be said to have knowledge, but not a printed (or on-line) text. A person can be said to 'have' (or have acquired) a concept, but a text cannot.

If a reader is unconvinced by this claim, we might consider a thought experiment – a reputable procedure in science, especially in circumstances where a real test is not feasible (Brown, 1991). Imagine the development of some kind of ultimate neutron-bomb that kills nearly the entire human race, but leaves most of the infrastructure intact. The philosopher Karl Popper (1979) discussed a similar scenario, suggesting that the knowledge in libraries would allow us to rebuild our technologically advanced society – that is, to rebuild relatively quickly, not by spending millennia repeating the slow work of iterative creation of culture as found in our actual history, with all its marginal increments, cul-de-sacs, and reversals. We could pick up on our current conceptualisation of potassium (see Chapter 5) with its isotopic composition, for example, from the off.

However, in my version of the thought experiment, we imagine that the only human survivors were a remote tribe of Amazonians who had avoided any contact with other societies or cultures, who once alone in the world found their population and range expanding, perhaps eventually (and one advantage of a thought experiment is that we are not restricted to the duration of a research council grant – so we could be talking many centuries) coming to occupy every continent. The world's libraries and other such resources potentially offer these survivors all the scientific and technological knowledge that had facilitated 'modern' society. Yet, these Amazonians lack the cultural resources to access it, having no knowledge of English, Portuguese, Spanish, or any other 'world' language.

If we really thought the knowledge was somehow stored in texts, then it should be possible for any physiologically modern humans to access it.

The reclusive Amazonian tribe are modern humans, just as capable as the rest of us, and indeed having their own highly evolved culture and language. By adulthood, members of the tribe are highly skilled in a wide range of activities that most readers of this book could not successfully undertake. Despite this, they will not simply be able to enter the world's libraries or patent offices, and start applying knowledge represented there to build nuclear power stations, mobile phones, and digital watches – or to produce antibiotics, kidney dialysis machines, or Sildenafil.

This is not a parallel situation to that of archaeologists trying to decode ancient inscriptions in an unknown language – where a version of the Rosetta stone would be useful. Archaeologists undertaking such work can assume that the inscriptions they are trying to decode largely represent entities they will be familiar with directly, or from their studies (*cf.* Whorf, 2012/1931). So, this symbol may represent a serpent, and that one a chariot, and this symbol seems to refer to some kind of supreme being living in the sky. The isolated Amazonians have no conceptual system in which to understand what a nuclear power station or a kidney dialysis machine is. Popper may well be right that even with these barriers (and assuming the Amazonians wished to take advantage of learning from the coded records) we might expect technological advances to be somewhat accelerated. Due to commonalities across human cognition and language (Chomsky, 1999), the survivors have some resources that may give them something of an advantage in decoding texts to the extent that there are human biases in how information is represented in cultural artefacts. Even so, this will not facilitate rebuilding advanced technological societies in a few generations.

The argument here is that knowledge cannot exist without a knower, so once those that knew were gone, along with those others who knew how to read the records, the knowledge is not around any longer. Knowledge cannot be stored in libraries, rather it can only be represented as something else. For knowledge 'transfer' we need a form of representation that is structured in a way that both the representer and the interpreter use. (There is clearly a useful comparison here to the use of ciphers and codebreaking.) We might argue that the texts contain *information*, but not *knowledge* (a distinction that will be developed in Chapter 11).

9.2.3 Texts Contain Only Representations of Concepts

Popper's conclusion, that given the libraries we could rebuild a technological society quickly, only applies, then, when there are survivors who retain the particular interpretive resources needed to decode the texts: as these texts are *representations of* particular knowers' knowledge and not, in themselves, stored knowledge. If, instead of the surviving indigenous population, the remains of human culture was discovered by archaeologists from an advanced alien civilisation then they might possibly do a better job of working out how to translate texts in human language (although they may not share human cognitive biases) – as presumably if they had the ability to get here to access human relics, then they would already have versions of the scientific ideas represented in the (human) scientific literature: but in this case they would not be acquiring new knowledge from our texts, but more recognising (versions of) their own current or historical scientific knowledge – just like the human archaeologists recognising the representation of a serpent or chariot.

The alien archaeologists are not like our Amazonian tribes-people seeking scientific knowledge in a post-apocalyptic world, but more like a classroom teacher carrying out formative assessment. The teacher asking a student what is meant by nucleophilic substitution is not seeking to learn about nucleophilic substitution, but to learn about what the student understands about nucleophilic substitution (Edwards and Mercer, 1987) – and that is only feasible because the teacher already knows about the topic.

The texts left by a defunct culture by themselves are not (and do not contain) knowledge (Taber, 2013b), and someone finding them without the means to interpret them would only have 'acquired knowledge' in much the same way as a banknote collector sourcing one of the fifty billion (5×10^{13}) Mark notes issues in Germany in 1923 (when a single US dollar would buy over four billion marks, or indeed four thousand billion marks in the American use of 'billion') can be said to have acquired 'a good deal of money'. That is, a high denomination banknote that no longer has legal status as a medium of exchange is not actually money in any functional sense. In a similar way, knowledge that has been represented in a text through a symbolic system (language) that no one alive can use, is no longer, in any useful sense, knowledge.

9.2.4 Criticisms of This Position

The position adopted here, then, is that concepts are an aspect of knowledge (see Chapter 4), and that knowledge is a property of a particular knower – and although texts can represent knowledge, they do not contain knowledge, and can only help a reader come to knowledge if they have access to the cultural tools to interpret the text. Chemical texts such as journal articles do not therefore of themselves 'contain' conceptual knowledge – and in any case there is a vast literature that, in relation to many concepts, we would not be able to interpret in a consistent way such that we could unproblematically de-code canonical chemical concepts from the texts.

This perspective on knowledge is not a position that everyone would accept as there are at least two well-supported positions inconsistent with this view (Collins, 2010). One approach does not tie knowledge to the mind but sees knowledge in terms of the potential for its practical application. On this basis, just as a hypothetical chemist might be said to have the knowledge to carry out some complex organic synthetic reaction, an acorn (though presumably mindless) may be said to have the knowledge to build an oak tree. As how we understand terms is based on convention, there is no inherent reason why this should not count as knowledge – after all much human knowledge is tacit. People who ride bicycles have the skill, the 'know-how' (procedural knowledge): but not usually the theory to explain the physics and biomechanics involved. In a similar way, an acorn has the 'know-how', given appropriate conditions (just as one cannot ride a bicycle without having access to a bicycle), to build an oak tree. However, the present book is concerned with the learning and teaching of concepts, and even if we feel an acorn should be considered knowledgable in some ways, it does not have any conceptual knowledge (see Chapter 4). In any case, in terms of the wider argument, if knowledge is about 'know-*how*', texts that we cannot interpret clearly and unambiguously, and that we have no inherent basis to choose between, do not offer us know-how.

The other issue concerns whether knowledge can be distributed, so that there can be knowledge that is not associated with any one individual personally, but that rather is understood as distributed across some kind of network. This may be a network of a person plus non-human entities (a notebook, a computer, *etc.*, – see Chapter 11), or networks of people. This latter perspective suggests that the real repository of scientific knowledge is not a canon of literature, but the scientific community itself.

9.3 Asking the Community of Chemists about Canonical Concepts

If texts, such as those found in the scientific literature, do not all represent current scientific knowledge (*cf.* Figure 6.1), then this suggests that there must be an arbiter of what still counts as scientific knowledge. This would presumably be the scientific community. If texts do not contain knowledge, but only *the representations of* knowledge, then perhaps the problem of needing to interpret representations is avoided by referring not to the text, but to scientists themselves. So, one alternative to assuming that scientific knowledge is located in the scientific literature (which the analysis above suggests is highly problematic) is to suggest instead that scientific knowledge is found in the community of scientists.

9.3.1 Who Are the Chemists Who Know?

There is a practical problem here, akin to knowing which research papers to refer to in order to find (representations of) canonical knowledge, which is, which scientists to ask. Would all chemists hold the canonical chemical concept {entropy}, and the canonical chemical concept {oxidation}, and the canonical chemical concept {nucleophilic substitution}, *etcetera*. If all chemists, or at least an *identifiable* cadre of reliable expert chemists, hold all the canonical concepts, then at least we know in principle who to ask.

Of course, chemistry is a very broad discipline, and different chemists have different specialisms, different areas of interest, and different levels of engagement with scholarship and research in different topics – so we may need a way to locate those who can be considered to hold particular canonical concepts. However, unfortunately, we might find that even when we identify supposed experts in particular topics, they do not always entirely agree on the concepts concerned – that they seem to hold somewhat distinct versions of chemical concepts (*e.g.*, {entropy}), leaving open the question which version should be considered canonical.

So, given the potential for diversity between supposed experts, it might be suggested that an academic discipline depends upon a community of scholars interacting to allow canonical knowledge to evolve. Then, the focus should not be on individuals, but on the relevant community: the community of professional chemical scholars, relying on the mechanisms that allow knowledge to be distributed and moderated by community processes. The focus should not be on the knowledge of a particular individual but on that consensual account that derives from the community as a whole.

9.3.2 Does Knowledge Need to Be Personal or Can It Be Distributed?

One common criticism of perspectives on knowledge that locate it (and seek it) in individual minds relates to the idea of distributed knowledge. Distributed knowledge has been defined as "knowledge that is 'distributed' among the members of the group" (Halpern and Moses, 1990, p. 549). This definition appears to refer to a group of individual people, but more recently there has been reference to 'knowledge bases' that can be "distributed across a range of technologies, actors, and industries" (Conceição *et al.*, 2003, p. 584), and to how organisations carrying out complex functions may rely upon a 'metacompetence' that is "somehow distributed in the network of agents that own part of the knowledge and collaborate" (p. 606).

At one level this is nothing more than the consequences of the idea of specialisation and division of labour: modern societies are based on the principle that instead of each of us doing all that is needed to survive (growing our own food, disposing of our own rubbish, building our own shelters, making our own tools...) it is more efficient if people develop different specialised skills and make economic arrangements to trade (Smith, 1827/1981). Firms (*i.e.*, companies, corporations) have been described as "distributed knowledge systems in a strong sense: they are decentered systems, lacking an overseeing 'mind' ... [such that] a firm's knowledge is the indeterminate outcome of individuals attempting to manage the inevitable tensions between normative expectations, dispositions, and local contexts" (Tsoukas, 1996, p. 11).

Cognition may be said to be situated such that the question of 'whether' someone has certain knowledge can only be answered contextually as the knowledge is in part a property of the person, something in their minds, something encoded in their brain; *and* also in part an affordance of the particular environment they are in. This may sound a little mystical to those not used to thinking in such a way:

> Knowledge, we suggest, similarly indexes the situation in which it arises and is used. The embedding circumstances efficiently provide essential parts of its structure and meaning. So knowledge, which comes coded by and connected to the activity and environment in which it is developed, is spread across its component parts, some of which are in the mind and some in the world much as the final picture on a jig-saw is spread across its component pieces. (Brown *et al.*, 1989, pp. 36–37)

I think this can be considered in different ways, and, in the end, it comes down to how we want to define a term such as knowledge. If chemists can modify a concept such as {acid} on the basis of convenience (as Chapter 6 suggests has been the case), then we can adopt the {knowledge} concept we think is most productive – that is, the version that we think offers the most useful way of discriminating between what it is helpful to think of as knowledge, and knowledge (*i.e.*, everything else).

Even within a single individual, we might think knowledge is inherently distributed within a cognitive system. Cognition is complex. I know the names of the first 20 elements (and a few more, but for purposes of the example I will stick to that), but when I am asleep I could not tell you them. Does that mean I do not know them at that time? The aspect of me that is capable of being consciously aware has no access to the information (unless I am dreaming about chemistry, perhaps), but I would not consider that a sufficient test. Taking failure to demonstrate knowledge because one is asleep as showing we do not have the knowledge seems an unhelpful way of defining knowledge, if we can demonstrate the knowledge when awake.

So, I am suggesting that to be considered to know certain things, we do not necessarily have to be perpetually demonstrating that knowledge, but should have the potential to do so under appropriate circumstances. When I am conscious and focusing on some unrelated challenging task my limited working memory would not allow me to access the element names: yet, if it became important to do so I could readily switch and access the knowledge needed from memory. I am happy that knowledge is distributed between the parts of my brain that can process it, and the parts that represent it in a more latent state for when I need to access it (perhaps the congenst, *cf.* Chapter 2). So that is a kind of 'distribution' of knowledge; but usually the idea of distributed knowledge is intended to go beyond that – to include some aspects outside of the person.

9.3.3 The Human Knower – Information Resource System

Now, whilst I am some way from forgetting the first 20 elements, I am at an age where I do readily 'forget' names, citations, and – indeed – much else.

The affordance of the internet means that rather than 'wracking my brain' waiting for a sought word to come into consciousness, I will often do a quick internet search. So, if I cannot think of the word that is most appropriate – I am pretty sure I 'know' just the right word, but I just cannot get it to mind right now: what is sometimes known as a tip-of-the-tongue experience (Brown and McNeill, 1966/1976) – I may type the word 'synonym' and a word that I think almost works for what I am writing, and see what is suggested. Often, I soon find a suggestion that I recognise as just the word I was looking for.

It is accepted that *recognising* (*e.g.*, 'which of these is a noble gas...?') tends to be easier than *recalling* (*e.g.*, 'name a noble gas?') – the stimulus allows memory to be searched more quickly, even when we cannot deduce the answer by eliminating other alternatives offered, as might be possible in a question such as:

- which of the following worked on nucleophilic substitution and helped introduce the aldol reaction:
Bach;
Beethoven;
Borodin;
Brahms;
Bruckner?

I imagine that if there was no internet, and I did not wish to go and find a thesaurus, I would recall the target word...in time. But I can wonder: 'what if' it never was to come to mind without recruiting external resources.

Consider a more clear-cut example. Imagine an author wanted to write about how chemical concepts have been developed. She or he might consider some target concepts. Perhaps one of these is {potassium} (see Chapter 5). The author might be aware that potassium was discovered in electrolysis experiments carried out by Davy, who first conceptualised potassium (and so formed a {potassium} concept). The author is also pretty sure that the {potassium} concept used by chemists today is different from that formed by Davy because a good deal more is known about the properties of the element potassium, much of which was based on theoretical and laboratory apparatus not available in Davy's time. However, the author is – shall we say – somewhat hazy about precisely what those differences might be.

The author's knowledge is limited. However, some time spent searching out and reading articles available in various academic sources provides more detail, enabling some specific examples to be discussed in a book chapter. Writing the chapter was made possible because the author was able to become part of a wider system (computer, internet link, university library gateway, publishers' servers) that collectively encompassed a wider knowledge base than that of the author himself.

This is a plausible scenario, and indeed is an authentic description of a real episode. This could be understood in terms of:

- author: limited knowledge located in this one person – insufficient for task in hand;
- author–computer–servers–*etc.* system: more extensive knowledge distributed across the system – facilitates task in hand.

Although I certainly recognise this way of thinking about the example, I would reject it when framed in precisely these terms. Given the way I understand conceptual knowledge, I do not think the system contained distributed knowledge, or any knowledge beyond that of the author alone, that allowed the task to be completed. Whilst recognising there was indeed a system with these different components, and without in anyway undermining the contribution of the non-human components of the system, I feel that a different description is needed.

In this preferred description, the writing of the chapter was made possible because the author developed his knowledge system (see Figure 9.1). He learned. All of the knowledge used in writing the chapter was – at the point of writing – the author's knowledge. Of course, that learning process was only possible due to access to suitable resources that supported the learning: both the texts themselves, and the information technology that afforded access to them. Some might consider this an example of distributed knowledge, but my own notion of knowledge is tied to a knower, and the only knower *directly* involved in this process was the author.

'Directly' is an important moderator here, as clearly the texts used as resources had their own authors, without whom the resources would not have

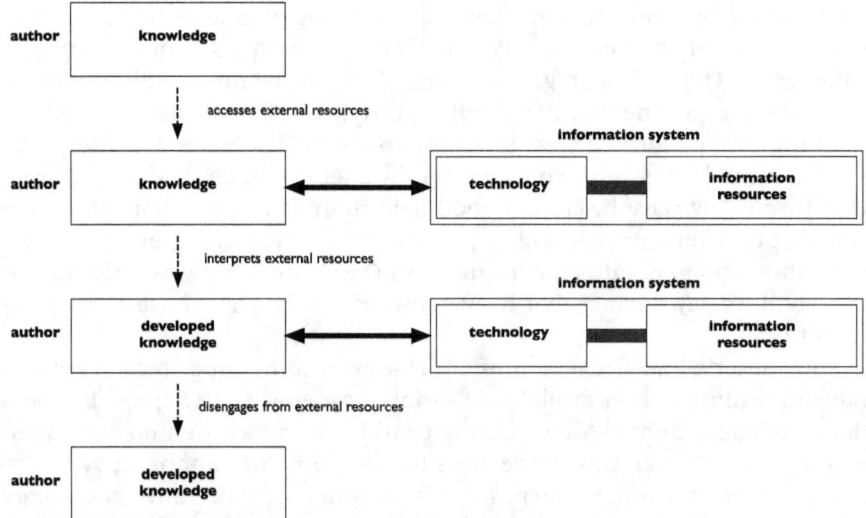

Figure 9.1 Developing knowledge by accessing external resources.

existed – but those authors were not consulted, and perhaps could not have been, so they were not (directly) part of the system. Once texts are produced, and disseminated, they exist as resources from which one can learn, even if the authors of the texts have gone senile, or have forgotten details of the work they were reporting, or have sadly died, or – even – changed their minds about what they reported in the text. (Scholars citing the philosopher Wittgenstein as a source will commonly refer to 'the early…' or 'the later Wittgenstein', as some of his most influential, posthumously published, ideas were inconsistent with the thinking set out in his *Tractatus Logico-Philosophicus* (Wittgenstein, 1922), the only book published during his lifetime.)

If we think of the human in terms of a cognitive system, as is sometimes suggested (Osborne and Wittrock, 1985), then we might think of the core of the system being the processing elements (working memory for conscious processing, but also all the other parts of the brain that are processing information), which is supported by perceptual systems that can input new information, and memory that is traditionally considered the place information is stored. This is a useful model for some purposes, but includes some important oversimplifications (see Chapter 13), which can be put aside for the moment. Then, the system may seem to be extended by notebooks, and calculators, and computers, and the internet – the sensory input is enhanced by new devices providing stimuli not otherwise directly available, and there is access to additional (extensive) memory stores, and supplementary processors.

9.3.4 The Importance of Understanding

This model works as long as we are thinking in terms of abstract information processing: of moving information about and storing it and undertaking algorithmic processing. What it ignores is meaning. Conceptual knowledge is not just information that is stored and accessed for use in algorithmic processing. Conceptual knowledge is understood – at least at some level (this theme is developed in Chapter 11). Learning about dative bonding is more than just adding a new category of chemical bond to a typology, and storing a list of characteristics and properties – but understanding how it arises, and how it is similar to, but distinct from other types of bonding. *Perhaps* digital technologies are potentially capable of giving meaning to information and understanding it, and machine learning is developing all the time, but at the moment there seems to be an important qualitative difference between what appears to be going on in a human, and in a computer that they are using.

9.3.5 Knowledge Distributed Across Networks of People

Similar arguments cannot be made against the view that knowledge is distributed across social networks – where people 'think together' and where 'two heads are better than one' and where two heads being networked can often (though not always) achieve more than two heads working in isolation. Many major scientific breakthroughs have involved teams: the Lavoisiers, the Curies; Meitner, Hahn and Strassmann; and so forth. High-energy physics depends on extensive collaborations that are partially grounded in political (funding) considerations, but only make progress due to the many experts working together. In particular, in this example, multidisciplinary teams bring different areas of knowledge: the physics, the engineering, the software design (Knorr Cetina, 1999). Many areas of research now depend on the expertise of different people – for example statisticians contribute critically to many areas of research where their own background in the substantive research topic may be quite limited.

Darwin is perhaps thought of as something of a scientific loner, avoiding attending too many scientific meetings or getting sidetracked by involvement in public affairs (although he did initiate a successful campaign to have Alfred Russel Wallace awarded a discretionary pension in recognition of his contribution to science), and largely working at home for much of his life (which is not to ignore some rather important and much celebrated field work early in his career). Yet, he relied on the expertise of a great many colleagues who he consulted – naturalists with different specialisms, anatomists, farmers, pigeon fanciers....

Breakthroughs can even happen when collaborations are sub-optimal. Crick and Watson worked well together and their personalities and areas of knowledge complemented, and this was important in the development of their DNA model (see Chapter 8). However, they also directly, if somewhat asymmetrically, collaborated with Wilkins (he shared the Nobel prize with them), and – more problematically – indirectly with Franklin and Gosling. They also consulted with people such as Chargaff (see Chapter 8) and in a sense (indirectly and unbeknown to him) with Linus Pauling through information provided by Peter Pauling, and had the advice of such esteemed colleagues as Perutz and Kendrew.

In any field, there will be a network of associations of people who work with each other, talk to each other, attend each other's presentations, and read each other's work. Each of the researchers have somewhat different knowledge of the field, and so it might be reasonably suggested (a) that the knowledge distributed across the network exceeds the knowledge of any one node (*i.e.*, researcher) and (b) that the existence of the network means that each researcher has access to the network and so, in a sense, to the knowledge distributed across it.

Whilst that is a perfectly reasonable account, this is not equivalent to suggesting that any member of the network personally knows everything that is known in the network, and neither, I would argue, that it makes sense to talk of the network (itself) knowing anything. If this were a system of linked computers, sharing common protocols for accessing, storing and transmitting information then we could talk of the cumulative information (sic, not knowledge, note) in the system, and consider each computer to have access to this full information set whilst remaining networked. That is not analogous, however, to a network of people with their individual 'stores' of knowledge that they understand to have meaning.

Accessing Chemical Concepts for Teaching and Learning 183

One thing lacking is the sharing of protocols that makes information transfer possible between computers with perfect fidelity. Each computer processes and stores information in the same way, so a network may act like a much larger computer with more processing power and storage. In the absence of telepathy, and much more advanced technologies than are currently possible, there is no analogous way of transferring thoughts from one mind such that they may be reproduced in perfect fidelity in other minds. (Given that the brain structure of an individual will always be unique, this may never be possible.) Rather, a more indirect process, that has to be mediated, needs to be employed.

9.3.6 Sharing Concepts Is a Process Involving Representation and Interpretation

If a scientist is considered to have some knowledge (say, some concept) then that cannot be directly transferred to a colleague. The scientist can only represent their ideas into talk, gesture, diagrams, writing, *etcetera*. Others then have to make sense of those representations (Taber, 2013b).

It may be challenging to represent some abstract ideas clearly in ways others will understand, and the process of making sense of a text is always an interpretation through existing knowledge and understanding (as will be explored in Chapter 11). Communicating ideas always then involves both a process of representation of something mental through a physical trace and a subsequent process of interpretation (see Figure 9.2). These processes are always likely to admit omissions and distortions such that communication between people is necessarily an imperfect process.

A well-crafted text will make sense to a colleague who shares background commitments and knowledge, and so will interpret the text much as intended. A baby would make no sense of a research report in a specialist chemistry journal text, nor would our hypothetical isolated Amazonians, nor any interstellar ethnographers who had not met our language. A colleague in the field, but working within a different research programme with different hard-core commitments; a scientist from a completely different specialism; and an undergraduate in the discipline; would be in intermediate positions.

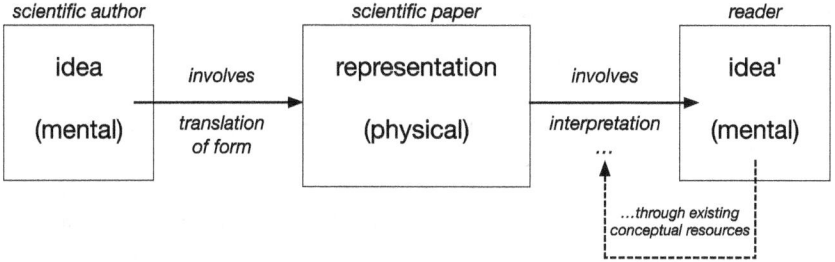

Figure 9.2 Scientific ideas, such as concepts, are represented in, but are not directly accessible through, media such as texts.

They would make some sense of the text, but are very unlikely to fully appreciate the nuances intended by the author (who by submitting to a specialist research journal is primarily writing for other researchers in the same area of work).

We have all experienced how, even when we read texts where we are part of the intended readership, it may be difficult to interpret meaning. Texts may appear confused or ambiguous or simply nonsensical. Sometimes we return to a text to find our initial take on the intended meaning seems to be wrong, and what is 'meant' is something quite different. We may misunderstand what someone says, and get 'the wrong end of the stick'. We may struggle with a text, and need to do some other reading or engage in some targeted conversations, after which it seems much less obscure.

There are of course many occasions when communication is effective and the author's or speaker's intended meanings are communicated sufficiently well for them to be understood pretty much as intended. Yet this is not automatic (in the way a file may be copied to another computer such that the copy is, in effect, identical to the source file), and depends upon the skills of the person making the representation and the person interpreting it. Such communication always depends upon the author working with assumptions (either explicitly, or tacitly) about the interpretive resources that the intended audience will have available.

The chemist changes the way she represents the ideas she wants to communicate when representing her ideas in a specialist research journal, or a review article for a wider pool of colleagues, or a piece intended for a broad professional audience (such as readers of 'Chemistry World'), or a lecture for undergraduates, or a talk to interested members of the public (such as at meetings of the British Science Association or the American Association for the Advancement of Science), or when invited to talk to children at the local primary school. Each audience is likely to bring different sets of interpretative resources, so the same representation is unlikely to make equal sense to members of such different target audiences. There is never meaning in the text, only meaning *represented* in the text that can be accessed if an appropriate interpretation of the text can be made.

9.3.7 Can We Avoid the Need to Interpret Representations?

To summarise the argument to this point:

- Individual people, such as chemists, conceptualise aspects of their experience to develop personal concepts.
- 'Sharing' of concepts requires one person to represent the concept (a mental representation abstracted from personal experience) in a totally different form (*e.g.*, talk, or an inscription), which then needs to be interpreted by a second person to allow them to make the representation meaningful.

Accessing Chemical Concepts for Teaching and Learning   185

- So, texts such as research papers only represent (do not 'contain') knowledge.
- The interpretation process relies on the resources that the person trying to make sense of the representation can bring to bear.
- So, 'sharing' of knowledge, such as a concept, cannot be assured – as the received meaning may not match the intended meaning.
- Therefore, the idea that there are canonical concepts that are shared by a community is suspect.
- Although knowledge may be distributed among people in the sense that a network allows (indirect) access to the knowledge of others, distributed knowledge is not a coherent, unitary body of knowledge that can be accessed in a straightforward way.

This argument assumes that there is not another means by which canonical knowledge can be shared. There is another possibility that has been long considered. Despite the arguments above about communication of technical concepts often being problematic, it is common experience that young children are taught 'this is a tree', or 'this is a dog', and generally acquire such concepts readily. This could be explained if people have some kind of intuitive direct access to knowledge that negates the need for others to represent their knowledge in a form that needs interpretation. This could explain how 'pointing and naming' works.

As discussed in Chapter 7, whilst some concepts in chemistry, so perhaps the concept {conical flasks}, refer to kinds of objects that can be readily pointed out, many chemical concepts, such as {reducing agent}, {ligand}, {unpaired electron}, {dative bond}, *etcetera*, are less 'ontologically obvious' from immediate unmediated experience of the world. In chemistry education, then, this direct access would require us to be able to intuit not just what is being referred to as a tree or a conical flask, but also what is being referred to as an acid, an oxidising agent, a transition state, a radioactive isotope, *etcetera*. It does not seem very likely that people would have such intuitive powers (and indeed much we teach in chemistry seems counter-intuitive to some learners) – and yet there is a very strong and well-established tradition of thought that does argue that humans are able to access knowledge in a more direct way and avoid the complications of communication discussed above. This tradition should then, for completeness, also be considered – as it could potentially offer another explanation for how people can share the same concepts.

9.4 Can We Access Canonical Chemical Concepts from a Non-material Realm of Ideas?

The final option suggested at the start of the chapter was that that canonical ideas have some kind of independent existence in their own right. This reflects a long-standing notion in philosophy that distinguishes a world (1) of objects, a world (2) of human subjective experience, and a third world (3) of ideas in themselves: what the philosopher of science, Karl Popper (1979, p. 106) referred to as "the world of objective contents of thought, especially of scientific and poetic thoughts and of works of art".

This relates to the idea of Platonic Forms. Plato suggested that the world as perceived was only a pale imitation of an underlying reality that was distinct from the material world. In the allegory of the cave, Plato told a story, supposedly originally due to Socrates, of prisoners born, and since birth held, in a cave – and who were chained so they could only face the wall of the cave. Behind them was a fire that they were unable to see. When there was movement between them and the fire, they would see shadows moving on the cave wall, but could not see the actual objects giving rise to the shadows. Plato implies that our experience of the world is like this, that our perception is limited and the phenomena we observe are mere shadows, or momentarily observed facets, of the actual real objects.

This idea has resonances in modern science. The apparent continuity of matter is considered to be underpinned by discontinuous particles. Centrifugal forces are 'fictitious' forces, only apparent due to adopting a particular frame of reference. Gravitation appears to be a force, but is 'actually' an effect of the geometry of space–time. However, science seeks more fundamental descriptions of the natural world, whereas Plato's Forms were considered to exist independently of the universe, outside of (or perhaps better, without regard to) space and time.

Plato's ideas actually link to some issues related to how learning is possible. Consider a hypothetical child who had not seen a tree – let's call him Emile. If Emile is taken for a walk and his adult companion points to a tree and suggests "that is a tree", Emile has to be able to identify what is being pointed out, and in particular to pick out the tree (not a leaf or a branch or a limb or the trunk) as a unitary object distinct from its surroundings (not the whole wood, or the tree plus the bluebells growing in its shade). In part, repetition and trial-and-error might be involved here, but there is an argument that humans are too good at this process for it to just be trial-and-error, as if they are born with some inbuilt bias to perceive the world in particular ways.

Another example might be Emily, a hypothetical student in a classroom where her teacher draws on the board a diagram to represent a cubic arrangement of ions in a crystal structure. The diagram will not be precise. Spacings will not be precise, angles will not be precise, and any attempt to reflect perspective will not be precise. If Emily tries to copy the diagram into her notes, she will be unable to precisely copy the diagram. The original representation will not quite reflect a cubic array, and nor will the copy – but each is likely to be distorted differently. Quite possibly, neither Emily nor her teacher are good draughtsmen, and the representations may seem to us to be obviously (differently) distorted versions of a cubic lattice.

Yet, there is an apparent mystery here: how is it obvious to Emily that the teacher's diagram is distorted. Unless Emily is an unusual young lady, a savant herself perhaps, the distortions in her own representation will not be attempts to copy the imperfections in the teacher's drawing, but rather her own different imperfections in attempting to draw the idealised lattice that the teacher imperfectly represented. Even if Emily has never seen representations of a cubic lattice before, it will be obvious to her (or perhaps, more likely, she will not even notice as she will have automatically, intuitively, appreciated the intended form) that there are flaws in the version offered by the teacher and she will (imperfectly) attempt to draw not the teacher's diagram, but the idea the teacher is attempting to represent.

There is a general problem here: that real spherical objects are not perfect spheres, and images of them are not perfectly circular; no real object has ever been a perfect equilateral triangle; no smooth surface has ever been perfectly smooth (as a scanning tunnelling microscope would reveal) – yet people seem to acquire concepts of spheres and circles, triangles and smoothness, of perfectly straight lines, and of pure substances, even if there are no true (pure) referents for these concepts in the material world.

Plato's solution was the existence of pure Forms that exist outside of the examples we might recognise in the world (Allen, 1959). In Plato's scheme, the soul is immortal, and has access to the knowledge of the Forms. So, a person is born with the potential to recognise beauty because the soul brings innate knowledge of the Form of beauty-in-itself. When a person dies (in this system) they forget the particulars of their life, but their reincarnated soul starts a new life again still having access to the Forms. Things and events in the real world are complex and may partake of the Form of beauty or justice or sphere, but the Forms are pure and comprise what is inherent in beauty, or justice, or being spherical. An artefact that is intended to be spherical will reflect the form of the (ideal) sphere as well as having imperfections contingent on its own material nature and history of production.

If one were to accept this, then one does not need to have ever seen a perfect tetrahedron to recognise that an imperfect representation of a tetrahedron is intended to represent that perfect Form. However, many (if not all) people today would reject notions of an immortal soul that gives its embodied person access to innate knowledge. This is also an account that should not be admissible as a hypothetical scientific explanation (even for a scientist who accepted reincarnation as part of their personal belief system), because science seeks natural mechanisms and explanations, and an immortal, immaterial, soul offers a supernatural explanation. That is, even if this account is true, it is not a scientific account, and scientists should seek a complementary, natural explanation (Taber, 2013a).

9.4.1 Do We Still Believe in Platonic Forms?

World 3 is not found in our material (World 1) universe, but is a completely separate realm. In Plato's philosophy, the Forms exist prior to (and are a kind of cause of, or pattern for) any of the material objects that can imperfectly reflect them. Popper (1979) populated World 3 with ideal forms, theories, conjectures, problems, and the like. So, the idea of a quadratic equation, and the idea of the formula for its roots, would exist in World 3, along with the ideas artists had in mind when creating their art. The form of Michelangelo's David exists in World 3, and the actual sculpture (that exists in World 1) is an attempt to realise this in material form.

In this model, the subjective knower has access to World 1 (containing tables and chairs and conical flasks and copper sulfate crystals and all other material things) by observation, and access to World 3 by rational thought. A great many people might have seen the Eiffel tower and have their own subjective (World 2) impression of it, but the real Eiffel tower exists in Paris, which is part of World 1. Analogously, countless thousands of students have learnt about Raoult's law and have their own subjective (World 2) understanding of it, but the real Raoult's law is (in this metaphysical system) found in World 3 (see Figure 9.3).

It is possible to buy material models of the Eiffel tower, and postcards depicting it, and read about it in books, but (as we all realise) these are not actually the Eiffel tower, but just representations of it. By analogy, then, we might suggest that it is possible to attend lectures on Raoult's law, find definitions and explanations of it in books, and references to its application in research reports, but these are not actually Raoult's law, but rather representations of it in World 1.

The idea of David was represented in marble by Michelangelo, and the idea of Raoult's law was represented by Raoult in an article published in the French scientific journal *Comptes Rendus*. Many reproductions of David have been produced in stone or plastic or other material forms, which indirectly represent the original vision of Michelangelo by being based on his statue. Many scientific texts and discourses have discussed Raoult's law as understood from his article, and the idea has been indirectly reproduced in the material of print media, computer files, patterns of disturbance of the air in lecture rooms, and so forth (Figure 9.4).

9.4.2 Does World 3 Really Exist?

Descartes famous suggestion that *je pense, donc je suis,* I think, therefore I am, takes personal mental experience as something that is self-evident, and from this makes an ontological deduction that there is a self (*i.e.*, that thinking implies a thinker), some kind of entity, thinking these thoughts. That does not, however, logically show the thoughts the thinker has represent any kind of external reality.

Some philosophers sometimes pose the question of whether we can be sure of the existence of World 1, the physical world, arguing that it could be an illusion or dream or virtual reality. The so called 'brain-in-a-vat' scenario (Putnam, 1981) posits that a brain could be detached from its body and

Accessing Chemical Concepts for Teaching and Learning 189

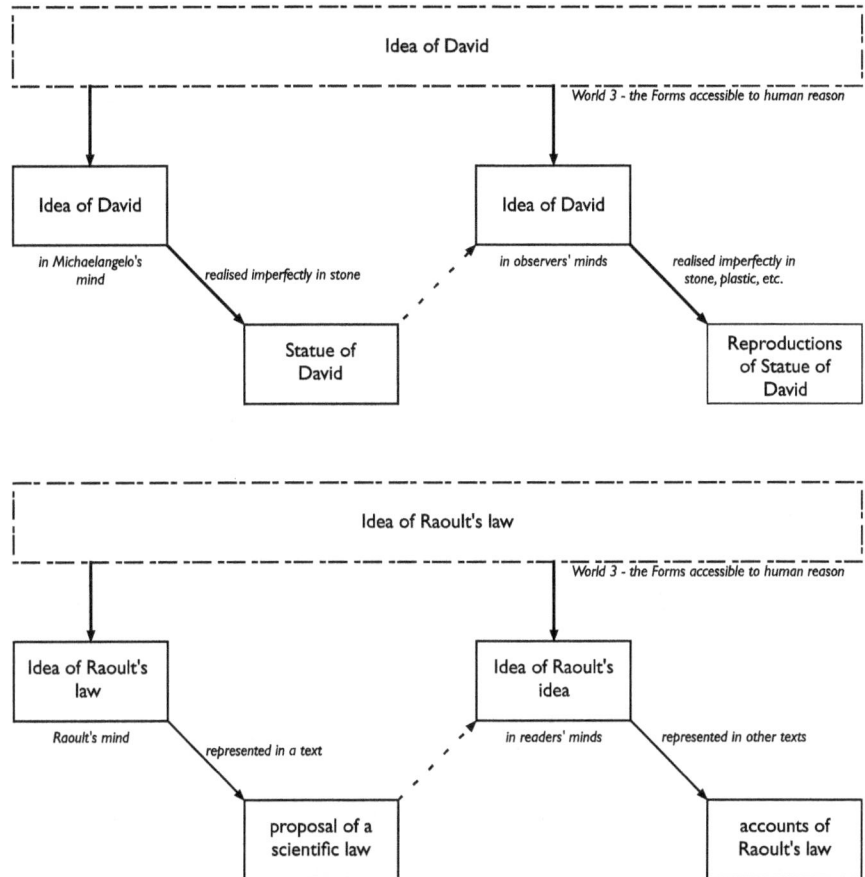

Figure 9.3 If thinking is able to directly access some realm of Forms then different people may access the same source for ideas, so that communication can be considered to be 'pointing to' ideas available to all rational beings.

placed in a bath of suitable nutrient fluid, and have all its external neural connections fed by synthetic circuits, such that the brain experienced what seemed a normal experience and was not aware of its actual status and location. This might be considered a parallel to the Turing test, which asks whether a person interacting with a conversant through an interface can be sure they are communicating with another person and not a sophisticated computer application.

That the current state of technology would not yet allow us to detach brains and immerse them in nutrient fluid, and connect them to totally immersive virtual reality is not the point – it is of course only intended as a Gedanken experiment or thought experiment. (It seems likely that immersion would not suffice, but there would need to be active pumping of the

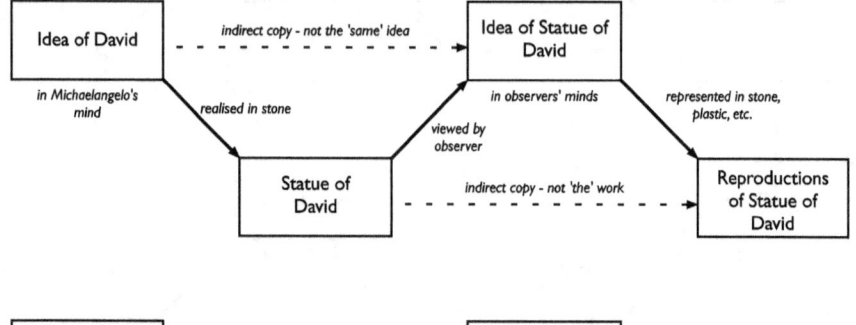
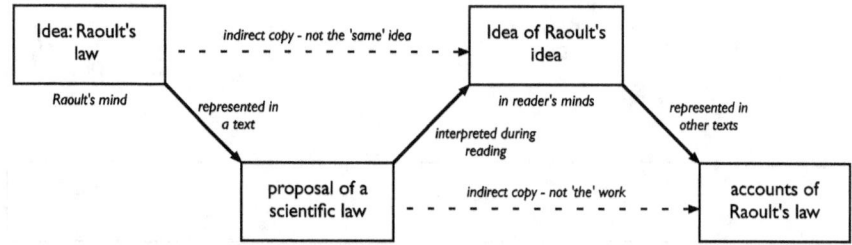

Figure 9.4 Without access to some common source of our ideas (see Figure 9.3), communication relies upon the ability of people to interpret representations (*e.g.*, in texts, works of art, *etc.*) of the original idea.

fluid through the brain's blood vessels, but that is a technical problem that in principle might be met: obtaining voluntary, informed, consent from participants might be a more serious impediment to a real experiment.) The brain in the vat would exist in a physical reality (the vat, the laboratory), but not the reality it 'knows'.

In everyday life, we generally operate as though there is an external reality that we experience and interact with. Yet, in the social sciences (including education), there are plenty of scholars who seem to claim that reality is not objectively out there but something created by their own minds or through social interaction among communities (Berger and Luckmann, 1991). One suspects that these scholars refer to bus or train timetables when planning their journeys *as if* they believe such timetables relate to something objective and external to themselves, but, in a sense, they are right that a belief that the world exists simply because we seem to perceive it is something of an assumption – albeit one that seems necessary for everyday functioning. Science gives us much reason to consider reality is not always as we directly perceive it (Gregory, 1997).

Certainly, the natural sciences, such as chemistry, seek to be objective. That is, science seeks knowledge of the world that is not relative only to the perspective or context of the particular observer, but that would apply to any observer able to replicate the conditions of the original observation. 'There was a nitrogen–oxygen atmosphere in the room' is a claim of objective knowledge that can be based on scientifically accepted measurement instruments and protocols. If such a statement is true, it is true for everyone.

By contrast, 'there was a very tense atmosphere in the room' is a subjective statement, relying on a personal perception that could be biased in many ways. The observer making such a statement might expect other observers to agree, but such agreement could not be assumed in the same way as expecting someone to corroborate that the gas sampled is indeed mainly a mixture of nitrogen with oxygen.

Science, then, presupposes that objective knowledge is possible. Scientists operate with certain metaphysical assumptions, assumptions that are not drawn from science, but rather need to be made *a priori* for science to be possible (Taber, 2013a). The existence of an objective reality (that different people exist in the same world, external to their own minds) and the possibility of humans having the potential to acquire reliable knowledge of that world, are (indeed, need to be) taken for granted in science. We do not usually notice these assumptions as they are not just taken for granted in science, but largely in everyday life as well. Scientists and members of the pubic more generally, wait at the bus stop in the expectation that there really are such thing as buses (an ontological claim about the world) and that they will recognise one when it arrives (an epistemological claim about the ability of humans to make sense of aspects of that world). The sceptical philosopher will join them in the queue, being pragmatic enough to appreciate that getting to the university to teach about our uncertainty in such matters relies on putting aside doubts about the existence of the bus, in the interim. (I suspect that many sceptical philosophers reserve the right not to be too sceptical on their own time.)

9.4.3 Objectivity in Educational Questions

I should perhaps have written above that the existence of an objective reality (that people exist in the same world, external to their own minds) and the possibility of humans acquiring reliable knowledge of that world are taken for granted in *the natural sciences*. However, these assumptions are sometimes much less secure in the social sciences. Chemistry is very much a natural science, but research *in chemistry education* often has more in common with the social sciences, suggesting that what we take for granted as scientists may not always be appropriate when working in education. Questions such as "was that a good lesson?", or "is she an effective teacher?", may provoke very different responses from different students in the same class. The researcher might pose the question in terms of a numerical scale, to measure an average response, but such an average might not be an authentic or meaningful representation of widely diverse perceptions: a class giving their teacher an average rating of 5 on a 9-point Likert scale could reflect a bimodal distribution where the individual students all rank the teacher at 1-3 or 7-9 reflecting very different experiences. One might imagine being given a box containing a number of fragile looking, unlabelled, bottles containing solutions. If you asked about the acidity of the liquids, how useful would it be to be told that, *on average*, they were pH7?

In chemistry education, we can be more objective about some claims than others. Class sizes; numbers of Bunsen burners, beakers, flasks, and the like, found in teaching laboratories; teacher qualifications; numbers of students choosing to specialise in chemistry studies: these things may be objectively measured with some precision, based on reasonable consensus about definitions and measurement techniques.

Student views, perceptions, attitudes and the like can be treated as if objective – but in doing so we produce an artefact that loses much of the qualitative nature of what we set out to investigate. That is not to say that numerical scales should be excluded, but if we are investigating personal experiences and opinions, we can only do these justice by collecting data that reflects the personal. That means people's own language – their own ways of expressing their views. (I imagine the reader has completed many questionnaires asking them to respond by selecting from a small number of closed response options. I doubt the reader has often been satisfied that these options actually provide adequate options to express how they would best answer those questions if given no restrictions.)

Of course, even if we do not question the existence of the world we might reasonably question whether it is quite as it often appears to us. Science has often shown otherwise. The issue of epistemology – how we come to sound knowledge of the world – is an important one. For example, chemists will happily look at a print-out from a spectrometer, and, from the appearance and location of peaks on the trace, infer the presence of structural features in molecules. That is some way from direct naive observation of the world and requires a theoretical chain of argument linking the observation (a curved line on a piece of paper that we see as a bump on a graph) with the inferred conclusion.

So, accepting that there is an objective external reality is not equivalent to believing it is possible to go beyond description and understand that reality (as science seeks to do). Science then relies on a number of basic assumptions that are prior to undertaking any empirical work – sometimes referred to as metaphysical assumptions. Natural science proceeds on the basis of the metaphysical assumptions that:

- there is an external reality;
- it is stable enough to make investigation of it worthwhile;
- it is knowable to some extent.

Some scientists are optimistic enough to suggest that science can and will fully understand the universe at some point – a view sometimes labelled 'scientism' (Stenmark, 1997). This requires a faith not only in the apparatus of science itself, but in the compatibility of the nature of the world with the potential of human cognition. Several centuries ago many practising scientists with strong Christian convictions adopted a position sometimes called 'natural theology' (Yeo, 1979), which assumed that humans could understand the world because God had created the world, and God had created people 'in God's image' so they could know God's creation. Indeed, for some of these scientists the study of the 'book of nature' (the creation) complemented the study of Holy Scripture as a form of religious practice.

Whilst some scientists would still adopt this position today, as least as a personal conviction, modern science is supposed to operate independently of reference to a supernatural realm, which makes it more difficult to justify an assumption that humans are, or will be, capable of fully understanding the natural world. (Despite this, oddly, it is sometimes scientists who are most keen to keep religion out of science who also seem most convinced that science will eventually allow humans to understand everything.) This is relevant to the concerns of this book because research suggests that often human cognition is *not* ideally suited to understanding scientific concepts (Wolpert, 1992), and that it commonly channels learners towards adopting alternative conceptions (*e.g.*, see Chapter 14).

Ultimately, we can only take a pragmatic approach in science. It is always possible that although we could understand some phenomenon well enough to predict and perhaps even control it, we might be actually completely misunderstanding its true nature. For example, if planning, and executing, international travel, much can be achieved by modelling the Earth as fixed in space, with its surface as stationary – as was commonly believed to be the case some centuries ago. Someone who believe this really was the true nature of things, would find their mental models allowed them to predict and control matters relating to travel fairly well.

That the earth is spinning on it axis, and following a complex trajectory that can be considered the sum of rotation around the centre of mass of the Earth–Moon system, rotation around the centre of mass of the Sun–Earth system, perturbation by other planets, the path of the solar system around the centre of the Milky Way galaxy, and the movement of that galaxy due to the influence of its local galaxies and the general expansion of space, can not only all be ignored, but introduce complications that would easily confuse matters. Sydney may be about 17 000 km from London if measured across the surface of the globe, but anyone travelling from London to Sydney ends up over 2 million km from their starting point within the solar system (*i.e.*, because Sydney, with the rest of the planet, has moved a long way during the trip)! Just because our ideas work, this does not imply we have a full understanding.

We find students sometimes seem to operate with their own alternative conceptions without spotting any incongruence between their experiences and their conceptualisations – even when their ideas are inconsistent with the chemical concepts taught (Taber, 1995). Scientists, historically, have also managed to make sense of their experiences using notions such as caloric or phlogiston or N-rays or cold fusion, even though it later became clear these conceptualisations of reality were flawed. It seems, then, that we can never know for sure if we have full or final understanding of some aspect of nature, and rather have to treat all scientific knowledge as conjectural and – in principle at least – provisional and open to correction (Popper, 1945/2011).

This is part of a modern post-positivist understanding of the nature of science that treats science as a source of reliable and robust, but not as proven-beyond-doubt, knowledge.

9.5 So Where Do We Find Chemical Concepts?

I suspect it seems obvious to many readers (as it once did to me) that a well-established discipline such as chemistry develops a set of canonical concepts, and that these are shared among its practitioners, publicly available in its literature, and so readily accessed by teachers to be passed on to learners. The present chapter has suggested that this may not actually be so. The chapter has discussed, and raised objections to, three possibilities:

- scientific knowledge such as canonical chemical concepts are found in texts, and, in particular, in the scientific literature;
- scientific knowledge such as canonical chemical concepts are to be found in the shared knowledge of the scientific community;
- scientific knowledge such as canonical chemical concepts have an independent existence and are accessible through rationale thought.

Concepts are mental entities, and knowledge can only be possessed by a knower. Sometimes we seem able to intuit the nature of things by pure reason (*e.g.*, we have moments of insight), but research into the learning sciences shows that this is certainly not assured, and indeed we may often have predispositions to understand the world in ways that science comes to question. Reasoning is essential to science, but application of reason alone is philosophy rather than science. Reasoning is a necessary, but not sufficient, criterion for developing scientific knowledge.

Scientists conceptualise the natural world based on an iterative process of observing, interpreting observations, forming theoretical accounts that suggest specific hypotheses for testing, and then undertaking further observations (*cf.* Figure 5.1). The concepts scientists form are mental abstractions from their interpretations of experience (including direct laboratory experiences, as well as experiences of reading each other's work, talking to colleagues, *etc.*, and the subsequent internal mental experience of reflecting on these various experiences), which they seek to share by representing them in words, figures, *etcetera*. So, the texts and other records do not contain concepts or other aspects of knowledge, but rather representations of someone's personal understandings in a form that needs to be interpreted so it can be understood by others.

Other scientists read the literature, and listen to conference talks, and so forth, and come to develop their own understandings of the work reported. They interpret the representations of chemical concepts in the literature through the interpretive resources they have available – their own conceptual frameworks, their own abstractions from experiences such as their own laboratory observations. Science is an objective process, so the system assumes that scientific reports should be comprehensible to others, and that differences in understanding will become apparent and can be discussed and tested. The analysis in this chapter is certainly not suggesting that science is a subjective activity. Science has checks and balances built in so as, as far as possible, to ensure that all in the community are sharing understandings and meanings – and that where there are discrepancies these are acknowledged and seen as a sign that further work is needed.

What the chapter does argue is that actually, to some extent, chemical concepts are indeed relative to individuals, because concepts are mental entities and so there is no perfect system of transferring ideas from one person to another. Communication is possible, and sometimes effective, but is not automatic. Moreover, science is an on ongoing enterprise: and so offers a moving target for the community of scientists who are inherently seeking to understand each other's ideas. Ideally, all chemists would share the same set of chemical concepts. Chemistry is a successful science in part because chemists have often been able to be effective in communicating ideas so others in the community understand them much as intended.

Yet this is not a perfect process, and the discipline of chemistry comprises a good many different specialisms with their own bespoke research programmes, such that most chemists are at best only part of one of many chemical research communities that actually share the components of a specific disciplinary matrix (Kuhn, 1974/1977). It is only within these different specialisms that chemists would be involved in the ongoing conversations that offer the (what has been termed) 'constant transactional calibration' (Bruner, 1987) that allows members to check their understandings against each other, and so ensure those concepts central to the research programme are (not perfectly and absolutely aligned, but at least) well aligned.

The scientific community more widely does not, strictly, share concepts as each person forms their own mental abstractions – so each scientist has a unique conceptual structure. We may think in terms of canonical chemical concepts, but what actually exists is a large community of chemists with somewhat overlapping expertise, working in a system that has mechanisms (such as peer review) that, for example, limit the extent to which members could use obviously idiosyncratic meanings for common terms, or apply a theory in a way counter to their peers, in published work.

In summary, then, there are mechanisms within science that tend to avoid chemists working in the same research programme forming substantially different conceptualisations within that active field of research without these becoming obvious and then subject to challenge. So, it is likely that, within such sub-disciplinary contexts, researchers operate with well-aligned (if not perfectly identical) concepts. Yet when considering a much wider frame – all chemists; or even all chemists and all members of the public interested in finding out about chemistry; the processes by which chemistry is communicated (*e.g.*, researchers representing their concepts in research reports that are interpreted by textbook authors who represent their own understandings in texts) are inevitably more open to the imperfections of human communication.

It seems:

- scientific knowledge such as canonical chemical concepts cannot be found in texts, as concepts are mental entities accessed by individual people, and texts only contain representations of personal concepts that need to be interpreted to be understood;
- scientific knowledge such as canonical chemical concepts are not to be found in the shared knowledge of the scientific community, as the community is highly fragmented, such that there is usually strong conceptual alignment within specific current research programme communities, but much weaker mechanisms to align concepts across the wider discipline and beyond;
- scientific knowledge such as canonical chemical concepts are not accessible through rational thought, such as forming intuitions about the world, as science is an iterative process of forming theoretical constructs that respond to, and motivate further, empirical observations, and this is subject to the limitations and inherent and learned biases of the human perceptual–cognitive apparatus.

None of this negates the worth of chemistry as a science or the value of chemistry education as a means to introduce people to an important and highly productive domain of scholarship, enquiry, and cultural achievement. It does, however, help explain something of why it can be so challenging to support students in mastering the conceptual content of chemistry.

References

Allen R. E., (1959), Anamnesis in Plato's "Meno and Phaedo", *Rev. Metaphys.*, **13**(1), 165–174.

Berger P. L. and Luckmann T., (1991), *The Social Construction of Reality: A Treatise in the Sociology of Knowledge*, London, UK: Penguin.

Brown J. R., (1991), *Laboratory of the Mind: Thought Experiments in the Natural Sciences*, London: Routledge.

Brown J. S., Collins A. and Duguid P., (1989), Situated Cognition and the Culture of Learning, *Educ. Res.*, **18**(1), 32–42.

Brown R. and McNeill D., (1966/1976), The 'tip-of-the-tongue' phenomenon, in Gardiner J. M. (ed.), *Readings in Human Memory*, London: Methuen & Company, pp. 243–255.

Bruner J. S., (1987), The transactional self, in Bruner J. and Haste H. (ed.), *Making Sense: The Child's Construction of the World*, London: Routledge, pp. 81–96.

Chomsky N., (1999), Form and meaning in natural languages, in Baghramian M. (ed.), *Modern Philosophy of Language*, Washington, DC: Counterpoint, pp. 294–308.

Collins H., (2010), *Tacit and Explicit Knowledge*, Chicago: The University of Chicago Press.

Conceição P., Heitor M. V. and Veloso F., (2003), Infrastructures, incentives, and institutions: Fostering distributed knowledge bases for the learning society, *Technol. Forecase. Soc.*, **70**(7), 583–617.

Edwards D. and Mercer N., (1987), *Common Knowledge: The Development of Understanding in the Classroom*, London: Routledge.

Gregory R. L., (1997), Knowledge in perception and illusion, *Philos. Trans. R. Soc. London*, **352**, 1121–1128.

Halpern J. Y. and Moses Y., (1990), Knowledge and common knowledge in a distributed environment, *J. Assoc. Comput. Mach.*, **37**(3), 549–587.

Kind V., (2013), A Degree Is Not Enough: A quantitative study of aspects of pre-service science teachers' chemistry content knowledge, *Int. J. Sci. Educ.*, **36**, 1–33.

Knorr Cetina K., (1999), *Epistemic Cultures: How the Sciences Make Knowledge*, Cambridge, Massachusetts: Harvard University Press.

Kuhn T. S., (1970), *The Structure of Scientific Revolutions*, 2nd edn, Chicago: University of Chicago.

Kuhn T. S., (1974/1977), Second thoughts on paradigms, in Kuhn T. S. (ed.), *The Essential Tension: Selected Studies in Scientific Tradition and Change*, Chicago: University of Chicago Press, pp. 293–319.

Latour B., (1987), *Science in Action*, Cambridge, Massachusetts: Harvard University Press.

Osborne R. J. and Wittrock M. C., (1985), The generative learning model and its implications for science education, *Stud. Sci. Educ.*, **12**, 59–87.

Popper K. R., (1945/2011), *The Open Society and Its Enemies*, London: Routledge.

Popper K. R., (1979), *Objective Knowledge: An Evolutionary Approach*, Revised edn, Oxford: Oxford University Press.

Putnam H., (1981), Brains in a Vat, in *Reason, Truth and History*, Cambridge: Cambridge University Press, pp. 1–21.

Shapin S., (1992), Why the public ought to understand science-in-the-making, *Public Understanding Sci.*, **1**(1), 27–30.

Smith A., (1827/1981), *An Inquiry into the Nature and Causes of the Wealth of Nations*, in Todd W. B. (ed.), Indianapolis: Liberty Classics.

Stenmark M., (1997), What is scientism?, *Religious Stud.*, **33**(01), 15–32.

Taber K. S., (1995), Development of Student Understanding: A Case Study of Stability and Lability in Cognitive Structure, *Res. Sci. Technol. Educ.*, **13**(1), 87–97.

Taber K. S., (2009), Learning from experience and teaching by example: reflecting upon personal learning experience to inform teaching practice, *J. Cambridge Stud.*, **4**(1), 82–91.

Taber K. S., (2013a), Conceptual frameworks, metaphysical commitments and worldviews: the challenge of reflecting the relationships between science and religion in science education, in Mansour N. and Wegerif R. (ed.), *Science Education for Diversity: Theory and Practice*, Dordrecht: Springer, pp. 151–177.

Taber K. S. (2013b), *Modelling Learners and Learning in Science Education: Developing Representations of Concepts, Conceptual Structure and Conceptual Change to Inform Teaching and Research*, Dordrecht: Springer.

Tsoukas H., (1996), The firm as a distributed knowledge system: A constructionist approach, *Strategic Manage. J.*, **17**(S2), 11–25.

Whorf B. L., (2012/1931), A Central Mexican inscription combining Mexican and Maya day signs. in Caroll J. B., Levinson S. C., and Lee P. (ed.), *Language, Thought, and Reality*, 2nd edn, Cambridge, Massachusetts: The MIT Press, pp. 55–64.

Wittgenstein L., (1922), *Tractatus Logico-philosophicus*, London: Kegan Paul, Trench, Trubner & Co., Ltd.

Wolpert L., (1992), *The Unnatural Nature of Science*, London: Faber & Faber.

Yeo R., (1979), William Whewell, natural theology and the philosophy of science in mid nineteenth century Britain, *Ann. Sci.*, **36**(5), 493–516.

CHAPTER 10

How Are Chemical Concepts Represented in the Curriculum?

10.1 The Curriculum

Curriculum is a major theme of study in education. There has been much attention, for example, to what is known as the 'hidden curriculum', or the implicit curriculum – that is how educational institutions may be teaching societal norms and attitudes that are not prescribed anywhere, but that are communicated through custom and practice, and so may be insidiously and inadvertently passed on (Cornbleth, 1984). This is an important topic, but for the purposes of this volume – with its concern with conceptual learning – the focus is on the formal chemistry curriculum.

The formal curriculum sets out what is to be learnt, and so what is to be taught. In some contexts, the curriculum will be constructed by the teacher. Someone offering a chemistry camp during the school holidays, perhaps as an enrichment experience for 'gifted' school children, or perhaps as a university's outreach activity to encourage students from disadvantaged communities to aspire to university study in the subject, might well have the luxury of deciding upon their own learning objectives, and designing curriculum accordingly.

Universities traditionally set their own curriculum, in which case there is likely to be a course team, required to account for their decisions to a hierarchy of committees, and delegating responsibility for determining precisely what is taught in particular lectures to the lecturers concerned. Even here, there may be a need to pay cognisance to external bodies. Industrial and other employers will have ideas about what they are looking

---

Advances in Chemistry Education Series No. 3
The Nature of the Chemical Concept: Re-constructing Chemical Knowledge in Teaching and Learning
By Keith S. Taber
© Keith S. Taber 2022
Published by the Royal Society of Chemistry, www.rsc.org

for in graduates, and learned societies, like the Royal Society of Chemistry, may have expectations about what needs to be included in a degree course that it will recognise as suitable for a degree labelled 'chemistry' (Royal Society of Chemistry, 2017).

In many school systems there is much less flexibility. Although it is not always the case, secondary-school (or high-school) chemistry teaching is often constrained by curriculum documents that are recognised in official government education policy and/or the syllabus documents that accredited examining bodies offer as specifications for what will be tested in their examinations. In the United States, the American Chemical Society has an Examinations Institute that offers examination services for introductory degree-level courses as well as high-school courses (Holme and Murphy, 2012). The chemistry curriculum may refer to particular attitudes and skills; as well as to areas of conceptual knowledge that may be tested, and so should be learnt – and so need to be taught. The curriculum may specify topics, theories, laws, models (see Chapter 8), and other more discrete concepts (see Chapter 7), and sometimes even specific examples.

10.2 Selection and Simplification

In constructing a curriculum, there are issues of selection and simplification. There is a need to decide what is included, and the level of treatment to be covered. Selection may be understood at two levels – deciding what falls under the remit of the subject of the course being taught, and then deciding what specific material should be included. The second stage here assumes the first: what is selected will be a sub-set of the potential curriculum defined by how the subject or topic is demarcated. Of course, the selection of material for inclusion in a curriculum assumes that it is clear what might be considered as a candidate for inclusion. That is, it might be taken for granted that the scope of chemistry is itself obvious.

10.2.1 Deciding What Counts as Chemistry

The educational theorist Basil Bernstein (Young, 2008) suggested that academic subjects varied in the extent to which they were understood either to have clear boundaries (strong classification) or to overlap with other subjects, without clear boundaries discriminating what 'belonged' under each subject (weak classification). Chemistry might be considered to be a subject with fairly strong classification – so a topic such as 'reactions of the halogens' is fairly obviously chemistry, whilst 'the causes of the industrial revolution' clearly belongs somewhere else. There are areas of somewhat weaker classification: there are topics that might be claimed by physics or chemistry, including some material – in thermodynamics, for example – that is taught in both subject areas. Generally, though, it may seem obvious what counts as chemistry. Classification in chemistry is also relatively strong in the sense that formal chemical discourse tends to be quite distinct from everyday talk – that may act as more of a barrier for novice learners, as the unfamiliar language highlights the subject's 'foreignness' and may make it seem more challenging to 'cross the border' into authentic disciplinary practice than may be experienced in some other academic subjects (Aikenhead, 1996). However, even in a subject such as chemistry, there might be questions over what should be included within the curriculum. Here I discuss two examples.

10.2.2 Chemistry in Its Social Context

In science education, going back a number of decades, there has been an argument that science needs to be taught in a social context. Before this, it was already common to teach about industrial applications and processes – the Haber process, for example – primarily to provide examples of where the chemistry was used and had economic importance. The science-in-a-social-context movement suggested teaching needed to go beyond simply discussing applications (Hunt, 1988). For example, students might be asked to consider the siting of a hypothetical new plant to process crude oil from an off-shore field. This might include economic considerations (cost of land), logistical issues (means of transporting products from the site to markets), environmental issues (disturbances of wildlife or areas of natural beauty – these days this would also include the wider consequences of burning fossil fuels), and so on. These are not strictly matters of chemistry, although they are issues that would be very important for some professional chemists who take up roles in industry or local government or the law. Including such material in a chemistry curriculum could be supported from a number of perspectives, including the importance of seeing how science interacts with society; the artificiality of disciplinary divisions in tackling real-world problems; and the need to engage and motivate those students who, strange as it may seem to some, do not find the chemistry itself intrinsically interesting.

More recently, there have been other arguments for this kind of approach. Given that most school students who study chemistry, and in some national systems also university students taking some chemistry courses (for example, if it is expected that arts and humanities students also take some science classes for breadth), are not going to use their chemistry professionally, then the aims of teaching the course should acknowledge the educational needs of all students, not just those wanting to be chemists. A science education for citizens – who will be consumers, voters, patients in health-care systems, *etcetera*, but mostly not professional chemists – needs to teach how science can be used to inform the decisions that people generally have to make when facing questions that have a science component. Often, such decisions cannot be decided by considering the science by itself, as they also involve value judgements (such has deciding whether to take a drug that is likely to prolong a terminally ill person's life, but has side effects the patient feels reduces the quality of that life). Issues that involve this combination of technical and conceptual knowledge, with the ability to consider different value-informed perspectives, are sometimes referred to as socioscientific issues (Zeidler, 2014). The argument is that not only is it important for people to appreciate the role science plays in such decision making, but also that educational experiences will prepare young people for facing such issues in the future by engaging them with socioscientific issues that present dilemmas of real-life complexity, where science has to be applied, but alongside other considerations.

10.2.3 Learning about the Nature of Chemistry

A second theme is the nature of science itself. My own recollection of my university chemistry degree programme is that it included a range of lecture courses on a diverse set of chemistry topics. I cannot, however, recall anywhere where there was any discussion of what chemistry was, and what made it a science. It seemed to be taken for granted that anyone signing up for a chemistry degree course would already appreciate these things, which itself apparently reflected a tacit understanding that the nature of chemistry was not problematic. Perhaps that is reasonable, as such issues should be considered in school science, so someone has a strong understanding of the nature of the discipline before they opt to study chemistry in higher education. My recollection of school science at the time, however, was that such matters were not explicitly treated there, either: perhaps the nature of science was not considered important or relevant, or maybe it was simply assumed that it would automatically become apparent through the process of learning some science.

Such an assumption may be true *to some extent*, but that may not always be a good thing. The way science is presented, and the language used, does communicate images of the nature of science. That is fine, as long as those images are ones we would consider appropriate. There are a number of areas where research suggests that many students have questionable ideas about science that have either been inferred from their science lessons, or have been maintained despite those lessons. This can include ideas such as (i) successful scientists are nearly always men; (ii) scientific breakthroughs are usually made by maverick geniuses who work independently of wider teams; (iii) scientific knowledge can be considered as certain and factual as it has been proven by passing experimental tests, or a corollary, (iv) where scientists publicly disagree, this is because no one has yet done the scientific experiments needed to confirm theories. For example, as there is public disagreement over the precise effects of anthropogenic inputs into the atmosphere on climate change, then (it is sometimes inferred, given that science produces unequivocal, absolute, proven knowledge) there is no firm scientific knowledge yet that can confidently inform public policy.

Earlier chapters discussed the nature of some of the conceptual entities met in chemistry courses – such as typologies (see Chapter 7), laws, theories, and models (see Chapter 8). Chapters 5 and 6 presented vignettes of the historical development of some chemical concepts (in particular, {acid} and {potassium}) as examples of how conceptual change occurs within a subject – such that, for example, to give a date to the discovery of potassium is necessarily oversimplistic because potassium, as we understand it in terms of our current chemical concept {potassium}, was not discovered at one point in time. Rather, the conceptual content (currently) associated with the label 'potassium' (*i.e.*, what we might think of as the content of the canonical {potassium} concept) has been refined over extended periods of time by different groups of chemists. The approach taken in those chapters (even if perhaps appearing simplistic to the historian of science, and naïve to the philosopher of science) is too detailed and sophisticated to be dropped into a school chemistry course – but might perhaps be suitable for including in a university chemistry course. Yet something of this essence of the nature of chemical concepts – and so the nature of chemistry as a science – should be appreciated by anyone completing a high-school chemistry course.

10.2.4 Chemistry Promoting Intellectual Development

Before moving on to focus on how chemical concepts are represented in the curriculum, I will address one further, related, point. A key aspect of my motivation in writing this book has been to set out some of the complexity of the subject of chemistry that faces novice learners of the subject, but that can come to be taken for granted by most chemists (including, if we are not careful, chemistry teachers). Conceptually, chemistry is a very complex subject because of the nuances that so many of our concepts have. I suspect that is a large part of my own attraction to the subject, and that may also be true of many readers of a volume such as this. Yet, this can be a source of frustration, rather than of fascination, for many learners if they are not sufficiently supported in meeting (and perhaps in a sense, initially even protected from) the many complications around so many chemical concepts.

Despite this, one strong rationale for including socioscientific issues and the nature of chemistry in teaching (and so inevitably engaging students with complexity and nuance) is that students can benefit intellectually from such challenges. One role for science teaching is to support the cognitive development of learners, that is, the development of their thinking skills. Research suggests that there is a common progression in such skills and that science education can provide suitable contexts for supporting development (Kuhn, 1999). A very simple summary of some of that research (certainly not doing it justice) is that

(a) young children do not differentiate the way the world is, from *accounts of* the way the world is (so being told by parents that a fat man in a red suit will be breaking into the house to deliver presents is just taken as a factual statement);

(b) later children make that distinction, but see matters in black and right: something is true, or not; one side in an argument is right, and the other side wrong;
(c) later still, young people come to appreciate that often things are not so simple, but that different perspectives may have merits even when in opposition (so there is a case for building the new off-shore wind farm in terms of supplying the national grid; but there is also a case that siting the wind turbines here will risk disrupting breeding grounds of an endangered species of bird), which leads to a shift to relativism such that right and wrong becomes irrelevant because decision making and evaluation are just down to opinion and personal preference;
(d) finally, people reach a point where they are able to adopt consistent, value-informed, positions, that they evaluate as being superior to other stances, whilst still acknowledging and understanding the merit of other positions, and appreciating their evaluation involves balancing a range of valid considerations.

Some of the early research in this area suggested that even elite undergraduate students at top colleges were often struggling to move beyond relativism (Perry, 1970), suggesting that (at least in the middle of the twentieth century) most school-leavers had not completed their intellectual development. The types of topics discussed above might well offer contexts for engaging in the kind of thinking that can support such development.

In a typical class, there will be a great diversity of learners – even if a school class is banded or set; or if a university class is only open to selected students. For classes to be genuinely educative the provision must offer all students suitable opportunities to develop. This means that all learners must both face challenge (as undemanding routine activity does not support development) and have sufficient support to be able to engage successfully (Taber and Riga, 2016). This is not easy for teachers who have to find ways to differentiate teaching for their classes. Some of the most gifted learners in chemistry classes may find lessons designed to be within the reach of their classmates do not really offer them great intellectual challenge – despite the thesis behind this book being that chemistry is conceptually a very complex subject!

In other words, just as the stereotype of a genius university professor who teaches in a way that only the most brilliant students can appreciate is a poor model for teaching most students, so the teacher who is very effective at making chemistry accessible to the weaker learners in a group, faces the danger of oversimplifying the subject for the more 'gifted'. This is not a dilemma that is easily solved – effective teaching is a highly nuanced and skilled actively. One approach is to start by designing teaching that is challenging for the most advanced learners in a class, and then consider how differentiation by support can enable other students to engage successfully (Taber, 2018).

10.2.5 Bounding the Chemistry Curriculum

So, we could judge that the content of a chemistry curriculum is just chemistry and a chemical education is about acquiring chemical knowledge and skills. Or, we may judge that the content also needs to include explicit teaching of the nature of chemistry as a science, as well as chemistry itself. (The analysis of chemical 'meta-concepts' in Chapter 8 suggests a strong case for teaching that is explicit about the nature of chemical knowledge: for example, the status of laws, theories, models, *etc.*) Also, we may feel that in a genuine chemistry education the chemistry curriculum needs to include relevant examples of socioscientific issues, where chemical knowledge has to be applied in the light of wider considerations.

10.3 Choosing the Chemistry

Even if a chemistry curriculum will only include the chemistry itself, devoid of its interactions with other disciplinary knowledge, it is inevitable that there will need to be some selection of, and within, topics. The content of chemistry, if judged in terms of the knowledge represented in the primary literature alone (which, as Chapter 9 suggests, is not an unproblematic source for identifying canonical chemical concepts), is too vast for a school chemistry course, or even a university chemistry degree, to include it all.

10.3.1 Breadth or Width?

A very common issue faced by those constructing a curriculum, concerns the balance between width and breadth of knowledge. A chemistry curriculum that looked to maximise breadth would seek to include material on all major areas of the subject, and all major topics within those areas. One argument for breadth, is that progression (from introductory courses to elective courses at school; from school to university study; from university to professional practice) may be compromised otherwise. If university entrance examinations test students on all major areas of chemistry, then students need to be prepared across those areas. If a degree course sends graduates into a wide range of careers in research, industry, environmental work, public health, and so forth, then it should prepare graduates regarding the chemistry they will most need in their future work (which will vary considerably).

The argument against too much breadth is that the thinner the content is spread, the less possibility there is for acquiring a deep understanding of the subject. The English National Curriculum for science, for example, has been criticised because the student experience was a constantly shifting carousel of topics (each with more new ideas to be studied) offering limited opportunities to ever become deeply engaged in any topic. Opportunities for substantive engagement are indicated if students are to meaningfully understand chemical topics. Focusing on a few areas in depth, leads to a deeper appreciation of the subject and so better provides the intellectual and disciplinary grounding needed to later be an effective learner of other chemistry that is met in a professional or other capacity. To some extent this is clearly the case – but most teachers would be uneasy taking this too far.

It would be possible to imagine a chemistry course that focused in great detail on Diels–Alder reaction chemistry. No doubt a great deal of chemistry could be considered when designing a course from such a starting point, but even so, if this was THE topic of the chemistry course, it would be considered unrepresentative of the subject more widely, and so inadequate as the basis for a chemistry curriculum.

So, in practice, there may be compromises – a curriculum may be devised to offer breadth, without needing to include all topics, with certain ideas or content areas from amongst those selected for inclusion identified for consideration in greater depth. As well as studying about the periodic table as a general theme, students may study in some level of detail the chemistry of – perhaps – the alkaline earth metals, or the halogens, and the first row of transition elements, but not other groups/periods. Perhaps in another course there will instead be a focus on the alkali metals and the nitrogen group. Some study of organic chemistry will be included, with the idea of homologous series given emphasis. Perhaps the alkenes and the alcohols will be discussed in some detail as examples: but alkanes, alkynes, ketones, *etcetera* given much less attention. There are clearly many possible permutations of such an approach.

10.3.2 Organising the Selection

There will need to be decisions about how to sequence and present the chosen material. For example, is it sensible to spend some time on basic ideas about particle models before proceeding to look at particular chemistry, or should these models be introduced as and when needed? Is it sensible to organise a course programme in terms of themes such as redox, and acids and bases – and within these topics include various material from different branches of chemistry? Alternatively, is it more sensible to teach according to the traditional distinctions between inorganic, organic and physical chemistry, and within that structure point out where connecting themes (such as redox) link the branches of the discipline?

Again, is it better to (i) teach about rate of reactions as an abstract topic early in a course, and then to revisit and reiterate the ideas that have been introduced in the context of different examples met later in the course; or is it more sensible to (ii) leave this topic till later in a course when one can revisit now familiar reactions to use as the contexts for thinking about rates? It may well be there is no reason to consider either of these approaches will be inherently better than the other, but any approach that seeks to emphasise the potential opportunities for linking and reinforcing ideas, and so seeks to integrate the subject knowledge as far as possible (*e.g.*, as either of options (i) and (ii) allow), is likely to be experienced by learners as more coherent, and to lead to learning that is more insightful and robust, than a teaching sequence that treats the chemistry curriculum as a succession of largely discrete units of teaching and learning.

It was argued in Chapter 2 that concepts inherently include associations with related concepts (the content of a concept can be modelled as series of propositions linking that concept to others), such that they can be considered as nodes in a complex conceptual net. This suggests that we have the choice of initially introducing a concept at a point where the learner does not have available many of the potentially related concepts making up the rich set of associations we wish to teach, in which case we later extend the teaching of the concept over time as we add new associations when we teach other material; or, we wait until all the related concepts are in place, and then teach the new concept when we can build up the richness the canonical concept has.

Needless to say, this decision has to be made on the basis of particular concepts: neither choice is available as a general strategy. If we were to decide to defer teaching the {base} concept to the end of a course and then develop it in the light of having already taught all the linked ideas we wanted students to have as part of their concept, then we cannot make the same choice about the concept {amines}; if we decide to teach about periodicity as our introductory topic before revisiting it in the context of other related concepts taught later in the course, then we are not able to make the same choice for the topic of atomic structure. Either we first teach about periodicity and later link it to atomic structure, or we decide to present atomic structure before periodicity, and later link it to periodicity.

10.3.3 Teaching Concepts in Stages

Some academic subjects can more readily be seen as being organised in a linear fashion than others. For example, mathematics is possibly the subject where it is most obvious that the subject matter is built up from particular starting points, and it is just not viable to teach some advanced topics without first teaching those from which they were developed. In contrast, in a course on literature, there may be no particular reason why, say, Kafka's 'The Trial' should be taught before, rather than after, Hesse's 'Steppenwolf'. Chemistry would seem to occupy an intermediate position in that it certainly involves the construction of more complex ideas upon more basic principles, but many of its ideas can only be fully appreciated in terms of others – indeed often others where the same consideration applies!

Learning chemistry then requires some degree of pedagogic bootstrapping, where one concept is partially taught, then used to support the teaching of other concepts, that in time allow the original concept to be revisited and better understood in terms of the others. For example, consider how one might teach the concepts {substance} and {chemical reaction}. If a substance is presented as something that cannot be broken down into anything simpler by chemical means (*i.e.*, by chemical reactions) then that aspect of {substance} is only meaningful to the extent that a learner *already* has a concept {chemical reaction} or {chemical change}. If a chemical reaction is presented as a process in which there are different substances before and after the change, then that facet of the {chemical reaction} concept is only meaningful to the extent that a learner *already* has a chemical concept {substance}.

Chemistry is then not inherently, at least not entirely, a linear subject where we can introduce a sequence of topics in a logical order such that each topic builds on, without substantially modifying understanding of, what has gone before. Certainly, there are prerequisite concepts needed for teaching some material (Herron *et al.*, 1977) – and some examples are suggested later in the chapter – but it is not possible to plan a whole course by an analysis based only on identifying prerequisite knowledge. There is no strict 'pedagogic Aufbau principle' (Taber, 2002) telling us what must happen in building up a teaching sequence: rather there are myriad potential routes, among which there are many that would be sensible approaches to introducing someone to the subject – providing suitable opportunities to link topics and concepts were taken.

Inevitably, although teaching needs to acknowledge the iterative way that many chemical concepts come to be understood – such that a better understanding of {concept (a)} developed using an understanding of {concept (b)} may through the more sophisticated understanding of {concept (a)} actually facilitate an even better understanding of {concept (b)} itself – whatever choices are made the concepts selected for teaching at the start of the schedule need to initially be presented in part, and then later elaborated in terms of concepts introduced subsequently. That is, on *introducing* a chemical concept, we will only be able to consider *some* of its conceptual 'content', through discussing *some* of the canonical associations (see Chapter 2). It may not matter so much whether we decide to first teach about periodicity and later link it to atomic structure, or to present atomic structure first, and later link it to periodicity, as long as we realise that we need to *continue teaching* the earlier concepts as we teach new concepts that are linked to them in the canonical version of the concept.

10.4 Keeping It Simple

So, one aspect of developing a curriculum, concerns deciding which topics are to be included, and another concerns the way the material ('content') that is selected is organised. However, the decision to include a concept in a curriculum is not *necessarily* a binary one. There is an asymmetry here. We may simply decide to exclude something: but we cannot simply decide to include material, without then making further choices.

For example, assuming we consider that the concept {hyperconjugation} is part of the conceptual content of chemistry (*i.e.*, it is included when we scope the range of canonical concepts in chemistry), we may well decide not to

How Are Chemical Concepts Represented in the Curriculum?

include it in an introductory chemistry course at secondary school level. Yet, whereas such a decision to exclude a concept is relatively straightforward, the same does not apply to the decision to include a concept. So, although we have sensibly decided not to teach 14-year-olds the concept of {hyperconjugation} (or {d-level splitting} or {antiaromaticity}, *etc.*), we may have decided we did wish to include in our curriculum the concepts, *inter alia*, {chemical bond}, {oxidation}, {rate of reaction} and {acid}. Whilst it is pretty clear how we do not teach the concept {hyperconjugation} (we do not mention it, we do not deliberately direct students to texts that refer to the concept, *etc.*) it is less obvious what it means to teach the concept {acid}. It was suggested in Chapter 6 that it may not even be clear what the canonical chemical concept {acid} actually comprises, but even when a concept seems unambiguous and to have (so to speak) tidy boundaries (or strong classification, to adopt the Bernstein terminology referred to above), it will not be possible to cover all potential aspects of the concept in a particular curriculum.

10.4.1 What Shall We Teach about the {Metal} Concept?

Let us take as an example a concept that is likely taught to some degree in most introductory chemistry curricula: {metal}. Put simply: we want to teach the chemical concept {metal}, and more importantly we would like the students to learn the chemical concept {metal}. We might break this question into considering (a) what is the canonical {metal} concept in chemistry, and (b) which aspect of the canonical concept do we wish students to learn – or, in other words, what counts as a creditable {metal} concept for students at this level?

The question of how we know what a canonical chemical concept actually is was met in Chapter 9, which raised some doubts about whether we can unequivocally characterise canonical concepts. Drawing on the discussion there, we might posit that:

(a) The canonical chemical concept {metal} comprises all the associated current knowledge claims found in the primary literature assuming that (i) we can easily decide what counts as chemical research; (ii) we can further decide which research reports are still considered to offer chemical knowledge, and have not been falsified or become obsolete; (iii) these accounts can be readily interpreted (more on this in the next chapter); and (iv) these accounts are mutually consistent such that there are no contradictions between claims in different papers meeting criteria (i) and (ii).

Or, we could instead suggest that,

(b) The canonical chemical concept {metal} comprises the ideas about metals that are consensual among the community of qualified chemists: assuming that (i) there is no problem identifying who counts as a member of this community; (ii) the community is in consensus; (iii) it is possible to identify this consensual content in some way.

As the reader will suspect (at least, if you have read Chapter 9) I am not wildly optimistic about either of these options as feasible and practical approaches to identifying canonical conceptual content, but we can put those concerns to one side for the moment to think about how we might proceed once (or if) we have judged that we had identified the canonical concept.

Perhaps, as a placeholder, we might put aside (a) and (b) and assume that;

(c) In practice, sometimes at least, a teacher or curriculum planner may substitute for the canonical chemical concept, their own concept.

In effect, the teacher can only ever teach from their own concept so what I am suggesting here is that instead of doing some specific research to support the planning of instruction, it is simply assumed that the teacher or curriculum developer, through their own education and the experiences that led to them becoming considered to have expertise, has already acquired 'the' concept. As the teacher only needs to have acquired the concept at the point of planning teaching, it would not be cheating (and indeed it may be appropriate) for them to revise their knowledge by consulting a textbook, searching the internet, talking to a colleague, *etc.*, but noting that, in practical terms, the 'updating' possible through such means inevitably falls short of what is strictly required under either (a) or (b) above.

So, what might be the content of the {metal} concept. Here are some suggestions.

The content of the concept {metal} includes examples of specific metals: sodium, magnesium, calcium, iron, copper… (that is, inherent to the concept {metal} is its set-member type relationship with other concepts such as {sodium}, so conceptions/propositions of the form 'sodium is a metal', *etc.*)

The concept includes some physical properties that metals commonly have – perhaps: lustre, sonority, relatively high thermal conductivity; relatively high electrical conductivity; ability to form alloys; crystalline form; ductility; malleability; … (that is, inherent to the concept {metal} is its associations in form of the typical characteristics that enable someone to discriminate with the concept: 'metals are conductors of electricity'; 'clean metal surfaces usually have lustre'; *etc.*)

The {metal} concept includes some chemical properties that metals generally have: they form basic or amphoteric oxides; they form salts with non-metals/acid radicals; *etcetera*.

The concept includes submicroscopic particle models explaining features of the observed properties (along the lines that the structure of metals is based on a regular lattice of atomic cores that allows some of the valence electrons to be delocalised).

10.4.2 What Details Shall We Teach about the {Metal} Concept?

I imagine many readers might want to suggest aspects they might wish to include that I have omitted. However, even without seeking to be comprehensive, my suggestions offer the potential for complications. Here are just a few of the questions that might need answering to refine our decision of what we need to teach, in order '*to teach the chemical concept {metal}*':

- If part of the {metal} concept consists of examples of metals (*e.g.* iron is a metal; copper is a metal; uranium is a metal, *etc.*) then the concept will be *incomplete* without the full list of metals – many of which are not usually met in introductory chemistry courses. If we assume we can do a decent job of teaching the {metal} concept without explicitly teaching 'X is a metal' for every X (is it essential for students to know samarium is a metal, if they otherwise have no concept {samarium}?), then how many examples do we explicitly mention (or direct students' attention towards in some other way), and which ones?
- If one of the properties we teach as being part of the {metal} concept is lustre, then how much detail do we give regarding when and why metals do not appear to have lustre, due to tarnishing of the surface. So, should the oxidation of aluminium be considered part of the concept to be taught, or patina on decorative metals, or verdigris on copper roofs?
- If we feel we need to teach about submicroscopic structure as part of the chemical concept {metal}, do we use a delocalised 'sea' of electrons metaphor? (Taber, 2003b) Do we need to explain conductivity in terms of a model of the overlapping of atomic orbitals forming molecular orbitals with extensive delocalisation (but where individual molecular orbitals will mostly not extend across the whole lattice)? Do we need to explain how these molecular orbitals may be considered to form partially filled energy bands such that thermal energy may be sufficient to allow electrons to move between different closely spaced levels? Do we need to consider how photons might interact with electrons in the metal to give rise to lustre? Do we need to consider models of resistance to explain why conductivity is not infinite and why different metals have different resistivities? And so on.
- As chemistry is a physical science, is it sufficient to consider metals as 'good conductors' or as having 'high conductivity'/'low resistivity', or is it important to include a quantitative criterion such as metals have conductivities of 10^6 S m^{-1} or more?
- In explaining why metals form salts with non-metals, to what extent should we link this with concepts such as {the periodic table}, {electronegativity}, {ionisation energy} ...?

I imagine the reader can see that this list of questions that are addressed explicitly, or implicitly, in modelling canonical knowledge in shaping a curriculum could be extended considerably. It is not being suggested this creates *an impediment* to teaching the concept, just that there is a good deal to be thought about in planning teaching.

It may also be fairly obvious that a similar sort of analysis can be carried out for any other chemistry concept, and that for some of the points that would be raised, as in this example, we would likely respond along the lines, 'no, we do not need to go into that at this level'. However, where we would draw that line (assuming 'we' might agree on where to draw it) would usually be different for 13–14-year-olds, and 16–17-year-olds, and again in a first-year undergraduate chemistry course primarily taken as a subsidiary or service subject, or again for a final-year honours chemistry degree lecture course.

In the latter case we would be teaching at a higher level, and so including more of the conceptual content, detail and nuance – but such final-year courses are usually taught by expert research chemists in their specialist areas, so that their own personal version of the chemical concept concerned likely has much greater content that that of a middle school teacher who is expected to teach content from across chemistry, indeed probably across school science (or possibly even across a whole gamut of STEM subjects). At whatever level someone is teaching chemistry, one might expect that the teacher's chemical knowledge would extend beyond the level of understanding that their students are being asked to develop. I intend that last statement to be normative in descriptive and also prescriptive senses: both that it would generally be found to be the case, and that students (and their parents or other sponsors) should be entitled to expect it would be the case.

10.4.3 The Spiral Curriculum

In 1959 the educational thinker Jerome Bruner chaired a ten-day conference set up by a committee of the US National Academy of Sciences to consider the teaching of science in schools, leading to a book 'The Process of Education' (Bruner, 1960). One of the ideas that Bruner popularised was that of a spiral curriculum.

In a linear curriculum, topics and ideas are arranged in a sequence that reflects the most logical order in which to meet them. So, a topic or concept may be analysed to identify prerequisite knowledge: to understand this (say, the concept {compound}) one has to use this (say, the concept {substance}) so we need to teach the concept {substance} before we teach the concept {compound}. There are many places in chemistry where this logic is important. So we might think:

- Before teaching the concept {ionisation energy} we need to teach the concept {atomic structure}.
- Before teaching the concept {dative bond} we need to teach the concept {covalent bond}.
- Before teaching the concept {keto–enol tautomerism} we need to teach the concepts {ketone}, {alkene} and {alcohol}.

- Before teaching the concept {variable oxidation state} we need to teach the concept {oxidation state}.
- Before teaching the concept {sp^3 hybridisation} we need to teach the concept {orbital}.

The reader can likely think of myriad other examples.

Yet, as discussed above, even though we can find many examples where we can only sensibly introduce one concept in terms of other ideas that need to be already understood, there is no overall simple linear order for teaching chemistry, as many concepts can only be partially taught in advance of teaching other concepts that they might be considered prerequisites of. Indeed, as we saw with {metal} above, even quite basic concepts become quite sophisticated when analysed in detail, and we seldom need or wish to introduce all their content at the same educational level. Teaching of these concepts needs a different approach from looking for a simple linear sequence of concepts. In a spiral curriculum, there is no attempt to find a single linear sequence for teaching topics or particular concepts, but rather the same topics and even concepts, are revisited at several points during learning.

We may ultimately wish students to acquire a sophisticated, nuanced understanding of some complex and subtle ideas. The nature of human learning seldom allows this to happen from a 'standing start', that is beginning with no relevant background in the topic. Student learning tends to be iterative and incremental (Taber, 2014b). Imagine that we wish students to understand atomic structure in terms of quantum numbers and energy levels, using the orbital approximation; and adopting models of electron 'clouds', appreciating Hund's rules, and the like. It would seem foolish to attempt to teach this topic in this depth to a typical class of, say, seven-year-olds, and some theories of cognitive development would suggest that such abstract ideas would not be accessible to children of this age (Piaget, 1970/1972).

Perhaps we see this as knowledge that we wish students completing upper secondary school elective courses at age 18 to have mastered. (We can call this 'target knowledge'.) If so, there is no need, and indeed little point, in seeking to teach this content to those seven-year-olds that we have decided are not able to make good sense of it. However, it would also be foolish to decide to completely leave teaching the topic until students get to 18, and then in quick succession to introduce the idea of particle models, introduce atoms, the nucleus, electrons, protons, neutrons, electron shells... and on to our target knowledge of quantum numbers and electron density patterns and the like. Rather we seek to provide a learning progression over time that builds towards the target knowledge.

Bruner (1960) also suggested that one could teach any topic to any child of any age in an intellectually honest manner. What he meant was that even if teaching a sophisticated scientific concept in its entirety to a young child seem futile, there would be an essence of the concept that could be effectively communicated – although not the full blown canonical concept. By 'intellectually honest', he implied that simplifications that do not retain the core essence of the canonical concept were not admissible. Even if this bold claim that *one can always find intellectually honest simplifications of any material suitable for any age of learner* was overstated, it has been seen as something teachers should adopt as an aspiration. We should avoid coming to such opinions as 'there is no way this class/student will ever understand hydrogen bonding' (or whatever), as it is the teacher's job to find a way to help learners grasp something of the essence of the concept. (How teachers might go about this is considered in Chapter 12.)

The principle of the spiral curriculum suggests then that we should not try to teach a detailed, nuanced account of atomic structure either by introducing it to very young students or leaving it till the age when we expect students to have mastered it. Rather we should introduce a simplified version of target knowledge early in the spiral curriculum, and then later revisit it in increasing sophistication and complexity. This is not inefficient, even if we have to commit teaching time to the same topic on a number of occasions, as revisiting a topic to build up an understanding through manageable chunks, viable 'learning quanta', reflects what is known about how human learning of complex material is necessarily an incremental and iterative process (Taber, 2014b).

So, the logic of the spiral curriculum reflects the mismatch between the complexity of some material we wish people to learn, and the way human learning tends to be limited by processing modest but manageable learning quanta; and, also, how learning only tends to become robust (so that it is readily accessed and applied) with periodic reinforcement. Brain changes indicative of learning do not happen all at once – but rather occur in stages, so that a learning 'event' is, in a sense, stretched over weeks and months even when triggered by teaching that occurs in one lesson or lecture.

In additional to these general arguments, that a spiral curriculum may better suit human cognitive processes (Johnstone, 2000); in a subject like chemistry, we have the additional complication (discussed above) that chemical concepts cannot be arranged in a linear sequence where each new concept is only taught after its prerequisite concepts have been learned: but that, rather, there is a need for an element of 'bootstrapping' where we use a partial understanding of one concept to help introduce another that in time will allow us to develop the first.

10.4.4 The Optimum Level of Simplification

So, if we are to teach redox to 17-year-olds, or acids to twelve-year-olds, or mass spectroscopy to second-year undergraduates, we need to think about how we represent the chemistry to that group of students in terms of their readiness to learn the material. Ideally, a teacher is able to plan teaching as a spiral curriculum (rather than only having responsibility for one turn of the spiral), but if not, as when a teacher is assigned to a class for only one year or to teach a single course that is part of a longer programme, we should at

How Are Chemical Concepts Represented in the Curriculum?

least have a decent knowledge of the helical pedagogic context within which our input occurs.

In general, our teaching of any concept will require a simplification of the canonical chemistry concept. As suggested above, the content of any chemical concept will be so extensive and nuanced that we would have to be selective just on the grounds that there would never be time to teach every detail. Usually, we will not only have to be selective about matters of the amount of detail (so we judge that it is not essential to mention every metallic element in teaching the {metals} concept, even if that is part of the full content of the canonical scientific concept), but in terms of sophistication and complexity.

So, for example, we may teach atomic structure in terms of shells, even though we know that is a simplification that limits full understanding, because this group of students are not ready to learn about the azimuthal quantum number. Rather than try and teach them something that they will likely fail to understand and learn, and so something that they will likely find frustrating, confusing, and off-putting; instead, we sensibly teach a model that is less sophisticated, but which the students have a good chance of finding meaningful (based on their existing understanding of relevant concepts), where they are likely to be successful in their learning and acquire ideas that they can usefully apply – even if only in a carefully selected sub-set of the problems and contexts that the canonical concept could be applied to. In these circumstances, simplification is necessary and sensible – as teachers tend to intuitively appreciate.

What should be avoided, however, is oversimplification. This does not just mean simplifying more than is needed by underestimating what particular groups of students are ready to engage with. Rather, this also links to Bruner's notion of presentations of concepts being intellectually honest. We can simplify concepts to such a degree that we are teaching something that no longer reflects the essence of the canonical concept. Students may be able to learn and apply the simplification to examples and contexts we suggest to them, but if the simplification is no longer an authentic reflection of the chemical concept there seems little value in this.

For example, students may be able to apply a teaching model that *ionic bonding is the transfer of an electron from a metal atom to a non-metal atom* to explain the formation of the ionic bond in NaCl: but if the model is flawed (as it is), and the explanation invalid (as it is – the process would be endothermic), and the learning leads to a conception of the ionic bond that is unhelpful in explaining the properties of NaCl (high melting temperate, solubility in polar solvents, conductivity of aqueous solution, *etc.*), then it may be better not to teach the concept at all, rather than oversimplify it in this way. (This example is not a 'straw man' set up for the purposes of the argument here, and will be further discussed below.)

So, the imperative is to seek the optimum level of simplification in teaching (Taber, 2000). This is when the concepts being presented are simplified enough to be meaningful and accessible to the student(s) concerned, without being oversimplified. In particular, an optimal simplification will provide a strong basis for later progression in learning to more sophisticated accounts. Any simplification that will act as a barrier to progression in learning should be avoided. It may not always be obvious what might count as a sub-optimal simplification in this sense, although the research into student learning in chemistry offers some examples.

The identification of optimal presentations of scientific ideas in teaching is itself a focus for scientific enquiry. Such enquiry can be informed by philosophical work, analysing the chemical concepts themselves in terms of their characteristics, affordances, core and peripheral features, links with other concepts, *etcetera*, but this needs to be related to studies of students' existing understanding, and of learners' responses to being taught with particular treatments of concepts. Like science more generally, progress is likely to depend upon an iteration of theoretical and empirical work.

10.4.5 Learning Progressions, Big Ideas and Threshold Concepts

Some of these considerations are being taken seriously in research in science education. One trend has been to seek to identify 'big ideas' in school science, around which the spiral curriculum can be planned and developed. For example, 'particles' was identified as one of five key science ideas recommended as foci for constructing teaching schemes for 11–14-year-olds in England (Key Stage 3 National Strategy, 2002). Such ideas can be used as the core theme for building learning progressions (LP). LP refer to the trajectories that students pass through in their learning (Duschl *et al.*, 2011) – and the term may either describe empirically observed LP or hypothetical LP (Alonzo and Gotwals, 2012). Such hypothetical LP take into account the learning demand (Leach and Scott, 2002) between the likely starting points of students and the desired learning outcomes at the end of that phase of teaching and learning – which ideally extend over multiple grade levels. The 'lower anchor', the assumed starting point of a LP, is informed by research into students' thinking – including any common alternative conceptions (see Chapter 14) – at the age the topic is introduced. The 'upper anchor' represents the target level of knowledge and understanding – so, what would count as creditable concepts for this level of education. The LP offers routes (ideally, more than one to allow for learner diversity) between these anchors in terms of intermediate conceptions.

Whilst terminology such as big ideas is commonly used in this context, a more sophisticated notion is of a threshold concept, which is a concept that is not simply important in the discipline, but actually has particular significance for understanding the subject. These have been characterised:

> 'conceptual gateways' or 'portals' that lead to a previously inaccessible, and initially perhaps 'troublesome', way of thinking about something. A new way of understanding, interpreting, or viewing something may thus emerge

– a transformed internal view of subject matter, subject landscape, or even world view. In attempting to characterise such conceptual gateways it was suggested...that they may be transformative (occasioning a significant shift in the perception of a subject), irreversible (unlikely to be forgotten, or unlearned only through considerable effort), and integrative (exposing the previously hidden interrelatedness of something). In addition they may also be troublesome and/or they may lead to troublesome knowledge for a variety of reasons. (Meyer and Land, 2005, pp. 373–374)

That is, threshold concepts in chemistry would not just be big ideas for chemistry, but of special importance in *learning* chemistry – and related to areas where there are tenacious alternative conceptions that have consequences for understanding large parts of the subject. These concepts may be problematic in ontological terms (as they may not fit intuitive typologies – see Chapter 7) or in epistemological terms (especially important in chemistry where students are asked to operate a good deal with models – see Chapter 8). So, we have seen earlier in the book, for example, that the chemical concept of {substance} is essential to making sense of chemistry as a discipline, and the key concept of {quanticles} (*i.e.*, the conjectured submicroscopic scale entities, that are often referred to as 'particles', but are quite unlike particles familiar to learners), may be challenging concepts for learners. Despite, for example, a very helpful treatment by Talanquer (2015) who discusses the concept of {atomicity} as an illustration, the notion of threshold concepts has not yet had the research attention it deserves in chemistry.

10.5 Curricular Models

There are a number of different kinds of models that are important in chemistry education. In Chapter 12, teaching models will be discussed. Mental models, that only exist inside the minds of individuals are also clearly important (Gentner and Stevens, 1983). Scientific models were discussed in Chapter 8. These are models proposed and used by scientists in their work. It was suggested that these are the basis of one class of concepts. Chemistry students are asked to learn about a range of scientific models, such as when modelling atomic structure in terms of concentric shells of electrons. It was suggested that students are expected to form a concept of a model such as {shell model of the atom} (although they may not use that terminology), which means they have the basis for discriminating this model from other models. This is not quite the same as saying they had learnt the model, as that would require more; but *part of* effectively learning the model would involve recognising this model as a distinct entity in its own right. It would be difficult to make sense of variously meeting atoms modelled as, say, tiny ball bearings, with a shell model, and with an orbital model, without forming distinct (even if linked) concepts of the different models.

Scientific models are candidates for being included in chemistry curricula – they may potentially be chosen when selecting from what is judged the content of chemistry at the scoping stage (see above). A student would be expected to not only recognise that a particular model was being invoked, but to understand and apply it. Of course, there are degrees of understanding that are possible, and different groups of students might be expected to be able to apply a model in different ranges of contexts, or to a different level of sophistication. Given what has been discussed earlier in this chapter, it is often not a matter of simply deciding that a scientific model will be excluded or included in the curriculum: if it is included, then there will be judgements to be made about the extent to which the full scientific model should be taught, or rather some simplification. A simplified version of a scientific model, set out as target knowledge for a group of learners, would be a curricular model.

Curricular models do not only relate to scientific models. The previous paragraph could, for example, be rewritten by substituting 'scientific theory' for 'scientific model'. So, a treatment of the {acids} concept set out as target knowledge for 13–14-year-olds would be a curricular model, as would the treatment of Raoult's law prescribed for teaching 16–17-year-olds, or the treatment of transition-metal complexes in a second-year degree course. These are models because they are selective simplifications of the canonical chemistry content (current canonical knowledge about acids, or Raoult's law, or transition-metal complexes) that limit the amount of detail/exemplification and usually involve omitting some of the complexity. A curriculum may contain models of particular scientific content that has been simplified and abridged with the intention of making the material suitable for teaching and learning in relation to particular groups of students. Ideally, curricular models should be honed towards, what is described above as, the optimal level of simplification.

So, it would be naive to consider that a curriculum document offers a selection of scientific content to be taught – certainly in school teaching, and to some extent in undergraduate education as well. Rather, such a document sets out re-conceptualisations of scientific knowledge adjusted to meet educational aims. To some extent the detail and sophistication of the current scientific account is likely to be necessarily compromised in terms of the teaching time available and the readiness of students to make good sense of the material, especially when it is judged that teaching a simplified account would be viable, but to present the topic in its full complexity would (currently) be futile or counterproductive.

The extent to which curricular models are detailed in documentation clearly varies. In some teaching contexts, curricular models are somewhat implicit. This is not problematic for teachers who are responsible for designing their own courses – as often happens with university lecturing where the teacher takes a topic, decides what should be taught and in what depth, and sets the examination questions that students will be asked to tackle. This may, however, be more of a problem when the curriculum specification was prepared by someone who is not the teacher. In these situations, a teacher teaches according to their understanding of what is expected, based on the presentation in the curriculum document. The teacher has to interpret the representations of someone else's conceptualisation of a suitable curricular model presented in documentation such as examination specifications (see Figure 9.2).

Perhaps it is only after students are examined by some external authority that the teacher receives an examiners' report that suggests that those marking examination scripts had a different understanding of what was being specified. Perhaps the teacher omitted some aspects that examiners expected, or used valuable teaching time to include some complications that the examiners considered unnecessary at the level being tested. (Of course, a teacher may deliberately teach something not examined on the principled grounds that in their view it is an important feature of the concepts being taught even though it is not specified. That is rather different from failing to correctly interpret the intentions of the person setting the curriculum or the interpretations of those examining student learning.)

Such misjudgments seem unfortunate, and perhaps explain the tendency in some educational contexts for the 'specifications' for an examination course (what use to be called a syllabus) to become highly detailed (specification, indeed), and accompanied by precise mark schemes for specimen and past papers (*i.e.*, examination papers from previous rounds of the examination).

This development (which I have seen at close hand in England where it has been quite extreme) inevitably replaces the potential unfairness of teachers having to guess at the level of treatment expected, with a tendency to encourage 'teaching-to-the-test'. So, there has been a historical shift as initially chemistry teachers were expected to teach chemistry, and initially examinations were designed to test the teaching of chemistry: but over time as target knowledge becomes more tightly defined and examiner's expectations more explicit, teachers focus less on teaching chemistry, and more on preparing students for the examination. To do otherwise would (assuming most other teachers are teaching-to-the-test) disadvantage one's own students compared with others sitting the same examinations, and could even be considered poor teaching by many students, parents, headteachers/principals, and other stake-holders.

The issues are slightly different in universities where the lecture course may (sometimes, but not always) be taught and examined by the same lecturer. I recall one lecturer suggesting to the class that if we forgot the rest of the material being taught in his lectures (I seem to recall the term used was, 'this bloody rubbish') we should make sure we remembered the Nernst equation. Another lecturer would pause in a lecture, very deliberately put a box around part of his notes on the chalk board (yes, this was before data projectors), and perhaps add an asterisk, before offering subtle hints such as 'if I was the person setting the examination question on this, which of course I am, I would think this could be something to base a question on in the exam'. In that context, where everyone taking the examination is given the same guidance (so only those absenting themselves from lectures would be disadvantaged), such comments could be seen as helping students prepare for an examination, or might alternatively be seen as a clever – if not particularly subtle – way of stressing the points the lecturer thought were most important to understanding the topic.

This treatment may all seem largely general, so it may be useful to consider an authentic example of how chemical knowledge becomes modelled as target knowledge in a curriculum context.

10.5.1 How Do Chemical Reactions Take Place in England?

As one example of a curricular model, consider the following extract from the official school science curriculum document for secondary age students (up to age 16), 'Subject content – Chemistry', in England (DFE, 2015):

> "chemical reactions take place in only three different ways:
>
> - proton transfer;
> - electron transfer;
> - electron sharing".

This statement does not suggest that students *only need to be taught* about three ways in which chemical reactions take place, but that there are "only three different ways [in which] chemical reactions take place". That is, it is not suggested that perhaps there are various ways that reactions take place, but English school children only need to learn about three that have been set up as target knowledge. Rather, the curriculum presents an explicit model of scientific knowledge.

The terms 'electron transfer' and 'proton transfer' are reasonably clear, but 'electron sharing' might be considered a little vaguer, even though it is a term widely used in chemical discourse. We might wonder (a) how do teachers present such sharing, and (b) does this reflect how professional chemists understand 'electron sharing'?

However, here I want to raise a different issue, which is whether these three categories exhaust the ways chemists consider chemical reactions to take place. We might consider precipitation reactions, such as

$$Pb(NO_3)_2(aq) + 2KI(aq) \rightarrow 2KNO_3(aq) + PbI_2(s)$$

I think this would count as a chemical reaction. The lead iodide precipitated is a chemical substance that was not present before the reaction, so this would seem to be an example of chemical change. Indeed, this is often used as an example of a reaction in school chemistry, along with those precipitation reactions used in inorganic analysis that produce new chemical substances such as silver chloride and barium sulfate.

These reactions involve discrete (initially hydrated) ions forming into crystal lattices. We might ask how this fits the statement that chemical reactions take place in only three different ways, that are proton transfer, electron transfer, and electron sharing. These reactions do not result from

How Are Chemical Concepts Represented in the Curriculum?

any of these mechanisms (certainly not in simple and straightforward ways as they are taught at this level). So, the curricular model of how reactions occur at the submicroscopic level, set out as target knowledge in the English school curriculum by the British government, excludes as chemical reactions some of the chemical reactions commonly studied at school level.

A common reaction studied in school science is the formation of sodium chloride by neutralisation:

$$HCl(aq) + NaOH(aq) \rightarrow NaCl(aq) + H_2O(l)$$

The product NaCl is initially formed in solution, and indeed in a sense (the sodium and chloride ions can be seen as spectator ions) the real reaction here is the formation of additional water molecules due to the equilibrium:

$$H_2O(l) \rightleftharpoons H^+(aq) + OH^-(aq)$$

If water is seen as the product, then this could be seen as proton transfer from a hydronium ion to the hydroxyl ion – yet I doubt that is the intention of the curriculum designers in terms of teaching this age group – or as the hydroxyl ion sharing a lone pair of electrons with the hydrogen ion. But, at this level, the focus of teaching neutralisation reactions is on the production of the salt, not of more water molecules. It is common in school science to allow the solution to evaporate (or even to heat the solution over a steam bath), so that solid crystals of the salt are formed. The reactants are two aqueous solutions, one an acid, one an alkali; and the product of interest is a solid salt. This is surely a chemical reaction. The English National Curriculum includes a reference to neutralisation reactions such as this one:

"Chemical reactions

- chemical reactions as the rearrangement of atoms;
- representing chemical reactions using formulae and using equations;
- combustion, thermal decomposition, oxidation and displacement reactions;
- defining acids and alkalis in terms of neutralisation reactions;
- the pH scale for measuring acidity/alkalinity; and indicators;
- reactions of acids with metals to produce a salt plus hydrogen;
- reactions of acids with alkalis to produce a salt plus water;
- what catalysts do".

Interestingly, the English National Curriculum specification here does not specify that a chemical reaction involves a new chemical substance being produced, but rather only offers a submicroscopic account: "chemical reactions as the rearrangement of atoms". By that account, my example (the formation of sodium chloride by neutralisation) is, strictly, *not* a chemical reaction as it concerns a rearrangement of ions.

The formation of sodium chloride by neutralisation followed by evaporation of the solvent, a simple laboratory procedure carried out by millions of students around the world, involves the formation of an ionic lattice by the interaction of ions that were previously in solution (from two distinct reagent solutions). New chemical bonds, which are normally characterised as ionic bonds, are formed, but there is no proton donation, electron donation, or electron sharing (unless these terms are interpreted in terms of subtle shifts in electron distributions about the ionic cores, which is not in keeping with the general level of treatment, and is not normally taught at this level in English schools) involved in the formation of the sodium chloride.

Nor are any atoms rearranged, as none of the species concerned are in the form of atoms: there are hydrated sodium ions, chloride ions, hydrogen ions and hydroxyl ions, plus water molecules. Few chemical reactions do involve atoms as such, as the only substances that commonly exist as discrete atoms are the inert gases, which have very limited chemical reactivity (and certainly no reactions commonly undertaken in school laboratories). So, according to the target knowledge expected of English school pupils, the formation of sodium chloride or other salts by neutralisation, just as precipitation reactions, do not seem to count as chemical reactions either in terms of the definition (what happens to atoms) nor in terms of the exclusive classes of types of reaction mechanism (proton donation / electron donation / electron sharing).

The English National Curriculum therefore seems to offer a poor curricular model of 'chemical reactions' (or at least a questionable curriculum model of an exclusive typology of reaction mechanisms – the only ways that reactions occur) as it seems the model offered of how reactions occur is inadequate to explain some of the reactions students will study. This may be a simplification of the chemistry, but it is not an optimal simplification, rather it is a curricular model that is simply inconsistent with current scientific models.

10.5.2 When Is an Atom, Not an Atom? (When Is an Ion an Atom?)

It might be argued that *in part* my criticism here is pedantic, and perhaps even indeed unfair, because I am distinguishing atoms from ions, and perhaps that is not intended in this curriculum. The same official government curriculum document, setting out target knowledge for students completing compulsory schooling in England (*i.e.*, at age 16) includes another subsection headed 'Atoms, elements and compounds' that include the phrase "differences between atoms, elements and compounds". I've reproduced the full subsection here:

"Atoms, elements and compounds

- a simple (Dalton) atomic model
- differences between atoms, elements and compounds

How Are Chemical Concepts Represented in the Curriculum? 223

- chemical symbols and formulae for elements and compounds
- conservation of mass changes of state and chemical reactions"
[(lack of) punctuation of the original retained]

The phrase "differences between atoms, elements and compounds" suggests (to this reader) two types of distinction. There is a major ontological distinction between (i) atoms that are conjectured theoretical entities of submicroscopic scale that are components of chemical theories and explanations, and (ii) elements and compounds, which are categories of pure substances (see Chapter 8). Of course, the curriculum document is meant as guidance to teachers who should be fully aware of this, even if school children themselves might find the statement a little bewildering.

However, an issue raised by the curricular model represented here is that although atoms are specified, there is no explicit mention of ions or molecules. This might seem an odd omission as very few materials students come across (in the laboratory or beyond) are in the form of atoms, and those that are (neon, helium, argon) do very little chemistry that is commonly included in school courses.

The substances lead iodide and sodium chloride discussed above do not contain atoms, and they are not produced in the laboratory from anything that does. Or, at least, not directly from anything that contains discrete atoms: sodium chloride could be made by burning sodium in chlorine (even if this is not a recommended practical activity for school children) but the chlorine is molecular, and the sodium begins as a metallic lattice of ions with a balancing quantity of (lattice/delocalised/free/conduction) electrons.

Many substances contain molecules – arguably, including the noble gases. Helium, neon and so forth are monatomic gases – they comprise of monatomic molecules – where the atom is the molecule. So, the term molecule might be more useful than the term atom. Here, I am assuming the {molecule} concept refers to the smallest discrete particle in a sample of a single substances made up of identical subunits (*i.e.*, so this would exclude the concept from encompassing sodium chloride, for example), so helium has molecules of a single atom. I am aware that molecules can be described in other, not necessarily consistent ways. So, one version of the {molecule} concept defines a molecule as a group of atoms joined by covalent bonds, by which definition the helium atom is not a molecule. Indeed, the ENC refers to "molecules (groups of atoms bonded together)".

The potential ambiguity here is not unique to this concept. As suggested earlier in this volume (see Chapter 9), it may be difficult to know sometimes which version of a concept can be considered canonical, or whether there are multiple versions in use, or perhaps there is a canonical concept that is fuzzy and shifts according to context (see, for example, Chapter 6).

10.5.3 Learning Progression between Educational Stages

The English National Curriculum is designed with cognisance to the principle of the spiral curriculum, discussed above. Content is presented under divisions for four 'key stages' (KS1-4) relating to students of different ages: KS1 : 5-7-year-olds; KS2 : 7-11-year-olds; KS3 : 11-14-year-olds; and KS4 : 14-16-year-olds. One of the phrases in this section of the English National Curriculum specifying what should be taught to school children aged 11-14 was "a simple (Dalton) atomic model". There are different atomic models that could be taught in school. It is common for secondary-school chemistry to use a shell model that can be readily linked to the periods of the period table for 'main group' (*i.e.*, s and p block) elements. In elective courses taken after the end of compulsory school, more advanced models are introduced involving orbitals and ideas about electron density and probability. There are other models, of course. For example, before the nuclear atom (of which the shell model can be seen as a variant or development) there was a so-called plum-pudding model that assumed the mass of the atoms was distributed through the atom, rather than nearly all concentrated at the centre. The plum pudding model is a scientific model, but not a current one (see Figure 6.1), and so would not be taught as a canonical chemical idea today.

The Dalton model predates any of these. Dalton's model itself borrowed an idea from the Greek Democritus – where atoms were indeed *atomos* (indivisible) and were thought of as perfectly hard and regular geometric bodies (Taber, 2003a). The atoms of the ancient world did not have nuclei or other structure. That is not to simply suggest that at that point in time such structure *had not been discovered* (as we might talk for example of potassium as it was characterised before it was known to have isotopes – see Chapter 5). Rather, homogenous structure was part of the essence of the {atomos} concept, so the modern {atom} concept is not merely a refinement of the ancient concept. (See Chapter 15 for discussion about when conceptual change in chemistry can be considered as concept refinement rather than a substitution of a distinct concept that may retain the same label, and possibly referent.)

Dalton's atomic theory was a major conceptual breakthrough and was certainly fertile in progressing chemical investigations as part of the iterative dialectic between theory and observation (see Figure 5.1). 'Dalton atoms' did not have clear structure. More pertinent to the point I have raised here, Dalton's atoms were not well distinguished from what we would today call molecules, and indeed what Dalton thought of as atoms might be better understood today as molecules.

If we take the English government's Department for Education's presentation of its chemistry curriculum at face value, then the concept of {atom} set out as target knowledge for English schoolchildren aged 11-14 is a concept that does not distinguish between atoms, molecules, and ions. Once we adopt this {atom} concept then the phrase "chemical reactions as the rearrangement of atoms" could be considered as less problematic as (in terms of more canonical concepts) it refers to "chemical reactions as the rearrangement of [submicroscopic particles such as atoms, molecules, and ions]". That reading would cover reactions such as the precipitation of lead iodide and the formation of sodium chloride by neutralisation, that otherwise have been excluded from being considered reactions in this curricular model.

That might seem to remove the problem I have highlighted, if the target knowledge of the atom that is prescribed for 11–14-year-olds (the Dalton {atom} concept) is also intended as target knowledge for 14–16-year-olds. However, a more sophisticated model is prescribed for 14–16-year-olds: "a simple model of the atom consisting of the nucleus and electrons, relative atomic mass, electronic charge and isotopes". We have also seen that according to this same curriculum document students are asked to learn about mechanisms involving proton transfer, electron transfer, and electron sharing. Other target knowledge statements in the same document require students to learn about "the shapes of molecules" and "when molecules collide".

It appears that there is some inconsistency in this official government curriculum document. The curriculum authority expects students to learn about reactions such as neutralisation, but also that all chemical reactions occur through three mechanisms (none of which apply to the types of precipitation reactions once referred to as double-decomposition reactions; and that only seem to relate to neutralisation reactions if the focus is on the equilibrium between water molecules and ions, not salt formation). Reactions are also conceptualised as being rearrangements of atoms that is consistent with the atomic model to be taught to younger (KS3) pupils, but not with the model used at this (KS4) age level where atoms, ions and molecules are clearly being distinguished.

The reason for focusing on this example in such detail is not to be critical of a particular flaw in a published curriculum (although to get something so basic to chemistry so confused in this way in a document published by a national government does seem extraordinary), but rather to reinforce a theme of the book: that the conceptual content of chemistry is complex, and teaching is far from straightforward. The {chemical reaction} concept is central in chemistry, so confusion and inconsistency about it in an official curriculum document is to be regretted.

The problem was raised shortly after the document was first made public (Taber, 2014a), but has not been addressed. Perhaps this means the English ministry's chemistry experts have a completely different concept of {chemical reaction} than the present author, and stand by their specification – or perhaps it is considered the model is 'good enough' for students of this age, and they leave it to teachers to address any questions raised by students who may notice that some apparently canonical reactions do not fit the model that they are taught. This example does show, however, that decisions about how to best simplify complex chemical concepts to produce curricular models for particular groups of students is not straightforward. As a teacher, interpreting a curricular model to inform a coherent teaching approach may be quite challenging.

References

Aikenhead G. S., (1996), Science Education: Border crossing into the sub-culture of science, *Stud. Sci. Educ.*, **27**(1), 1–52.

Alonzo A. C. and Gotwals A. W. (ed.), (2012), *Learning Progressions in Science: Current Challenges and Future Directions*, Rotterdam: Sense Publishers.

Bruner J. S., (1960), *The Process of Education*, New York: Vintage Books.

Cornbleth C., (1984), Beyond Hidden Curriculum? *J. Curriculum Stud.*, **16**(1), 29–36.

DFE. (2015). National curriculum in England: science programmes of study. Retrieved from https://www.gov.uk/government/publications/national-curriculum-in-england-science-programmes-of-study.

Duschl R. A., Maeng S. and Sezen A., (2011), Learning progressions and teaching sequences: a review and analysis, *Stud. Sci. Educ.*, **47**(2), 123–182.

Gentner D. and Stevens A. L. (ed.), (1983), *Mental Models*, Hillsdale, New Jersey: Lawrence Erlbaum Associates.

Herron J. D., Cantu L., Ward R. and Srinivasan V., (1977), Problems associated with concept analysis, *Sci. Educ.*, **61**(2), 185–199.

Holme T. and Murphy K., (2012), The ACS Exams Institute Undergraduate Chemistry Anchoring Concepts Content Map I: General Chemistry, *J. Chem. Educ.*, **89**(6), 721–723.

Hunt A., (1988), SATIS approaches to STS, *Int. J. Sci. Educ.*, **10**(4), 409–420.

Johnstone A. H., (2000), Teaching of Chemistry – logical or psychological? *Chem. Educ. Res. Pract.*, **1**(1), 9–15.

Key Stage 3 National Strategy, (2002), *Framework for Teaching Science: Years 7, 8 and 9*, London: Department for Education and Skills.

Kuhn D., (1999), A Developmental Model of Critical Thinking, *Educ. Res.*, **28**(2), 16–46.

Leach J. and Scott P., (2002), Designing and evaluating science teaching sequences: an approach drawing upon the concept of learning demand and a social constructivist perspective on learning, *Stud. Sci. Educ.*, **38**, 115–142.

Meyer J. H. F. and Land R., (2005), Threshold concepts and troublesome knowledge (2): Epistemological considerations and a conceptual framework for teaching and learning, *Higher Educ.*, **49**(3), 373–388.

Perry W. G., (1970), *Forms of Intellectual and Ethical Development in the College Years: A Scheme*, New York: Holt, Rinehart & Winston.

Piaget J., (1970/1972), *The Principles of Genetic Epistemology*, Mays W. (Trans.), London: Routledge & Kegan Paul.

Royal Society of Chemistry, (2017), *Accreditation of Degree Programmes*, Cambridge: Royal Society of Chemistry.

Taber K. S., (2000), Finding the optimum level of simplification: the case of teaching about heat and temperature, *Phys. Educ.*, **35**(5), 320–325.

Taber K. S., (2002), *Misconceptions re-conceived: why the effective teacher pays heed to the aufbau principle of learning*. Paper presented at the What does a chemistry teacher need to know? Virtual asynchronous RSC conference. Retrieved from: http://www.leeds.ac.uk/educol/documents/00003627.htm, 29th December 2018.

Taber K. S., (2003a), The atom in the chemistry curriculum: fundamental concept, teaching model or epistemological obstacle? *Found. Chem.*, 5(1), 43–84.

Taber K. S., (2003b), Mediating mental models of metals: acknowledging the priority of the learner's prior learning, *Sci. Educ.*, 87, 732–758.

Taber K. S., (2014a), *Ignoring research and getting the science wrong*. educationinchemistry blog, 6th May 2014. Retrieved from http://www.rsc.org/blogs/eic/2014/05/ignoring-research-and-getting-science-wrong, 29th December 2018.

Taber K. S., (2014b), *Student Thinking and Learning in Science: Perspectives on the Nature and Development of Learners' Ideas*, New York: Routledge.

Taber K. S., (2018), *Masterclass in Science Education: Transforming Teaching and Learning*, London: Bloomsbury.

Taber K. S. and Riga F., (2016), From each according to her capabilities; to each according to her needs: fully including the gifted in school science education, in Markic S. and Abels S. (ed.), *Science Education Towards Inclusion*, New York: Nova Publishers, pp. 195–219.

Talanquer V., (2015), Threshold Concepts in Chemistry: The Critical Role of Implicit Schemas, *J. Chem. Educ.*, 92(1), 3–9.

Young M., (2008), From Constructivism to Realism in the Sociology of the Curriculum, *Rev. Res. Educ.*, 32(1), 1–28.

Zeidler D. L., (2014), Socioscientific issues as a curriculum emphasis: theory, research, and practice, in Lederman N. G. and Abell S. K. (ed.), *Handbook of Research on Science Education* (vol. 2), New York: Routledge, pp. 697–726.

CHAPTER 11

How Are Chemical Concepts Communicated?

This chapter seeks to consider what it means to communicate something, and in particular how to communicate something like a concept. This will involve making a distinction between (what will be termed here) as *information* and *understanding*. It will be assumed that in teaching and learning, the kind of communication especially valued involves more than communicating information and is more centrally about developing understandings. The background to the discussion here concerns claims made earlier in the book about the nature of concepts (see Chapter 2): that is, that a concept is a mental entity (or perhaps better, a mental attribute).

The programme for this chapter moves through three phases. To start with, (i), the notion of concept (the '{concept} concept' in effect) being used in this book is revisited through the claim that concepts must be considered as subjective rather than objective. This is certainly not intended to fundamentally question the nature of science as an objective activity, as long as this is understood in terms of scientists aspiring to, and adopting procedures intended to, maximise objectivity; rather than assuming scientists *can assume*, or *assure*, objectivity. Then, (ii), there is a consideration of the importance of understanding, both in science, and in the teaching and learning of science – and this will be related to the nature of concepts, and the teaching and learning of canonical concepts. Then, (iii), the process of communicating concepts is examined from the perspective developed in relation to *concepts and understanding*.

11.1 Objectivity and Subjectivity

Earlier in the book, concepts were characterised as mental entities (see Chapter 2), and it was also suggested that canonical concepts were something of a useful fiction (see Chapter 9). Here, I will build on this idea by considering the nature of concepts in relation to the descriptors 'objective' and 'subjective', in order to develop a perspective on the way concepts may be communicated.

I am taking objectivity to refer to the situation where something is considered in a way that is accessible to (and checkable by) others. As I write this, I have music playing from two loudspeakers on window sills either side of my position. The window sills are at approximately 45° angles to the open French doors directly in front of the table on which I have placed my computer keyboard. The speakers are each something like 0.75 m from my head. This is an objective description as it is in principle open to being confirmed (or disconfirmed) by other observers. There is no one else in the house. In principle, I am observable to a neighbour on a very tall ladder, or a very low flying hang-glider pilot, or someone parachuting into my back garden, or someone sending a camera on a drone over my hedge, and so on – but no one seems to be taking these potential opportunities right now. I cannot be certain (even if I can consider it extremely unlikely) that I am not being observed by an alien scientist in orbit with access to technology that allows high-definition images of ground-level activity in good resolution and fidelity; nor that the intelligence services, judging me as a threat to the social order (after all, I criticised the English National Curriculum in the previous chapter), have not planted a tiny spy-camera in some nook or cranny in the room. But there does not need to be an independent observer present to make something I report observable *in principle* to others.

The description is verifiable as I could invite another observer in to check the situation or I could make a record by taking some photographs or video recordings (and audio recording to demonstrate music is playing). I could increase the detail of the account by using a ruler and a protractor to measure quantities I have estimated, and my measurements could in principle be checked by someone else using these, or their own, instruments – instruments that would be made with reference to standards such that they would enable the same readings (within experimental error) as would be obtained by other protractors or rules.

11.1.1 Objectivity in Science

Science seeks objectivity. Science examines features of the natural world that are, in principle, available to observation to others. The 'in principle' qualification is important. Some observations made with the first electron microscope, or the first mass spectrometer, or with the Large Hadron Collider, or with the Hubble space telescope, were (and in the latter cases, are) not available to anyone who did not have access to those instruments. Yet, these are still objective observations to the extent that there is a technical account of how they may be made so that it is *in principle* possible for someone else to independently repeat the work and see if they obtain comparable observations. Perhaps, for example, someone with sufficient funds could (that is, in principle) build and launch their own replica of the Hubble Space Telescope to repeat some of its observations – although perhaps not an absolute replica, but a copy with the primary mirror ground to the intended shape within the expected tolerance (NASA, 1990), so it might perform to the original specification without requiring repairs *in situ*. If a naturalist returns home from an expedition to a remote and inhospitable jungle and reports a new species of bird, few would have the skills or resources to go and check the observation. Yet this is an objective observation as long as the report offers sufficient detail of how to locate the site of the original observation and any other details needed to, in principle, repeat the observation.

In practice, laboratory and fieldwork undertaken in science often relies on a great deal of tacit knowledge (Polanyi, 1962). So, it has been found that the reproducibility of many scientific experiments is much more problematic that a naive account of science would suggest (Collins, 1992): the accounts of technical developments often leave out key information because it involves subtle aspects of procedure that the reporting scientists not only do not appreciate are important, but actually do not even realise they are implementing. Just as we learn to climb stairs without being conscious of the actual sequence of muscular actions we are taking, scientists can acquire tacit knowledge of how to adjust scientific instruments (or how to prepare specimens or reagents, *etc.*) that they are not consciously aware of.

Due to the potential of the human organism for implicit learning, feedback allows the refinement of technique without this being a conscious process. In a similar way, the ability to spot fossil fragments on a beach, or distinguish specimens of superficially similar species, can only be partially specified – as some of the ability involves learning that is outside of deliberate control and examination. There are then limits to just *how* objective science can actually be in practice (at least as practiced by human organisms): but it is, *in principle*, an objective activity, as it concerns phenomena that are objectively available for observation, and that are observed and reported in systematic ways to allow observation to be as reproducible as possible.

Although my account of my location whilst writing was objective in the sense of describing features that could be checked, in principle, it fell short of being the kind of phenomenon science focuses on, in that it was one-off historical event. (That is, it is historical as a unique event in time, rather than in the sense of actually being of any interest to historians.) If I *was* being observed by an alien anthropologist in orbit who was exploring a research question along the lines 'How do earth scholars set up their immediate environment when writing manuscripts' then such an observation might be considered scientific, assuming it is treated as one datum among many in a survey adopting a suitable sampling frame – but, of itself, it is merely an individual biographical anecdote.

11.1.2 Subjective Reports

As I write this paragraph, I am listening to a very beautiful piece of music. It is a short instrumental piece for oboe and string ensemble – short, as in

4 min 44 s according to my computer. As I am playing a copy of a commercially released recording, I can easily offer objective details that would allow someone else to identify the track (it is entitled 'Prelude: Song of the Gulls (Rehearsal Take)'). I can also offer a biographical anecdote (again) that in principle could be objectively confirmed – that is, that a version of this piece was used as background music on my wedding video.

What is less objective, is my aesthetic judgement that this is 'a very beautiful piece of music' (even if it might be inferred likely that at least one other person agreed with me, given its selection for use for the wedding video soundtrack). What I see when I describe the position of the loudspeakers is detail available (in principle) to other observers. What I hear when I listen to music could be described in terms of tempo or key – these are matters agreed by convention and so somewhat objective and that may even be formally represented in a score – but my feelings as I listen to the music, and my descriptions of its aesthetic qualities are subjective.

Someone could confirm or challenge my report of the duration of the recording, and subject to agreement on standards and instrumentation for measurement (as is always needed in scientific work) there should, within measurement error (the scope of which would itself would be part of the agreed standard), be an objective, right, result. Someone else, an independent observer, might agree with my aesthetic evaluation of the piece, or they might feel the piece is trite and derivative, or schmaltzy, or whatever, but this does not (and cannot) confirm or challenge my judgement – as this is just another subjective evaluation. Of course, I could be influenced by someone else's opinion, and learn to change my taste, and even to hear the music differently in future. Be that as it may, right now my judgement is my subjective entitlement, and nothing anyone else can say, or do, changes that – or makes it right or wrong.

Not only is my evaluation that the piece is 'beautiful' subjective, but my communication of this evaluation is in subjective language. If other observers were unsure what I meant by reporting that the duration of the track was $4'44''$, as they themselves (perhaps those hypothetical aliens again) measured time in Andromedan fortnights, then it would be possible for someone to provide them with a reference clock that would allow them to find out how long $4'44''$ actually was in their own units. If they were unsure what I meant by 'beautiful' then there is no independent reference – they could be shown a sunset, or a flower, Michelangelo's David, Millais' painting of The Martyr of Solway, or a clip of Ingrid Bergman in Gaslight, or... (as this is a subjective matter, the reader is invited to suggest their own options here), but there is of course no assurance that someone else would have subjective experiences that relate to those that might lead to me using the descriptor 'beautiful'. This is reflected in an exchange in Douglas Adam's "The Hitchhiker's guide to the Galaxy" stories between displaced earth-man Arthur Dent and the ('paranoid') android Marvin. Arthur reflects, "I lived on a beautiful planet once". Marvin asks, "Did it have oceans?"; to which Arthur replies: "Oh yes; great big rolling oceans". Marvin responds: "I hate oceans".

11.2 Are Concepts Subjective or Objective?

Having established what is meant by the terms objective and subjective as they will be used here, I turn to the question of whether concepts should be considered subjective or objective. Gilbert and Watts pointed out long ago that there was ambiguity in how the term concept is commonly used, as 'concept' "is applied with equal facility to both an individual's psychological, personal, knowledge structure and to the organisation of public knowledge systems" (Gilbert and Watts, 1983, pp. 64–65). Yet, these two meanings of concept refer to very different things: concepts as personal, individual and psychological, and concepts as public, formal, and institutionalised. It seems that it would be useful if different terms were used for these very different referents. Gilbert and Watts suggested the term conception be employed for the former meaning: "that 'conception' be used to focus on the personalised theorising and hypothesising of individuals" (p. 69). In earlier chapters I have used conception as some aspect of a person's conceptualisations (fitting with Gilbert and Watts), *but* have considered such conceptions reflect facets of people's concepts. Instead, we might preface the term concept to suggest there is a distinction here between *personal* concepts and *canonical* concepts.

11.2.1 Personal Concepts and Canonical Concepts

Personal concepts are those mental entities discussed in Chapter 2: aspects of personal knowledge represented in brains and applied in cognition – such as when identifying acids, calculating lattice enthalpies, explaining the relative stability of benzene, predicting expected peaks in the visible–UV spectrum of an unfamiliar organic compound, and so forth. The concepts are only directly available to the individual using them, and indeed to the extent that conceptual knowledge can be tacit (see above), may sometimes be drawn upon without becoming explicit. Someone else may draw inferences about a person's concepts from their behaviour, especially their talk and writing (Taber, 2013), but if the person decides not to express their ideas, then the nature of their concepts remains private.

Despite this, personal concepts seem ontologically to be on a sounder ground than do canonical concepts. Each of us has our own experience of thinking with concepts, and when we talk to others there seems strong evidence that they also apply concepts, and often we can detect whether aspects of their concepts are consistent with our own or not (which of course is what teachers are asking during assessment). By comparison, we might naively expect canonical concepts should be more visible: they are supposedly shared across a community, and subject to consensual agreement, and are linked to a whole discipline. They are supposedly public, as aspects of what is termed public knowledge (Ziman, 1968), unlike the private nature of personal concepts.

Yet, I have already suggested this may not really be the case. I implied earlier that canonical concepts are useful fictions. Canonical concepts are important referents – teaching chemistry relies on them – but are actually hard to pin down. In Chapter 9, the possible repositories for these public concepts were discussed in some detail (and so I will not repeat all that here). Neither the chemical research literature nor the community of chemists (or even a combination of the two) can provide unambiguous access to canonical concepts. If concepts can only exist in minds, and if concepts cannot be directly copied from mind to mind (but rather have to be represented and then reconstructed by interpretation), and if every mind is unique, then there is nowhere where we can find the canonical chemical concepts {acid}, {orbital}, {oxidation}, {square planar complex ion} or whatever.

The other possibility is that canonical concepts exist somewhere outside of time and space, in some non-material world of ideas, that human intuition allows us to access as we can recognise the true essence of things. Somehow, we do know what is meant by a spherical shape, without ever seeing a perfect sphere, and we learn to recognise examples of trees even though no two trees are identical, and indeed trees can vary considerably. Somehow, we seem to 'get' the essence of 'sphere' and 'tree'. We should be wary though here though. For thousands of years it seemed pretty obvious that different types of living thing were fundamentally distinct, and now science suggests that species have evolved incrementally and iteratively from a common ancestor. Along the way, as discussed in Chapter 7, species blurred into each other.

If we can recognise a tiger and distinguish it from a lion then this shows we can form {tiger} and {lion} concepts that 'work' given the world we find ourselves in: but not that there are 'tiger' and 'lion' essences that are eternal and that we can intuitively recognise in some way. We can usually effectively discriminate specimens of lion, and of tiger, from each other, and from other entities that are neither tiger nor lion, given the range of entities that exist in the world today. That said, few of us have the expertise to examine an early stage foetus and confidently determine if it is lion or tiger (or something else). Were we to be taken back in time many thousands of years the concepts we have developed living in this present time may not clearly help us decide whether a creature alive then was actually a specimen of tiger or was rather a specimen of an ancestral species much like a tiger.

If someone's {tiger} concept is intuiting an essence, it is not accessing some eternal essence, but rather a historically contingent one. Darwin, surely one of the greatest naturalists of all time, undertook a study of barnacles. Not expecting the extreme sexual dimorphism demonstrated by one species, he apparently initially discarded a good many male specimens that he thought were merely parasites on the females. The dwarf males live on the body of the female, but are not technically considered parasites as they provide sperm – indeed in some species the tiny males are reduced to be little more than testes (Vollrath, 1998). (There is surely potential for a radical feminist allegory to be constructed from this natural phenomenon.) If there was a barnacle essence, then Darwin's considerable biological intuition did not access it to recognise the tiny male barnacles. The conclusion drawn here is that unless one does believe in the reality of World 3 (see Chapter 9), as a non-material depository of ideas, that is somehow accessible to rational thought, then public knowledge, and more specifically, canonical concepts, cannot be definitively found anywhere, because *they do not actually exist*.

What do actually exist, are personal concepts in individual minds, that cannot be transferred directly between minds, but that have to be communicated indirectly (by representation of one person's ideas in some kind of language using material media, and then interpretation of these traces of a person's thinking through the language and other interpretive resources available to another person) with the potential for miscommunication (see Chapter 9). I have stressed this conclusion because of its pertinence to educational processes such as the construction of curriculum, teacher development, planning teaching sequences and presentations, and classroom teaching.

11.2.2 Canonical Concepts Are as Useful as Ideal Gases

Despite rejecting the existence of canonical concepts, I am not suggesting *the notion* of canonical concepts (the concept {canonical concepts}) is worthless. It is possible to draw an analogy here with the notion of an ideal gas. The {ideal gas} concept is very useful, and the {ideal gas law} concept (see Chapter 8) is often applied fruitfully. That there are no ideal gases neither prevents us having a concept {ideal gas} nor making good use of it, just as people may have, and find uses for, concepts such as {unicorn} or {Maxwell's demon}.

The concept {ideal gas} should not lead us astray because we know it is an ideal (there is a strong clue in the name). Real gases are not ideal gases. The ideal gas law never strictly applies to real gases, but real gases may approximate ideal gas behaviour, and as long as we acknowledge this, and consider when real gases will (and will not) closely approximate ideal behaviour, then the idea of an ideal gas can be used fruitfully. We need to consider the concept {canonical concept} in a similar way. We want to teach canonical concepts, or suitable simplifications of them. If there are no canonical concepts we must substitute alternatives that act as, suitable, good-enough, approximations.

11.2.3 Chemical Education Without a Focus on Concepts

One response to the analysis offered in this volume might be that I am following a fundamental error in terms of how chemical education is often understood. I start the book by arguing that chemistry is a highly conceptual discipline, and so that chemistry teaching and learning is heavily (if not entirely) about teaching and learning concepts, but then I offer an analysis of *the concepts of chemistry* and conclude they do not exist.

Certainly, concepts exist, as mental entities, or affordances, which are enacted in minds. That is, in the minds of people, including in the minds of chemists, chemistry teachers, and chemistry students – but these are all personal concepts. The canonical concepts that are supposed to provide the 'content' of the subject, and that are the reference points for what we should teach, and for evaluating conceptual learning in students, are nowhere to be found – unless we consider they live in a kind of 'ideas heaven' outside of space–time, that is neither material nor mental, but that we somehow are able to access through some mysterious process (see Chapter 9).

11.2.4 Induction into a Community of Practice

A logical response might be that given that chemistry teaching and learning does occur, and can often (if certainly not always) seem to be effective, then the premise that chemistry education is largely about learning the (seemingly non-existent) canonical conceptual content of the subject must be mistaken. An alternative perspective might support an argument that chemistry should be understood as a set of practices, and that chemistry education is induction into those practices (Olitsky, 2007). Practices are observable, unlike concepts, and so can be a more objective basis for thinking about education. Practices are activities, but this can certainly include practices that we normally associate with higher-level cognition. So, a practice that might be valued in chemistry could be something like, say, assigning molecular geometries to molecules. There can be canonical assignments, and conventional agreements of which examples should be included in education at different levels – a more limited set in secondary/high school, a more expansive set in undergraduate courses. Students can be judged – objectively – on their ability to assign molecular geometries – and if they demonstrate that they have acquired this practice effectively, then that is what matters.

It might be asked whether this is any more than a change in language that might make no substantive difference to our analysis. The practices still need to be represented in curriculum and teaching, but if they are defined in terms of specific behavioural objectives then it becomes possible to be much more explicit about whether they have been achieved. So, if the aim is that students can assign molecular geometries, then the practice can be operationally defined, and more objectively evaluated. This contrasts with an aim that students learn scientific concepts (which only exist in an ideal sense) where it is only possible to test this indirectly in terms of behaviours that can be considered to *reflect the activation of a personal concept sufficiently aligned with the target concept* to lead to the desired behaviour.

It would be possible to proceed according to such an approach. One could list the practices of chemistry (determining melting temperatures; calculating lattice enthalpies; interpreting mass spectra, whatever), and decide which should be taught (at what level of skill) at different levels in school and university curricula. Students would be judged only according to how well they demonstrate the prescribed practices. Teachers would be evaluated according to how well their students show progression in developing competence in the assigned practices.

In such a regime, students may be tested by being asked such questions as:

Is Na_2O an acidic, amphoteric, basic, or neutral oxide?
What shape would a molecule of ammonia (NH_3) be?
What kind of bonding is found in each of NaCl, HCl, Cl_2?
Which of the following formulae can represent compounds that are unsaturated? CH_4, C_2H_4, C_2H_6, C_6H_6?

and so forth, but they would not be asked to 'explain your thinking' or 'give reasons for your answer'. If a student could give 'correct' responses to these and all the other questions based on the practices they were being taught, then that student would be judged successful in having mastered the prescribed practices. In a sense, in practice [sic], that might be fine. If teachers managed to successfully induct students into the practices that were identified and specified as being the curriculum for chemistry courses, then that could be considered successful chemistry education.

11.2.5 Learning as a Black Box

Such a regime could treat learning as a black box. Students learn to assign molecular geometries (and so forth), and if they do so successfully, it does not matter how. Just as infants learn to walk, and people learn to ride bicycles, and are judged by those outcomes, without being expected to explain how they achieve this, so it could be with chemistry. The Olympic gold medallist, or the winner of a literature prize, is judged on their achievement regardless of their ability to offer a high-quality account of how they managed that achievement. The politician who gets the largest number of votes cast for her is elected, regardless of how well she can explain her success. The company salesman-of-the-month has to sell more product than rivals, not provide an account of how that came about (and indeed may well suspect that repeating the feat in future months may be more likely if they do not offer a detailed account to their peers). And so forth. We could think of chemistry education as a similar kind of enterprise, where what counts is effective practice.

Most of us are perfectly happy with such a basis for evaluation in many aspects of our lives. As long as behavioural outcomes are as desired, we are not usually concerned with how the black box works: the postal service has a procedure for sorting mail so that it nearly always it gets delivered to the intended address; entering numbers into a telephone key pad nearly always puts a call through to the number entered; supermarkets generally have procedures to ensure that there is usually stock on the shelves; sports federations have a procedure for organising fixtures so that teams are not required to play in two different places on the same day, and that by the end of the season each team has played a certain schedule of matches (*e.g.*, against each other team in the league, home and away). The lift (elevator) stops on the floors relating to the numbered buttons pressed. The chemical stores manager has a system for making sure that common reagents and consumables are always available for the labs. The search engine returns potential hits relating to our search terms. The coffee machine in the staffroom provides a range of options of beverage at the press of a button. As curious people, we may sometimes give thought to some of the procedures, mechanisms, and algorithms, involved in such achievements, but as busy people we often tend to be satisfied with the desired outcome without feeling the need to delay satisfaction till we are also provided with an account of how it was achieved.

There are two obvious reservations that may occur to readers. For one thing I am equating chemistry, an academic discipline, with more practical matters like delivering the post and scheduling sports fixtures. An approach based on practical outcomes alone seems antithetical to the nature of chemistry as a science – science seeks to understand the world, and is not just about being able to predict things correctly (which is in effect what the kind of successful practice described here could allow) – even if one criterion for evaluating understanding might be the ability to predict.

Science is based on a dialectic between theory and empirical observation (*i.e.*, observations motivate theoretical conjectures, which motivate testing through further observations, *etcetera* – see Figure 5.1), whereas the application of science just to achieve practical ends is something else – technology. However, if the graduates of such an education are actually successful in the practices of chemistry (including those where tacit knowledge has been recognised as important) then perhaps they might be considered to demonstrate a functional kind of understanding – as they will understand chemistry well enough to successfully practise it.

The second reservation is that even if we did only care about a student correctly assigning molecular geometry (*etc.*) rather than being able to explain how they were successful doing so, we might suspect that usually students are only going to be successful on non-trivial practices (*i.e.*, where they cannot simply rote learn a limited set of examples) if they do indeed have an understanding of what they are doing. We want students to understand principles they can apply to novel examples, and not just learn a set of standard examples they might be tested on. (That 'we' certainly includes this author, and I would imagine many readers of this volume, even if, as suggested in Chapter 10, the tendency to over-specify curriculum found in some countries increasingly encourages teaching- and learning-to-the-test.) So, understanding (relating to internal, mental states), and not just externally observable practices, is important to chemistry education.

11.3 Information and Understanding

The position adopted here, then, is that chemistry education is not just about inducting learners into practices that are to be judged directly in terms of behavioural outcomes, but also centrally about helping students develop a conceptual understanding of the subject – an understanding of chemical ideas. Yet, here we are faced with the problem already discussed earlier that whatever we may want to assess, we can only do so by observing behaviours. We cannot see students' concepts, but can only draw inferences about them by observing what they say, write, draw, do, *etcetera*. The same problem arises when discussing 'understanding'. We cannot experience someone else's understanding – we can only evaluate it indirectly in terms of how they can represent their understanding in forms we can access and interpret. A useful distinction to make might be between the notions of information and understanding.

11.3.1 Information in Chemistry

I am taking information as something objective. This does not mean information is necessarily true, there can be misinformation. However, information is potentially open to corroboration or refutation. A spectrum produced by a spectrometer would count as information in this sense. It could be described objectively as it is of material form (there is a peak on a chart or screen with a maximum at such-and-such wavenumber, with an area of whatever). The spectrum offers objective information, although its veracity of course depends upon the careful design, engineering and calibration of the machine that produced it. If there was a flaw in the application of theory in designing the spectrometer, or in the construction of the device, or in its calibration, then the information produced may have limited value, but it is still objective in the sense that the output is in a material form, and so potentially open to different observers.

Books and computers store information, in the sense that whatever information is recorded in them can be read out. Consider the statement that "the alpha crystal phase of tantalum has a body-centred cubic structure". This information could appear in books (such as this one), and be represented on a computer (such as the laptop I am using to write this passage) in a form that can be recovered in true fidelity. In that computer file the information will be in the form of a string of binary digits, and in this book the information will be in the form of a graphic pattern that can be unambiguously read as letters of the Latin alphabet.

The systems we have for storing and reproducing information are such that they can be highly accurate – because what is being stored, information, is unambiguous and so can be unambiguously coded. It is important to note here that the information concerns the statement "the alpha crystal phase of tantalum has a body-centred cubic structure", and not, say, the distinct statement "tantalum has a phase, denoted alpha, with crystal structure body-centred cubic". This other statement would be represented in my computer with a different binary string, and on the page with a different graphical pattern. So, in that sense, the 'information' is different in these two cases.

11.3.2 Understanding and Meaning

Now, it might be countered that actually the informational content of these two statements ("the alpha crystal phase of tantalum has a body-centred cubic structure", and "tantalum has a phase, denoted alpha, with crystal structure body-centred cubic") is entirely the same – that there is no difference in what is being communicated in the two cases. Yet, I am using 'information' in the sense of what is objective, raw data – and in terms of raw data these two patterns of letters (and so the binary strings used to represent them on my computer hard-drive) are clearly different, as they do not match. Yes, there are many of the same symbols, just in a different order, but if that were enough to make them the same information, then by that criterion the two strings: '£97 531' and '£13 579' (with exactly the same set of symbols, just re-sequenced) contain the same information. If any reader thinks that is indeed the case, then please get in touch – I would like you to lend me £97 531, and I promise to immediately pay back the full £13 579.

We can also show this by considering another sentence: "the structure of body-centred cubic alpha phase crystal has a tantalum". This is a different string of the same symbols, but clearly does not contain the same information (as can be seen even more clearly in the more extreme example: "–   aaaaaaaabbcccccddeeeeefhhhhhilllmnnnoopprssssrrrttttttuuuuyy"). So, if we are seduced by the idea that the two strings "the alpha crystal phase of tantalum has a body-centred cubic structure" and "tantalum has a phase, denoted alpha, with crystal structure body-centred cubic" should be judged to have the same information then we are making such an evaluation based on something other than the information itself. Indeed, we are not comparing the actual information at all, but rather the meaning we can make of it.

I am then making a distinction here between information, as something that can objectively exist in the world, and any meaning we may associate with that information. Our hypothetical spectrum is an objective record of information – but inferring meaning from it depends upon interpreting it in a particular theoretical context in order to construe it as meaningful.

We can see that the information '£97 531' has a different meaning to the information '£13 579' (or that "chemistry is a more useful discipline than necromancy" has a different meaning than the alternative sequence "necromancy is a more useful discipline than chemistry"), so it is intuitively clear that the information content must be different. However, what about:

1. "the alpha crystal phase of tantalum has a body-centred cubic structure";
2. "tantalum has a phase, denoted alpha, with crystal structure body-centred cubic";
3. "the structure of body-centred cubic alpha phase crystal has a tantalum".

I suspect most readers would readily agree that the information in string 1 is not equivalent to the information in string 3, probably arguing that 1 has a clear meaning and 3 does not, and that whatever meaning 3 might have, it certainly does not have the same meaning as 1. String 2, however, seems to mean the same as string 1, so is it in effect the same 'information'? Not in the sense I use information here, as the information has to be totally objective, in the sense that it could be agreed upon by any observer *without assuming common frameworks for interpretation*. A reader may suggest that, in this case, any competent reader of English can tell that 1 is equivalent to 2 – but my argument is that English cannot be assumed as a basis for deciding, as it is not a universal framework available to any objective observer. So, if we consider:

1. "the alpha crystal phase of tantalum has a body-centred cubic structure";
2. "tantalum has a phase, denoted alpha, with crystal structure body-centred cubic";
4. "the alpha crystal phase of tantalum has a body-centred cubic structure";

then in principle a non-English speaker, a simple computerised system for comparing strings (of the kind that generates strings of symbols on webpages that users are asked to copy to ensure they are not robots), an illiterate medieval peasant, or a passing alien anthropologist, can all determine that 1 and 4 have the same information, but 2 is different.

So, strings 1 and 4 have objectively the same information content, and string 2 is different, although someone with the appropriate resources to interpret the strings (*i.e.*, having learnt the necessary symbolic language) might judge that the information in 2 has *the same meaning* as that in 1 and 4. This, however, relies on something other than the objective information presented, requiring, in addition, an understanding of what the information means within a particular interpretive context.

This might be more obvious if we take another example:

5. "the magnitude of the first ionisation enthalpy of sodium is greater than the magnitude of the first electron affinity of chlorine",
6. "more energy is required to remove one electron from an atom of sodium, than is released when a chlorine atom attracts an electron to become a Cl^- ion".

In terms of strings of symbols, it is very clear that 5 and 6 are quite different, so in the sense of 'information' being adopted here, these two strings have quite different information content. One only has to look at the first symbol in each string to see they are not identical in terms of information. However, we might argue that they in effect have the same meaning – so are largely equivalent.

My argument is that the objective information content is quite different (they are very different strings of symbols, and would clearly be quite distinct to our non-English speaker/simple computerised system/illiterate peasant/ passing alien anthropologist) and that determination of their equivalent meaning depends upon subjective understanding. I will just highlight two points here.

Technically, strings 5 and 6 are (once interpreted, decoded) about rather different things. String 5 (in principle) concerns properties of macroscopic samples of materials as ionisation enthalpy and electron affinity refer to energy changes measured when one mole of sodium atoms or chlorine atoms undergo changes. Sting 6, however, is about single atomic entities – a sodium atom, an electron, a chlorine atom, another (or perhaps, the same) electron. Now, to the chemist, this is an important distinction, but one that can immediately be seen to be unproblematic here (that is, in this context) because of the way we can scale back and forth between moles and molecules (see Chapter 8). Even if the strict meanings of the two sentences are quite different, we understand that what happens at one level can be taken *to imply* what happens at the other level: an understanding that relies upon an abstract theoretical, conceptual framework.

A slightly more trivial issue concerns terminology. However, this may not be so trivial in particular cases of interpreting the information content. Consider a secondary-level student who has successfully completed an introductory course in chemistry, but not a more advanced course. In my own teaching experience (in the English system) this might be a 16-year-old who has studied chemistry to the general school leaving certificate level (currently, in England, the General Certificate of Secondary Education). The student would likely understand the string "more energy is required to remove one electron from an atom of sodium, than is released when a chlorine atom attracts an electron to become a Cl^- ion". Such a student's depth of understanding may be limited, so their understanding is likely not very nuanced, but this string would probably be meaningful to the learner. The same learner may attempt to make sense of "the magnitude of the first ionisation enthalpy of sodium is greater than the magnitude of the first electron affinity of chlorine", but for most learners at this stage in their chemistry education it is unlikely they could understand the information to the same extent as string 6. As chemists, we might prefer string 5 as it is more formal, in more technical language, and phrased in a way that fits without our valuing the ability to shift between macroscopic and microscopic descriptions; whereas string 6 is more informally phrased.

So, if we consider presenting these two strings, 5 and 6, to a range of individuals, we might expect the interpretation they place on the information provided will be quite different. The learner just starting out their first chemistry course would likely make little sense of either string; the 16-year-old just discussed might understand one string reasonably well, but not the other; a somewhat more advanced secondary/high school student might understand both sentences without fully appreciating how they may be considered chemically equivalent, where a graduate chemist used to making effortless mental shifts between the submicro- and macro-levels, and appreciating how statements at one level have implications at the other, might in effect decide they have much the same meaning – they can be understood in the same terms. If information content were to be considered in terms of how that information is understood, it would be a subjective issue, and the information content of the strings would be very different for different people. Here, however, I am taking information as objective content, and so strings 5 and 6 have different information content, even if, for some people, with particular interpretive mental resources, they can be understood in much the same way.

One might consider one last example string here:

7. "♦≈♏ ☺●☐≈☺ ♏☐✇·♦☺● ☐≈☺·♏ ☐✗ ♦☺■♦☺●♦○ ≈☺· ☺ ♌☐♎✇♍♏■♦☐♏♎ ♏♦♌✳♏ ··♦☐♦♏♦♦☐♏."

Objectively, this clearly is not the same as

1. "the alpha crystal phase of tantalum has a body-centred cubic structure".

However, string 7 was produced by simply replacing symbols in the original string 1, with alternative symbols, according to a simple algorithmic substitution rule. Anyone who has the cipher, who knows the code, could translate 7 into 1, and might feel that they can then claim the two strings include the same information – but this claim relies upon the possession of something external to the strings – some kind of framework for interpreting string 7 to make it meaningful. Objectively, strings 1 and 7 contain different information. In the same way "CuSO$_4$" and "copper sulfate" are quite different in terms of *information*, even if those people who also have possession of the relevant 'code' might interpret the information so as to understand both strings as having the same *meaning*.

11.3.3 An Analogy with Data (and Evidence)

The particular distinction I am using here, then, has a strong analogy with another distinction very familiar to chemists and other scientists (see Figure 11.1). This is the distinction between data and evidence. Data is certainly used as evidence in science, but data is not automatically evidence, and indeed is not of itself evidence. Data (*cf.* information) is objective, and has no meaning. When an argument is made, to make a scientific claim, data is interpreted within a particular conceptual/ analytical framework. It becomes presented as evidence (*cf.* understanding) only within that larger structure of argument. Placing the data within such a framework is part of the meaning-making of the scientist(s)

How Are Chemical Concepts Communicated?

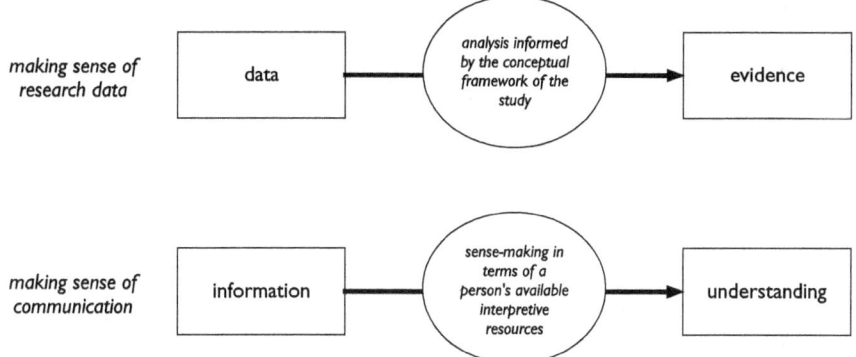

Figure 11.1 Just as research data has to be understood within a particular framework for it to be considered evidence, information has to be interpreted in a process of meaning-making.

making the report, and is intended to have the same meaning for other scientists reading their work. Sometimes, of course, other scientists do not interpret the data as intended, and do not share the intended meaning, as they do not consider the data to be evidence for the claim being made. The data is objective, and open to various interpretations. Evidence is not just data, but data interpreted in a particular frame. Similarly, information is only understood when it is meaningful to someone, and information is not meaning, but is only assigned meaning when it is understood within a particular conceptualisation.

It seems then, ironically perhaps, that if we wish chemistry education to be an authentic reflection of *the practices* of the community of chemists, we need to consider teaching as intended to bring about the ability to understand chemical ideas, and not simply the ability to apply them as if 'black boxes'. Practices of chemistry involve making meaning of chemical information to build arguments and explanations – and that requires understanding concepts.

11.4 Understanding as Subjective

If information is something objective, something open to observation by any observer, then understanding is something subjective, only available to an individual. It is part of the subjective experience of someone. Indeed, understanding may be associated with feelings as well as with cognition. When we have moments of inspiration and realise something, there is a feeling (Brock, 2015). When we are struggling to make sense of something, but feel (sic) we do not understand it, then we may be disconcerted, frustrated, and so forth. We may feel we understand something, only on closer analysis to find this was too optimistic. One person may feel they understand something on a basis that another person feels is inadequate or mistaken.

Many school students (and, indeed, more advanced students) are content that they understand why chemical reactions occur in terms of atoms seeking and attaining octets or full electron shells (Taber, 1998) – though this is not sensible from a chemical perspective.

11.4.1 Understanding and Explanation

Although understanding may be accompanied by, and even evaluated, in terms of feelings, it tends to be justified (to ourselves, and others) in terms of explanations. If we understand something, then we can offer an explanation that we judge as a sound basis for our understanding. So, we might explain the anomalously high boiling temperature of water in terms of the strength/extent of hydrogen bonding in water compared with other hydrides or the alcohol series (*i.e.* taking water as the $n=0$ member of the homologous series $C_nH_{2n+2}O/C_nH_{2n+1}OH$ when compared with actual alcohols). For some students, the explanation might be simply water has a high degree of hydrogen bonding, and this leads to an anomalously high boiling temperature – and they feel (sic) this is a satisfactory and sufficient explanation. Another student may not be satisfied with this explanation, and needs to extend it to consider how boiling temperature reflects attractive forces between molecules, which to a first approximation depend upon the van der Waals forces that tend to be greater with larger molecules (with more potential for distorting the electron distribution), so that boiling temperature depends upon molecular mass. From here, it may be argued that in the case of water the hydrogen bonding provides an additional force supplementary to the van der Waals' forces, so the simple relationship with molecular mass does not apply.

Alternatively, it could instead be argued that as hydrogen bonding means there is association between supposedly discrete molecules in liquid-phase water, water should not be considered to have molecular mass 18 when considering its boiling temperature, but rather the effective molecular mass is higher, and so the boiling temperature is considerably higher than expected for a substance with molecular mass 18. (These two lines of explanation can be considered (understood to be) two alternative ways of understanding the same phenomenon that are complementary rather than being in opposition.)

Would this be a full understanding? Well, no, as water is a polar molecule, and so its dipole moment is relevant when compared with a non-polar or less polar molecule: a molecule that is polar but that could not form hydrogen bonds would also show some increase in boiling temperature over what might be expected purely on the basis of (nominal, unassociated) molecular mass. Clearly a range of concepts are drawn upon here: {hydrogen bonding}, {van der Waals' forces}, {boiling temperature}, {molecular mass}, {electron distribution}, {polarity}... and each of these can be understood to varying extents (and, indeed, in various non-canonical ways).

11.4.2 Meaningful and Rote Learning

One commonly used idea, proposed by Ausubel (2000), is the distinction between rote and meaningful learning. This is often presented as a dichotomy, but is better understood as being a matter of degree. We can imagine three particular extreme situations, when a student experiences an attempt to teach them something.

(a) the student has no way of comprehending the teaching – and nothing is retained;
(b) the student recognises a familiar symbolic language in which the teaching occurs, but cannot make any further sense of it;
(c) the student is able to make sense of the teaching in terms of their existing conceptual resources.

These may be taken as reference situations, as surely most real situations of teaching tend to fall somewhere between these.

Situation (a) is not usually discussed in any detail as it should not occur in teaching. Examples might be:

- A teacher talking in a language the student is not familiar with.
- A teacher talking with such strong dialect or accent that the student cannot make out the words intended.
- A teacher presenting a pie chart, mass spectrum, reaction profile, *etc.*, without explanation (assuming it is familiar) to a student who has never met that form of representation.

We might all recognise some approximations to these situations: many readers will have attended lectures by visiting overseas academics and perhaps had no idea what the person was actually saying, even if they were nominally speaking in our own language. The graphical examples may be somewhat less convincing. A pie-chart may seem self-evident unless the students has virtually no exposure to any graphical kinds of representations. Anyone who understands a line graph could make something (not necessary what is intended, note) of a reaction profile as it uses some of the same graphical conventions (see Chapter 12). In principle, however, we can recognise such a situation as potentially possible, even if rare in extreme form.

Rote learning is learning without understanding. The learner has access to the symbol system in which teaching is presented but does not assign any meaning to the symbols. Consider a ten-year-old school pupil asked to learn "Na: $1s^2\ 2s^2\ 2p^6\ 3s^1$". This is a string of letters, upper or lower case, and numbers, some in superscript. A typical ten-year-old should have the resources to *access the information* here by recognising the symbols used. Assuming the ten-year-old has no interpretive resources that allow them to make any sense of this string, it might be challenging to learn it accurately.

However, it is certainly possible. The child can practise and test themselves, and given sufficient effort could learn the string – and later reproduce it. Learning has taken place. This would be rote learning, however, like learning lines of poetry in an unfamiliar language.

Readers of this book might find it difficult to put themselves in the situation of that ten-year-old, as for them the string "Na: $1s^2\ 2s^2\ 2p^6\ 3s^1$" is not just information – it conveys meaning because they have available interpretive resources to make sense of it, to understand the information within a wider context such that it is not just information, but *meaningful* information. (It is not impossible that "Na: $1s^2\ 2s^2\ 2p^6\ 3s^1$" could be some meteorologist's coded record of observations of cloud cover and type observed over the town of Northallerton over a four-day period – but even if that is what the information was meant to represent, it is likely a chemist would make quite different meaning from the information.)

Ausubel talked of meaningful learning occurring where a learner had available the conceptual resources to link to the learning, *and* was actually able to make such links (*i.e.*, at some level the learner recognises how the information can be understood in terms of existing conceptual resources).

So then, here I am distinguishing three reference situations:

(a) teaching is experienced as noise, as the learner is unable to make out any signal that can be perceived as information;
(b) teaching is recognised as providing information, as the learner recognises the symbol system in which teaching is presented and so can detect a signal, but that information is not meaningful to the learner as they cannot decode it;
(c) teaching is understood as the learner not only recognises information presented, but interprets the information in a meaningful way (which may, or may not, reflect the teacher's intended meaning).

Probably most situations are intermediate. The listener recognises some, but not all, of the words that a person with a strong dialect or accent says. The student looking at a mass spectrum for the first time, without it being explained, recognises something familiar in the representation from experience of working with graphs, but does not appreciate fully what can be inferred from the spectrum.

The student who has never heard of the term 'transition element' or 'transition metal' and who is told that 'niobium is a transition element' understands that niobium is an element, but does not understand what 'transition' is meant to mean. (They might well interpret the information to produce meaning – perhaps a transition element is one that is between a metal and non-metal; or an element in the process of reacting to become a compound; or one element produced from another in a radioactive process?) The student who has never heard of the terms 'transition element' or 'transition metal' and who is told that 'niobium is a transition metal' understands that niobium is a metal but as (they know that) alloys – bronze, steel, *etc.* – are commonly described as metals may not appreciate from that information that niobium is an element.

A student who already has in place a concept of transition element may see the statements 'niobium is a transition element' and 'niobium is a transition metal' as equivalent – objectively those statements contain different information, but when that information is understood within a particular conceptual framing the information is understood to have much the same meaning. So the statement 'niobium is in the d-block' can be understood to entail that niobium is an element, and a metal, and indeed a transition element/metal, but that is not entailed in the information itself, as it is not unpacked purely from 'niobium is in the d-block', but rather is constructed by relating 'niobium is in the d-block' to prior conceptual learning that makes the statement meaningful to a particular person. Consider the following six statements:

(a) (i) niobium is a transition element; (ii) niobium is a transition metal; (iii) niobium is in the d-block;
(b) (i) propene is an alkane; (ii) propene is an alkene; (iii) propene is an alkyne.

Observation tells us that the three statements (b, i–iii) are very close in term of their *information* content, whereas (a, i–iii) show more distinctions. Yet, to someone who has the appropriate conceptual resources (*i.e.*, chemists) to meaningfully interrogate (a) and (b) it is clear that (a, i–iii) are similar in *meaning*, whilst (b, i–iii) are contrary alternatives, where logically only one could be correct.

11.5 So, How Are Chemical Concepts Communicated?

If we understood concepts as no more than strings of symbols, as a form of information, then there is no problem regarding how they might be communicated. The communicator carefully presents some information through its representation in a symbol system that is shared with those to be communicated with. The communicator and communicatee need access to a common system of symbols (a language) and there needs to be a clear channel of communication – the student has remembered their eyeglasses, the teachers speaks clearly and sufficiently loudly, display screens are not obscured by bright lights from windows or lamps 'bleaching' them when seen from the viewpoint of the students, the classroom is not amidst a building site where a lecture hall is being noisily demolished to improve facilities for conference catering, *etcetera*.

However, it has been argued here that concepts have to be seen as subjective, as personal mental entities, that are understood within a wider conceptual structure. So, if a teacher has a concept (be that {acid}, {potassium}, {combustion}, or {antiaromaticity}, or whatever), then that concept is in effect the activation of a node within a network of meaningful associations – and, moreover, a somewhat unique, idiosyncratic, network of meaningful associations. If that is what needs to be communicated, then clearly the task is somewhat more complex than communicating information. Indeed, it seems likely that the task of reproducing a network of such associations in high fidelity from one unique conceptual context – or mind – to another is simply not feasible.

In that absolute sense, then, concepts cannot be communicated. This conclusion will likely not surprise those who have carefully considered the analysis and examples presented in this chapter – or those who have been introduced to constructivist notions of learning. Indeed, anyone who has taken seriously the argument in earlier chapters that canonical chemical concepts do not really exist, might well have suspected from the outset of this chapter that its challenge, 'how are chemical concepts communicated?', was going to prove futile. After all, it is rather difficult for a teacher to pass on something that she has never herself possessed.

11.5.1 Authoritative Concepts and the Impression of Canonical Concepts

However, of course, in another sense concepts are (reasonably effectively) communicated all the time. Concepts such as 'combustion', 'neutralisation', 'oxidation', *etcetera*, are *sufficiently* shared among chemists that much of the time chemistry experts can operate *as though there are* canonical concepts that those in the community 'share'. So, we can say that professional chemists share a range of canonical concepts as long as we are careful to critique what we mean by 'share'. Just as we might say that in covalent bonds atoms share electrons, as long as we are careful to treat this use of 'share' as metaphorical, as a shorthand for something other than common ownership in the way a couple may share their home.

In answer to the question of how concepts may be communicated among chemists, then, the response is that this has to be understood in a much wider context than the episode of 'communication' itself. A chemist imagines some new possibility not previously discussed in the chemical community to make sense of some empirical pattern (perhaps, *cf.* Chapter 5, a new element 'potassium' in response to observing minute globules appearing in moist potash) and seeks to share this conceptualisation.

Consider the first person who imagined that bubbles commonly appear in tap water when salt is added because dissolved gases come out of solution as the salt dissolves. This chemist conceptualises this 'salting out' and communicates the idea to other chemists by the usual means – informal discussion, conference talks, journal articles. They can readily make sense of this idea because they have appropriate interpretative resources relating to the nature of electrolytes, solutions, gas solubility in water, *etcetera*. Other chemists have conceptual networks with nodes with much the same set of labels ('solution', 'gas', 'salt', 'dissolve', 'competition', *etc.*) as the chemist proposing the idea, and those nodes have sets of associations that to a large degree (if never exactly) overlap with that of the proposer and with their other peers. Communication is effective because the information in the communication can be interpreted within a conceptual context that is aligned (although not entirely replicated) between different experts in the community. That is, communication can be effective to the extent that the information in the communication can be interpreted within a conceptual context that is sufficiently aligned between experts in the community.

I am here introducing a new notion into this discussion: that of an authoritative concept. Canonical concepts may not strictly exist, and it is argued here that any actual concept is someone's personal concept, and that all accounts or traces of concepts that can be objectively found in the material world must be representations of people's personal concepts. Yet, the idea of a canonical chemical concept – the {canonical concept} concept – does useful work for us because within an expert community it is often possible to proceed *as if* there are canonical concepts because there is sufficient alignment between the pertinent personal concepts of different experts to proceed as though each of them have identical concepts. Even if, occasionally, inevitably given the account presented here, this assumption breaks down, it still applies much of the time.

There are of course levels of expertise: we might assume that, generally, graduate chemists share a well-enough aligned {element} concept to allow communication that assumes there is a shared understanding of the term. Yet other concepts may be more specialised ({endosomolytic polymer nanoparticles}; {homopropargylic amines}: {Knoevenagel condensation}; {molecular docking}; {nanoflowers}; {spiny metal–phenolic coordination crystals}; {sesterterpenoids}; {synergistic proteome modulations}; {5,6,12,13,19,20-hexaimidazolium-ethoxy-cyclotriveratrylene}; *etc.*), and we might then limit 'expertise' to research chemists, or even those working in a particular field. That is important, but does not undermine the general principle – and of course those writing or talking about chemistry are usually aware when they are writing for, or talking to, a specialist audience within a research field, or for a wider community of chemists (or scientists more generally, or indeed the wider public).

Given this proviso, that the term needs to be understood contextually, then we can consider there is a class of (personal) concepts that we might label 'authoritative' in relation to people's expertise. If the canonical concept is a useful fiction, an ideal, then we might define the authoritative concept as being a personal concept of an expert that is sufficiently aligned with the personal concepts of other members of their expert community that it can be treated *as if* it is a canonical concept.

Communication between experts, such as between chemists, can assume that those communicating have authoritative chemical concepts so that concepts may be readily represented (using standard terms, such that the information follows recognised conventions) and will usually be understood much as intended. This allows communication between experts to be largely (if not always) unproblematic, and to generally occur without substantive misunderstandings (or indeed failures to make any sense of what is being communicated).

Consider an example highlighted earlier in the book. In Chapter 6 the reader was told that "according to the perspective developed by Brønsted and Lowry, acids and bases are understood in terms of reactions where one substance (the acid) donates a proton, that another substance (the base) accepts". Yet, in Chapter 7, it was suggested this statement is strictly inappropriate. Whether this phrasing is considered as adequate and succinct, or confused and incorrect, depends on how the information presented is interpreted. Anyone who read the statement completely literally (a confused novice learner, perhaps) would not understand the intended meaning. An expert has no problem understanding the intended meaning as the interpretive resources of an experienced chemist (automatically) provide the unmarked switches between a description of bench chemistry and the theoretical model of what is happening at the level of individual molecules or ions. The literal meaning, that (just) one proton is donated by a substance during a reaction, is simply not noticed as this is not a viable interpretation – so the reader's sense-making interprets the statement in a meaningful and chemically sensible way. This raises the question of how we should respond to a student who offered such a definition in an examination: do we regard this highly, as demonstrating a shorthand of language reflecting a fluency of thinking in switching between levels of explanation – or see it as a confused and clearly inaccurate response? If the latter, we should be very careful in our own use of language with our students.

11.6 Concluding Comments

This analysis then raises two key issues for education.

First, how does the process of communicating conceptual material work in teaching, given that by the nature of being novices, students do not have conceptual structures that share the attributes of the expert, and so do not (to a sufficient degree, unlike experts) align with those commonalities 'shared' among the community of chemists. That is, learners do not yet have authoritative concepts that allow them to readily understand information presented in the conventional form suitable for experts.

Secondly, how did the situation arise that professional chemists, generally at least, have sufficient alignment in their personal conceptual networks for effective communication: such that a new conceptual understanding imagined by one chemist can be represented in the form of information that can, usually, be interpreted much as intended – to the extent that for most practical purposes we can consider the concept itself to be communicated and shared. That is, how did experts acquire authoritative concepts – how did they develop such community-recognised expertise?

These two questions are linked, of course, as the answer to the second depends on the first: effective communication between chemists depends upon their effective induction into expertise through chemistry education. That this works, shows how successful teaching can be. This does not, however, negate the problem of how to effectively communicate concepts, as

(i) this suggests the *immediate*, proximate, cause of effective communication of chemistry concepts between experts may appear to be a brief episode of listening or reading, but only because such episodes are essentially supported by prior learning that often extends back over many years;
(ii) the community of professional chemists is just a small sub-set of those who enter chemistry education, and many of those studying chemistry in our schools and colleges are not successful enough to proceed towards becoming experts and members of the professional community, even if they would choose to do so.

So, among those individuals where chemistry education has been very effective (so they have developed a large repertoire of highly connected authoritative concepts that are generally well aligned with those of their peers) the sharing of conceptual understandings about chemistry can be achieved by representing these ideas through framing information in conventional forms that others in the community can (usually) readily interpret much as intended. Yet, teachers are charged with sharing conceptual understanding with precisely those lacking a large repertoire of highly connected authoritative concepts that are generally well aligned with those of experts. So, it is worth considering how teachers actually carry out this difficult work in classrooms. This will be the focus of Chapter 12.

References

Ausubel D. P., (2000), *The Acquisition and Retention of Knowledge: A Cognitive View*, Dordrecht: Kluwer Academic Publishers.

Brock R., (2015), Intuition and insight: two concepts that illuminate the tacit in science education, *Stud. Sci. Educ.*, **51**(2), 127–167.

Collins H., (1992), *Changing Order: Replication and Induction in Scientific Practice*, Chicago: University of Chicago Press.

Gilbert J. K. and Watts D. M., (1983), Concepts, misconceptions and alternative conceptions: changing perspectives in science education, *Stud. Sci. Educ.*, **10**(1), 61–98.

NASA, (1990), The Hubble Space Telescope optical systems failure report. Retrieved from Washington, DC: https://ntrs.nasa.gov/archive/nasa/casi.ntrs.nasa.gov/19910003124.pdf 29th December 2018.

Olitsky S., (2007), Promoting student engagement in science: Interaction rituals and the pursuit of a community of practice, *J. Res. Sci. Teach.*, **44**(1), 33–56.

Polanyi M., (1962), *Personal Knowledge: Towards a Post-critical Philosophy*, Corrected version edn, Chicago: University of Chicago Press.

Taber K. S., (1998), An alternative conceptual framework from chemistry education, *Int. J. Sci. Educ.*, **20**(5), 597–608.

Taber K. S., (2013), *Modelling Learners and Learning in Science Education: Developing Representations of Concepts, Conceptual Structure and Conceptual Change to Inform Teaching and Research*, Dordrecht: Springer.

Vollrath F., (1998), Dwarf males, *Trends Ecol. Evol.*, **13**(4), 159–163.

Ziman J., (1968), *Public Knowledge: An Essay Concerning the Social Dimension of Science*, Cambridge: Cambridge University Press.

CHAPTER 12

How Are Chemical Concepts Represented in Teaching?

It was suggested in the previous chapter that, by their very nature, concepts (as personal and subjective mental entities, being the activation of conceptual structures represented in an individual's brain) cannot be *directly* communicated. The 'sharing' of concepts therefore relies on clear communication of something else, information (symbolic strings, such as: 'chromium exhibits multiple oxidations states', '$C_6H_5NH_3Cl$', 'damıtma', 硫黄の花, *etc.*). This information is understood by the communicatee (the 'receiver' of information) through making sense of the information in terms of existing conceptual networks that are able to act as interpretive resources. The information can be understood (much) as intended, as long as these conceptual resources are sufficiently aligned with those conceptual frameworks employed by the communicator (*i.e.*, 'transmitter' of information) when shaping the communication. That is, information may be transmitted between people (this is a technical issue – *e.g.*, if the communicator is in Brisbane and the communicatee is in Montreal then a channel is needed to allow the information to be clearly transmitted), but communication of *ideas* depends not only on the information itself, but a sufficiently shared interpretive conceptual framework in which that information is understood – and that relies on applying mental apparatus, rather than information technology.

The reader may spot here an analogy to the description of how science progresses, discussed earlier in the book, where it was pointed out how measuring the properties of pure samples of substances is only possible once it is possible to determine the identify and purity of substances, which in turn depended on having characterised the pure substance by having measured some of its properties. Similar challenges arise in establishing

---

Advances in Chemistry Education Series No. 3
The Nature of the Chemical Concept: Re-constructing Chemical Knowledge in Teaching and Learning
By Keith S. Taber
© Keith S. Taber 2022
Published by the Royal Society of Chemistry, www.rsc.org

an effective measurement scale: to develop a new temperature scale we need to employ existing independent ways of measuring temperature (see Chapter 7). This may be related to the learning paradox, suggested by Socrates, asking how we can learn something new without already knowing it well enough to know when we have successfully learnt it. In a similar way, coming to share (much the) same concepts between people depends upon those people already having some common concepts. In both developing scientific knowledge, and in coming to share conceptual frameworks, an iterative approach needs to be adopted where initially approximate and imprecise starting points are slowly refined and built upon.

This need for a degree of existing alignment between concepts to facilitate shared meaning is a general requirement for the communication of ideas – it applies to everyday conversation, it applies to scientific communication between expert chemists, and, of particular interest in the context of this volume, it applies to teaching where a teacher communicates with students. Within a community of experts, sufficient alignment usually exists to support generally effective communication of ideas. The personal concepts of experts are authoritative concepts – although idiosyncratic in nuance, they are sufficiently similar from one expert to another that they are represented (in writing, talking, *etc.*) in ways that other members of the community would generally recognise (and can assume) as canonical. Developing authoritative concepts relies on processes of induction into a community of experts that allow different people to develop sufficiently aligned conceptual frameworks (Kuhn, 1970). Those processes of induction imply an extensive educational process. By definition, most learners in chemistry classes are relative novices that lack such a well-aligned set of concepts.

12.1 Communicating and Teaching

Teaching, then, involves a teacher of an area of expertise, such as chemistry, working with learners who, at the outset at least, lack a level of conceptual alignment to make good sense of much of the material to be taught. At the early stages of chemical education, this lack of alignment includes not having any version of many of the relevant concepts. Despite this, much of the time, students – or at least some of them – come to learn concepts that (based on the indirect evidence available to an examiner) are judged *sufficiently close to* the target knowledge, curricular models of authoritative concepts (see Chapter 10), that the students *are considered to* have learnt or mastered some of 'the' concepts of chemistry. We might for shorthand say that *these students have effectively learnt 'the' concept* of combustion, or *'the' concept* of tautomerism, or whatever, but should be aware that the definite article here (the 'the') is a little misleading (see Chapter 7). A student learns 'a' concept of combustion (*etc.*) that appears sufficiently matched to selective aspects of the examiner's {combustion} concept to be treated as 'the' chemical concept for particular purposes.

12.1.1 Teaching as Moving Students Towards Accredited Conceptualisations

Just as we may consider the personal concepts of an expert to be authoritative when they appear to others in the community of experts to be canonical, we might consider that (some) students come to hold concepts that are sufficiently aligned with target knowledge to be evaluated positively by teachers and examiners. If the student concerned is a PhD candidate in chemistry, and they are judged to have passed the doctoral examination, then we might consider this to mean they have acquired authoritative concepts in their field of research.

By comparison, however, if the context is a class of 14-year-olds taking a school test, we would not consider that such students are being asked to show they hold authoritative concepts. Perhaps, in such a context, target knowledge of acids is limited to their action on universal indicator and the nature of their reactions with alkalis, metal oxides and metal carbonates. A student who is successful in the test produces responses to test questions that are judged as showing that they have acquired a concept {acid} that sufficiently matches such particular target knowledge in the curriculum. Yet their concept may not include any associations relating to pH or hydrogen ions, and indeed they may have no {pH} or {hydrogen ion} concepts within their conceptual structures with which to link their {acid} concept as part of a network of concepts.

In general, then, when learners are judged (through the way they apply their concepts when representing their thinking), this is not in terms of whether they seem to have acquired authoritative concepts that would appear canonical to members of an expert community of chemists, but rather they are being judged as having developed concepts that meet more limited standards relating to how chemical knowledge is represented in the curriculum at their level of study (a matter discussed in Chapter 10). We might describe personal concepts evaluated positively in this way by a suitable authority (the teacher, or external examiners through a formal test or examination) as being 'accredited'.

At the highest levels of chemical education, accredited concepts become synonymous with authoritative concepts. However, at earlier stages this need not be so. Many concepts developed in school-level chemistry, and indeed sometimes in undergraduate-level chemistry, may be considered accredited whilst being deficient compared with authoritative concepts. Accredited concepts would (indeed, should) usually be expected to be canonical in the particular sense of being consistent with, even if not as sophisticated or extensive as, authoritative concepts. In practice, this may not always be so. As one example, a student may be judged to have an accredited version of the {ionic bond} concept when their personal concept is based around the idea that an ionic bond is the transfer of an electron from an isolated metal atom to an isolated non-metal atom (Taber, 1997). This would be considered an accredited concept if that notion of an ionic bond is set out as target knowledge (see the discussion of curricular models in Chapter 10), even though such a notion is non-canonical and is inconsistent with an expert's authoritative concept.

Ideally, a chemistry teacher's own concepts should be authoritative; and, at the highest levels of the education system (*e.g.*, if the 'teacher' is the doctoral supervisor, or the post-doc's mentor), it is important this is so. It is unreasonable to expect this at all levels of the system. The primary school teacher introducing students to basic ideas in chemistry *could* be a PhD in chemistry who would be considered a member of the expert community of chemists. A small number of primary school teachers might be this well qualified in chemistry, but clearly the vast majority are not. Most will not be chemistry graduates. Many will not have taken any advanced elective courses in chemistry at secondary/high school level.

The minimum necessary to support students in acquiring accredited concepts is that the teacher of chemistry should at least hold accredited concepts (at the level they are teaching) themselves. Without this, it is difficult to see how they will support learners in acquiring accredited concepts, or indeed evaluate student learning (and so their own teaching). The once fashionable idea that teachers are facilitators of learning, and may effectively support students in learning things they themselves do not know, may have some logic in certain learning contexts: however, in teaching an academic subject like chemistry it is difficult to see how someone who does not have a good knowledge and understanding is going to facilitate much learning beyond the trivial level of helping students study by logging them onto the classroom computer or directing them to the library. Effective teaching of chemical concepts requires teachers to monitor the learning demand (Leach and Scott, 2002) – the 'gap' between students' current understanding and the target knowledge – in order to inform ongoing teaching, and clearly that is not possible for someone who does not themselves have a good grasp of what is to be learned.

Sensibly, teachers should have concepts that reach beyond the level of what would be accredited status at the level they are teaching, further towards being authoritative concepts: as the effective teacher attempts to teach not just to bring students to the expected standards ('teaching to the test'), but to do so in a way that puts in place sound foundations for further progression in learning. As an example, a school teacher expected to teach an introductory class about oxidation in terms of reaction with oxygen, and who's own personal concept was no more sophisticated, is not well placed to prepare the learners for how this concept will be developed later in schooling in terms of electron transfers or shifts in oxidation state. It is not suggested that a teacher charged with teaching oxidation as reaction with oxygen should complicate matters for students by also teaching other definitions/models, but it would be counterproductive if the teaching implied that {oxidation} was a unitary concept that would not later need to be expanded; or if such a teacher responded to a student who commented on having read that oxidation was a loss of electrons by rejecting this and treating it as a student failing to take on board teaching.

In particular, a teacher who did not appreciate something of the nature of chemistry as a science, and, in particular, the status of the typologies and models that chemists use to make sense of the subject (see Chapters 7 and 8), and treated chemical ideas as definitive knowledge, things that chemists had discovered to be just the way things are, is likely to teach in a way that is unhelpful. As explored earlier in the book (see the treatment of the {acid} concept in Chapter 6) chemistry relies on a good many imaginative constructions that chemists find useful (*e.g.*, an acid as proton donor) but that cannot be treated as simple 'facts' about the natural world.

A teacher who has a naive understanding of the nature of chemistry and teaches (for example) that electrons have planetary type orbits in shells around the atomic nucleus as if this is a fact proved by science, rather than one model that fits with some of, but not all of, the empirical evidence, is teaching in a distorted way. Arguably, it is less important if chemistry teachers are always aware of how the concepts they are teaching are deficient in relation to current authoritative concepts of expert chemists, than that their teaching presentations are informed by an appreciation that chemical concepts are human constructions that are valuable in offering useful accounts of nature, but that will often need to be developed and complemented as students, or indeed chemistry itself, makes progress.

That is, teaching based on a limited knowledge of chemical concepts supported by a sophisticated epistemological framing of the nature of chemical ideas (Taber, 2010) is likely to give students a better basis for later progression than an extensive knowledge of concepts well beyond the level taught, but underpinned by a naive positivism about the status of those ideas.

12.1.2 The Communication of Information Is Necessary but Not Sufficient for Teaching

What allows effective communication is, then, two-fold. First, it is a shared set of symbolic resources for representing information (along with the practical issues referred to in the previous chapter relating to a clear communication channel: if the lecturer has been told the class is in lecture theatre 2 and the class have been sent to tutorial room 7, then this will be a practical impediment to effective communication). 'Information' is here used in the sense developed in Chapter 11 – as something objectively available to observation (such as a text), rather than any meaning that someone might associate with the information (how they subjectively understand the text – which can vary from person to person). Much of the symbolic system will be common within a particular social context (*e.g.*, everyday English or Spanish or Chinese or some other natural language; common conventions of graphical representation; *etc.*), although some may be specialist conventions within the area to be taught, and so needs to be learned within the chemistry classroom context (Taber, 2009).

Secondly, the process also relies upon sufficient interpretive resources at the learner's ('receiver') end of the communication to understand the information in the communication – not just in a meaningful way, but in a way that has *similar meaning* to that intended by the teacher (or another communicator). So, what is actually directly communicated is information, but the framing of information has to be judged to allow it to be understood as intended. The teacher can control the presentation of the information (speak clearly, write legibly, project text in a clear large font, carefully label diagrams, *etc.*), but does not have direct control over the interpretive resources that might be brought to bear to make sense of it. Yet chemistry teaching, a process to support developing conceptual understanding rather than simply passing on information about chemistry (see Chapter 11), relies upon the teacher engaging those interpretive resources available at the 'receiving' end of the information stream.

Concepts are communicated, then, by presenting information in such a way that the information can be interpreted so that it is understood, and, moreover, understood in particular ways. Communication about chemical concepts between professional colleagues may (at least much of the time) be based on an assumption that there is sufficient alignment of interpretive resources such that a clear presentation of the right information suffices for meaningful understanding along the lines intended. Teaching those not yet inducted into expertise in the subject is more challenging, and therefore requires the teacher to frame information in ways designed to engage intended interpretations, which in turn requires the teacher to have a good mental model of the interpretive resources the learners have available to make sense of teaching.

12.1.3 Lecturing

Some readers may well appreciate the relevance, here, of the criticisms that were once very commonly made of lecturers (but one hopes are much rarer these days). A caricature of lecturing might compare it with teaching (*i.e.*, what critics of the lecture might characterise as 'real' or 'proper' teaching), where the presentation of a good teacher is designed as much with the particular learners in mind as the material to be taught. A pedagogically poor (even if informationally rich) lecture is designed purely in terms of the presentation of information – albeit perhaps information that had been carefully selected, well sequenced, and clearly phrased. The lecturer's job, in this caricature, is to make sure that the right information is clearly presented, but it is the students' responsibility to interpret that information and make sense of it.

Learners certainly do have responsibilities (they should arrive for lectures on time, having undertaken any recommended reading or pre-lecture task, be suitably rested and alert, and they should pay careful attention, *etc.*) – but if they do not have the interpretive resources to make sense of teaching, or do not recognise how relevant prior learning links to teaching in order to make sense of it (Taber, 2005), then such tactics as having had a good night's sleep, sitting at the front of the room, and concentrating as hard as possible, will not be sufficient to convert the stream of information from a lecturer lacking pedagogic expertise into meaningful learning.

There has been much research and scholarship into teaching chemistry in university settings (indeed for some years the Royal Society of Chemistry published a journal with this focus called *University Chemistry Education*, that then became incorporated into *Chemistry Education Research and Practice*), and this caricature is relevant to many fewer university lectures than might once have been the case. Despite this, in many countries pedagogic preparation for teaching in a university is still somewhat minimal compared with the usual expectations for teaching in a school, so developing pedagogic knowledge and skills may often largely rely on the lecturer's initiative. Moreover, if the claims made in some research papers about what is typical in many classrooms are accurate (Taber, Forthcoming), then in some countries school teaching itself may often reflect the caricature of lecturing as communicating information.

12.1.4 The Non-linearity of Teaching and Learning

It is important to appreciate here that the situation is particularly complex because teaching chemistry is not, and cannot be, a matter of the linear treatment of a succession of concepts (see Chapter 10). Conceptual analysis is very important in teaching to ensure that when teaching one concept, one does not assume one can present it in terms of other concepts that have not yet been taught. Things would be more straightforward if chemistry had been a subject that could be developed linearly. That would mean that one could start from one concept – say {substance} – and teach this concept (by presenting examples and non-examples, by offering a definition and explanation in everyday terms, using non-technical concepts already available to all the learners, *etc.*) Then this would allow us to teach, say, the concept {element}, based purely on using the concept {substance} along with demonstrations, examples, and reference to common knowledge all students have from outside the subject. This might allow us to teach the concept {compound}, relying only on the concepts {substance} and {element}, as well as common non-technical resources. Perhaps we can then move onto {reaction}; and so forth. Eventually, somewhere in the sequence, would come {dipole moment}, {enthalpy of neutralisation}, {aromaticity}, {the inductive effect}, and all the rest – each positioned to be taught drawing upon those concepts earlier in the sequence.

If chemistry was like this, we could analyse any concept – {combustion}, {enthalpy}, {neutralisation}, {orbital}, {solvent}, {substitution}, {unsaturated}, *etcetera* – and, for each, produce a list of those concepts we need to have in place before we teach that concept (Herron *et al.*, 1977). Then it would simply be a matter of finding a logical linear order to work through.

That is not to suggest teaching the abstract ideas would be easy, but this would certainly make it less challenging than it is.

Yet, as discussed earlier in the book (see Chapter 10), chemical concepts do not fall into a simple sequence where each concept can be sufficiently understood in terms of those that precede it and can be taught earlier in the sequence. Chemical concepts can often be understood at different levels of sophistication, so one cannot seek to teach a concept such as {metal} or {acid} or {oxidation} as a once-and-for-all shift from ignorance to canonical understanding. Indeed, sometimes there is a need for 'bootstrapping' as analysis of two particular concepts may well suggest that for a full understanding of each of them, the other is a prerequisite concept – which is one reason why it is appropriate that accredited concepts (those judged to match target knowledge in the curriculum) are often considerably impoverished compared to authoritative concepts (those that experts operate with). Someone cannot really understand what a chemical reaction is until they have mastered the chemical {substance} concept, but a good understanding of what is meant by a substance relies on understanding what is meant by a substance being potentially changed into another substance by chemical means, which in effect requires some form of understanding akin to a {reaction} concept.

12.2 Framing Information to Allow Meaningful Understanding

The argument being made here then is that no matter how skilled a teacher (be that a university lecturer, or a school or college teacher) at selecting and sequencing material and presenting clear information – based on a good understanding of the chemistry to be taught – this does not ensure that learners will understand that information in the way intended. Teaching designed to support conceptual understanding and development therefore has to ensure that the information presented can be interpreted appropriately by the learners, something that is entirely dependent upon both their having suitable resources to make sense of the information, and on them making those links between prior learning and the new information when teaching occurs. The former criterion does not assure the latter: it is quite common for learners to fail to bring to mind available prerequisite learning that the teacher assumes is ('obviously') relevant to the new information being presented.

This therefore raises a number of issues widely explored in educational theories and research. Teachers needed to develop what has been termed pedagogic content knowledge or PCK (Kind, 2009), a form of synthetic knowledge that builds upon their knowledge of chemistry, and of general pedagogic principles (Taber, 2018a), to develop expertise in teaching particular chemical ideas to specific groups of learners. PCK would be informed, *inter alia*, through research that characterises how learners understand chemical ideas, including the alternative conceptions they commonly hold (see Chapter 14).

12.2.1 Modelling the Other Person's Sense-making

Anyone who communicates with another person makes assumptions (if not deliberately, then implicitly) about the interpretive resources the other person has available for making sense of that communication. Assuming a genuine intention to communicate, then speaking in English assumes that the listener speaks English; sending a letter assumes the recipient is literate; talking of iron as a transition element assumes the listener has a suitable {transition element} concept to understand the intended meaning. Without such a concept, and relying only on a knowledge of the language, all that can be inferred is that iron is being categorised into some kind of set or other.

In effect, there is a modelling process here. In everyday communication, this is largely implicit and automatic. When parents are with a young child they will automatically adjust their language as they switch between talking to each other and to the child. They use different vocabulary according to whom they are addressing, and also shift the level of complexity of sentence structure used. Indeed, they do something more complex than this – when the parents are talking to each other, they make these adjustments, again effortlessly, according to when they wish the child to be an audience for the conversation. So, people automatically employ mental models of how their communication will be understood when framing that communication. This process is fallible, especially when dealing with people that are not known personally. We might ask someone at the bus stop the time, only to then realise from their incomprehension that they are not an English speaker. We might avoid technical terminology in a conversation with someone (perhaps because it seems reasonable to presume that most taxi drivers, say, will not know what ascorbic acid is), only to realise, given their response, that we have underestimated their knowledge of a topic. Despite such occasional misjudgements, modelling the way our communications will be understood to inform how we should best frame such communication is largely effective in everyday life.

As this is something people learn to do automatically, teachers have these abilities like anyone else. A chemistry teacher telling an anecdote of how a student nearly set fire to the laboratory will frame it somewhat differently when talking to another chemistry teacher, rather than when telling the story to teachers from other curriculum areas. The teacher will talk about a student's learning difficulties differently when talking to the student's parents from when consulting an educational psychologist with specialist knowledge – and indeed the psychologist will moderate their own language to avoid use of some terms they would use with other psychologists because they are talking to a teacher. A teacher does not need to be told that they need to adjust their presentation of a topic to be taught to both eleven-year-olds and 18-year-olds in the school – or one that is being presented as part of a first-year undergraduate lecture course, and also a research symposium for post-graduate students and other academics – or to interested members of the general public as an outreach activity.

However, although this process occurs automatically, this does not imply that teachers will automatically optimise their presentation of topics. Effective teaching requires more than relying on the implicit processes by which we normally make these adjustments in everyday situations. Students often misunderstand teaching, and sometimes can make little sense of it, as the interpretive resources necessary to make good sense of the information presented are either absent, or are not engaged because relevant prior learning is not recognised as being pertinent. This seems to be very common in teaching at all levels, and shows that (unlike in most everyday conversations) this modelling of how teaching will be interpreted, used by teachers to frame the presentation of information so it is understood as intended, can often fail.

This is understandable for two reasons, First, it is much easier to make adjustments in everyday conversations when communication falls short, as usually we are talking with one person at a time in a symmetrical situation (that is, they can do as much of the telling). This rather different to talking to a class of 25 14-year-olds or in a lecture room to 150 undergraduates.

Secondly, most everyday conversations are about transferring information to be interpreted within frames where the available interpretive resources are inherently adequate. We discuss medical issues with the qualified doctor, buy a bus ticket from the trained bus operator, talk to a shop assistant employed in a particular shop when we want to buy something in that shop, and so forth. In these situations, we can normally presume that we are communicating in a context where both partners in the discourse share the relevant background knowledge.

The family doctor does not know in advance of the consultation that the patient has been suffering for some weeks from a cough that does not seem be clearing – but the doctor obviously does have a detailed and sophisticated {cough} concept that has associations with other concepts relating to the respiratory tract, diseases of the lungs, micro-organisms, antibiotics, *etcetera*, so that the patient does not first have to explain to the doctor what is meant by having a cough, and why this might be something worth raising in a consultation between a doctor and a patient. The patient does not need to remind the doctor that coughs can indicate disease, or that viral coughs normally clear after some days, or that the lungs are important organs of gaseous exchange (which is important for maintaining cell respiration, and so cell viability, and so forth), to help the doctor interpret the information about the long-standing cough and so understand why this is a concern for the patient. A teacher introducing a concept such as {steric hinderance} or {transition state} is in a rather different situation.

So, teaching a conceptual subject such as chemistry is very different from much everyday talk, because the aim of the communication is not just to pass on information that can be understood within an existing conceptual

How Are Chemical Concepts Represented in Teaching? 263

framework, but to actually build and develop the conceptual frameworks themselves. There is perhaps an analogy here with feeding someone. A person can be fed by providing them with a balanced set of foodstuffs. But imagine feeding someone who had not developed their digestive system, and needed nutrition to construct the digestive system as well as to maintain metabolism. Although offering them some fruits, vegetables and nuts might provide an adequate supply of the nutrients they needed, they would not be able to deconstruct these to extract those nutrients. In a sense, teachers need to pre-digest the conceptual 'food' they provide to learners as we are not only feeding learners information, but simultaneously developing their ability to digest that information (see Figure 12.1). (If this analogy seems a little

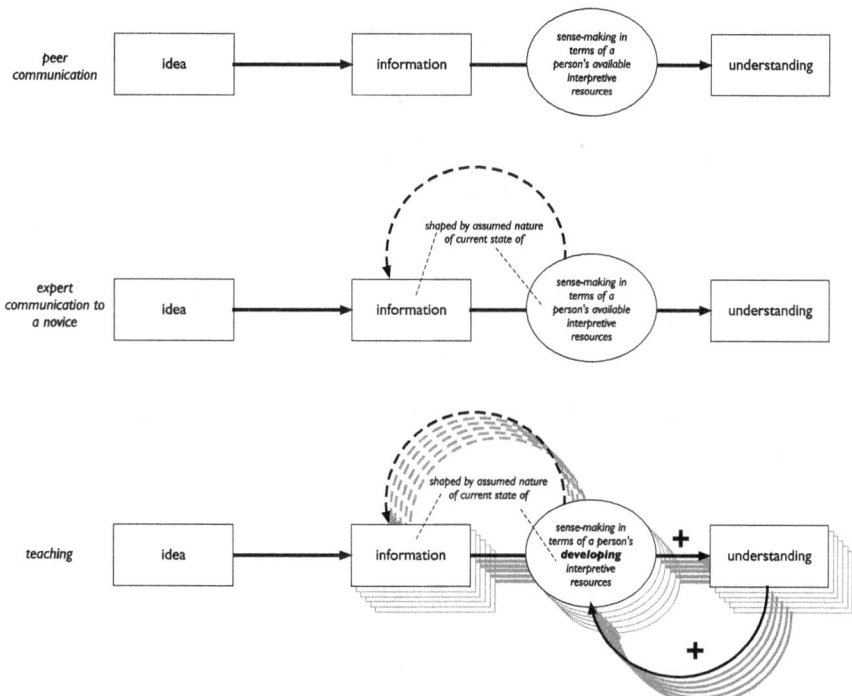

Figure 12.1 When communicating to someone with the same level of expertise (in some domain) we can usually assume they will understand the way we express ourselves. When communicating as an expert to a novice we need to frame the communication in a form that the novice is able to understand given their more limited level of knowledge. Teaching often requires helping learners understand something they do not yet have the resources to make full sense of. Teaching is therefore often iterative, as first we have to develop the interpretative resources until the learners are ready to understand what they are expected to learn. This relies on a positive feedback cycle (indicated by the two + symbols) where further developed interpretative resources can support more sophisticated understanding, which provide more powerful resources for interpreting further information, which. ...

forced, because the image of feeding someone who had not developed their digestive system seems somewhat bizarre and unlikely, it is worth noting that we all started life without a digestive system, and had to be fed pre-digested nutrients until we developed one.)

Developing sufficiently detailed models of the interpretative resources learners have available to make sense of teaching is not easy, and it is unlikely teachers can ever build a well-matched model without feedback from testing it in operation whilst under development. Certainly, the teacher needs to start somewhere (using resources such as published research studies, or maybe administering a pre-test), but then to be constantly testing out, and so iteratively updating, the mental model used to frame teaching.

Teachers then need to refine their mental models of the interpretive resources that the particular students in their classes can bring to bear, and to update those models, as teaching develops. By the very nature of teaching, the interpretive resources that learners have available for making sense of teaching are 'moving targets'. Even if a teacher has well judged the learners' starting points, if the teacher is doing a good job the students' conceptual frameworks will be changing and this needs to be taken into account in subsequent teaching. This suggests that teaching involves not only careful conceptual analysis of the material to be taught (to identify the necessary prerequisite learning, and sensible teaching sequences), and diagnostic assessment to check the starting points (*e.g.*, prior learning and existing relevant alternative conceptions) of particular groups of students, but also formative assessment – that is ongoing assessment that monitors how students are interpreting teaching, and how this is informing their developing understanding. That, in turn, means effective teaching has a dialogic aspect (Mercer, 1995; Scott, 2007) – involving the voices of the students to give expression to their thinking – rather than just being a one-way channel of communication (Taber, 2014).

12.2.2 Making Unfamiliar Concepts Familiar

A key aspect of teaching may be characterised as *making the unfamiliar familiar* (Taber, 2002). The mentality of doing research is sometimes described in a similar way (although it may need to start by *making the familiar unfamiliar*) and indeed effective teaching is much like research. It certainly involves an ongoing iterative process of observing students, forming hypotheses about their developing understanding, and testing those hypotheses. A teachers' role is to help learners become familiar with entities they are not yet familiar with (where those entities may be material things like conical flasks, or abstract things like Beethoven's Eroica symphony or the {transition state} concept). The most direct way of making the unfamiliar familiar is to provide experience of that entity to allow learners to familiarise themselves with it. Students can be shown conical flasks, told they are conical flasks, allowed to handle them, and set tasks that require the use of such flasks (all the time referring to them as conical flasks). They will

How Are Chemical Concepts Represented in Teaching?

develop a {canonical flask} concept, an abstraction that allows them to recognise other examples of conical flasks (see Chapter 2), and come to be able to use their {canonical flask} concept in appropriate ways (*e.g.*, when asked to put 50 ml of a solution in a conical flask, they do not pour the solution into a beaker or a round-bottomed flask).

To some extent, similar tactics work with more abstract entities – such as playing a recording of the Eroica to help learners become familiar with it. It is not so easy to show and point to a transition state as it is to a conical flask. If one was explaining the {transition state} concept to a hypothetical chemistry expert with a good set of authoritative chemistry concepts, but who had somehow (unlikely as this would be) managed to never meet the {transition state} concept before, the task would be simple enough. One could show the expert a reaction profile and point to the maximum and say something like 'the transient structural state that the reacting molecular complex passes through at this point would be the transition state'. That information, in the context of the interpretive resources available to the expert (including sophisticated chemical concepts) should do the job: the concept, and its significance, should be clear. This is in effect how scientists do communicate when they moot a new concept: it can usually be unambiguously and clearly defined in terms of other related concepts using language and other shared symbolic systems of meaning-making.

If teaching the concept of transition state to an advanced high-school student, it would be wise to refine this approach and add additional support (perhaps 'scaffolding' in educational terminology) to support the interpretive processes. Even if students have met reaction profiles before, it would be important to reinforce the nature of the representation and what was represented, in particular in terms of the changes molecules undergo during reactions, and the energy dimension. It would be useful to check students are thinking of the energy state as related to stability. It might also be sensible to check that students have a notion of the kind of time-scales likely to be involved in such changes, as well, perhaps, as how the relationship between the initial and final endpoints link to thermodynamic considerations, and how reactions can be seen as equilibria, as well as pointing out the aspect of the profile that relates to activation energy, and how this is important to kinetics. Our unlikely expert (who, somehow, had never come across the notion of a transition state) knows all that, and moreover likely readily brings it all to mind spontaneously on being shown a reaction profile. Well-consolidated learning (*i.e.*, expert knowledge) tends to be well integrated in memory, and can often be accessed and 'loaded' into working memory as if a single chunk of prior learning.

However, even if students have been taught all of this background, many will not likely readily call it to mind unless it has previously been regularly reinforced and rehearsed so that it has become well consolidated. More likely, this background knowledge will make substantive demands on working memory as it will be treated as a series of discrete chunks of prior learning, rather than accessed as a whole. The teacher cannot assume that students have learned what has previously been taught well enough to effortlessly call upon it, and nor can the teacher assume that what is obviously relevant to the teacher or another expert will be automatically noticed by a learner.

When the expert is told which point on the reaction profile relates to the transition state, she will readily appreciate that a finger pointing near the peak indeed refers to that maximum. A teacher knows that needs to be made explicit to the learners, and that the 'obvious' fact that this point on the profile is an energy maximum needs to be made explicit in case some learners have lost track of the dimension (because they have limited working memory capacity, and even if the teacher mentioned this among other points just a minute earlier, it may no longer be actively within the locus of conscious awareness).

Even more critically, the implications of this maximum need to be explored. The mental simulation of a reaction mechanism that may come readily to an expert is not likely to be readily available to the learners, and in particular the key ontological difference between a transition-state structure and the structures of the reactant and product molecules is likely to be missed. Indeed, the very definiteness of drawings of molecular structures (including those of transition states) may act as an unintentional and unhelpful cue that these are stable entities.

We might consider here the difference between an intermediate structure, which is in principle, and often in practice, isolatable, and a transition state. The conventions of the usual representations of drawing a reaction profile may suggest that an intermediate and a transition state are similar as they both occur at extrema on the profile rather than the terminals where the reactants and products are found. In some classes, some students will spontaneously have an insight into the significance of the position of the transition state as they will construct the meaning by integrating the new information with their background knowledge: students who tend to have such insights when their peers need things set out for them one step at a time may in some contexts be referred to as gifted learners.

Teachers will need to 'scaffold' most other learners to get to this point. A scaffold is a temporary support that allows a learner to succeed in some task that is beyond their current capacity unaided (Taber, 2018b). This derives from a theory of learning and development proposed by Vygotsky (1978). Scaffolds have to be well judged, as if the teacher is too ambitious then – although the learner may appear to have learned whilst the support is present – the learner does not move beyond mediated (supported) learning and can no longer succeed later when the scaffolding is not available. Ideally scaffolding is 'faded' at a rate that matches the students' developing capacity to demonstrate competence.

This may seem rather abstract to readers new to these ideas. So, consider the following two scenarios. The first scenario adopts the caricature of lecturing referred to above, and the second scenario concerns a more pedagogically informed strategy a teacher might adopt.

12.2.2.1 Scenario 1

In a lecture, the teacher says something like:

> This point of the reaction profile [indicating a diagram] represents the position of the transition state. As you see, this is the maximum of the profile. We can draw structures that relate to this point. So, for example, in an S_N2 reaction the transition state would have both the leaving group and the attacking nucleophile partially bonded, one group forming a bond, the other breaking away. Of course, as this is the structure at the maximum it is the least stable structure during the whole reaction, making it impossible to isolate. That's different from an intermediate, which is a structure formed on a local minimum on a reaction profile, as these can be stable enough to isolate. For example, if we rapidly cool a reaction mixture we can sometimes detect an intermediate by spectroscopic analysis. We cannot isolate transition states due to their instability. However, we can see the structure must pass through this point, and it is useful to know what this structure would be if we are to imagine the reaction mechanism.

Six months later, students might be presented with a question on an examination paper along the lines:

> (a) Explain what is meant by the term transition state. (b) Draw a reaction profile and label the point associated with a transition state. (c) (i) Give an example of a reaction where the mechanism is understood well enough to characterise the hypothetical transition state. (ii) Write a reaction equation for this reaction and sketch the structure of the transition state as well as the structures of the reactants and products. (Label each sketch.) (d) Explain why it is not possible to isolate the transition state in the actual reaction.

The examination question tests what has been presented – but the responsibility is on students to have made good notes, make sure they understood the teaching, and to have learned the material.

12.2.2.2 Scenario 2

A school teacher might develop some teaching resources and set group work, but could also lead the class through an exposition of the material. However, when compared with the hypothetical lecture, this would be more dialogic (where more than one voice, *i.e.* perspective, is considered), based on something like a Socratic dialogue, rather than a one-way stream of information (Scott, 1998). Such a dialogue could be quite involved, so here I abbreviate a little and leave readers to 'fill in the blanks'.

> Who remembers what this type of diagram is called...good, and can anyone tell me what it actually shows. That's right... [*not assuming*

students readily recall having seen the diagram, or, more importantly, what it is intended to represent]

So, let's just check how well we recall this. I have some labels here – 'reactant', 'product', 'activation energy', and 'enthalpy of reaction'. Can you make a quick sketch of the diagram and add the labels where you think they should go. Then compare your answer with the people either side of you. Ask them to explain what they have done if you are not sure – see if you agree.... [*waiting till this has been done*]... Okay, who is going to suggest how I should label my diagram, which label goes where. ...

Someone in another class told me that the reaction profile was just a graph showing energy – what do you think, is that a fair point? [*mooting an idea from another party of uncertain status, so the class do not know if this is being presented as an authoritative idea or not, and so they are invited to critique it rather than just accept it*] ...Can anyone suggest why a reaction profile is not just an example of a graph – can you suggest any differences [*it is important in teaching to highlight neutral and negative features of analogies and other comparison used, as well as the positive mappings*] ...

Okay, I'm going to add a new label we have not seen before. I'm going to label this point here – can you see what I'm pointing to. [*Not assuming students appreciate which point is being indicated*] ... Can anyone describe this point – how might we describe this to someone down the telephone who could not see where I am pointing? [*Both checking there is a shared referent, and engaging a different modality – students can see the diagram, but are being asked to link this with a verbal description. Thinking about how to describe the point means the students are actively processing rather than passively listening: and research suggests that associating verbal and imagistic information supports recall of learning.*]

We label this point the transition state. Perhaps you should all add this label to your sketches as well. ... I want you to be nosey now, have a quick look at what your neighbours have done, and check you think they have put the label in exactly the right point so they will be able to remember what they have labelled when they look at their lecture notes on Saturday evening to review the week's work (as I am sure you all look forward to Saturday evening as a special time for reviewing all your notes.) ...

Now remember this line reflects the molecules reorganising as the reaction occurs, so a point on the line represents a stage in that process. If we could film the reaction at a molecular level – using a magic camera able to pick out the details of individual molecules and able to catch really, really [*exaggerated stress for emphasis*], quick changes, then we could play it back frame-by-frame and each frame would be what the molecules are like at one point on the line. [*The teacher uses a kind of thought experiment to help students imagine the chemistry by referring to something expected to be familiar to the learners: a movie that consists of a series of stills.*]

Now the shape here, around this point we are going to label the transition state [*using some repetition to reinforce the point*] reminds me of what might happen if I flicked something like a ball up in the air, and then caught it. Imagine I had a ball of my dirty socks here on the bench – don't worry it's just pretend, you are all safe from that, thank goodness – and I flicked it up, and then let it land on the bench. Look here I've got something we can pretend is my ball of socks, that the technician sorted for me. Imagine we are filming this – here goes – okay no surprise there, the ball went up, then it came down. We can pretend we've got a really fast state-of-the-art camera that costs a fortune: so, what would we see if we pinned up print-outs of all the frames in order? What would correspond to our transition state in those stills? Do you think we could get an image of that point, at least near enough, within a tiny fraction of a second?

Now someone in one of last year's classes, I think they were trying to be clever, told me that we could not get an image of the ball at the very top of the flight. They said that that was impossible, because the ball goes up, and then it comes down, and it does not stop at the top so we could never get a picture of it there. Does that make any sense? ... [*letting students discuss the point – again this is mooted as an idea from someone who may or may not have an authoritative view*]. In fact, this 'clever clogs' suggested that the ball never existed at the top of the flight as if you calculated how long it was there the answer was one divided by infinity seconds, which is zero seconds. So apparently the ball passes from going up to coming down without ever being at the top? What do you think? Is that clever, or just trying too hard to be clever, and just silly? ...

I think the argument was that I could never get a picture at the precise top of the trajectory – only ever either just before whilst the ball was still moving up, or just after, whilst starting to move down, as even if I could take a frame every millisecond the point where the ball should be at the exact top of the trajectory was always between frames...if the best I can do is get a picture say 0.000 5 s before the ball stopped moving up, and another one 0.000 5 s after it starts moving down, does that mean there is not a point where the ball has stopped moving up and it's not yet moving down?...[*Here, the teacher is drawing an analogy between the abstract maxima on the reaction profile and another situation that is more familiar and that can be directly demonstrated with a simple prop. The teacher assumes that as students can see the 'ball', and indeed toss balls in the air themselves, they might be persuaded that the ball passes through a point at the top of the parabola even if it is difficult to capture that moment. Then between going up and coming down the ball must be doing neither, even if only for an infinitesimal time. Similarly, to get from reactants to products the system passes through a transition state even if that structure cannot be isolated.*] ...

Look, here I've got a marble track – actually it's just some plastic tracking they use for hanging curtains, but it works well with marbles. You will see it's shaped a bit like our reaction profile. Who likes going on those rides at fairgrounds designed to make you feel sick or scared? Is anyone in the class good at those rides – able to enjoy them without losing control of bodily functions? ... Good, come out here Mohammed, I have a challenge for you. This is a track for the kind of marbles who enjoy those kinds of thrills rather than the delights of doing chemistry – adrenaline junkie marbles perhaps? [*Here, anthropomorphism is being used, as marbles are treated like sentient beings with human ways of experiencing the world, as a means of developing a narrative that draws upon learners' own experiences.*]

Now I have Mohammed, our volunteer, who is an expert at fair rides. I want you to do something really simple. I want you to send this marble down our reaction profile shaped track, but not so hard that it flies off – can you do that. Can you move the marble through the profile so that is gets to the other end? Okay let's see. ...

Good, but that was just a practice run, because the real task is actually just a little bit more difficult than that. This time I do not want the reaction to go to the end. I want you to push the marble just hard enough to get to the top of this peak here, I know they give silly names to these kinds of rides at fairgrounds, so let's call the bump 'mount activation' and the very top 'peak transition state'. So, can you give the marble just enough energy to get to the peak, but not so much that is goes on to the end. I will be very generous – I'll let you have three attempts. I suppose if you succeed I should let you have a prize, like they do at the fair – perhaps I'll let the class off any homework this week if you can get the marble to stop on top of peak transition state... [*that most students will immediately see that it is not going to be possible to get the marble to stop at the maximum does not matter here: actually having someone try it out highlights the point by offering a concrete, material, analogue of the reaction profile. Attempts to succeed will only reinforce the principle that the maximum represents an unstable state.*]

... I thought you were an expert? Does anyone else think they might be able to do a better job and become a hero by helping the class avoid homework this week? ...

This is clearly a fictional account, and as this book is not a work of fiction, I will cease the account at this point. The teacher has more work to do in this lesson, and if the teaching is to be fruitful will find opportunities to return to and reinforce learning of this material over subsequent lessons – ideally over a period of weeks and months as that will support students in consolidating learning in memory.

I wonder what readers make of this account (as this will depend upon different reader's own experiences and so the interpretive resources they bring to make sense of the information in the text). My fictional teacher clearly has something of a jokey style and not all teachers would feel comfortable using humour as a pedagogic device – and indeed this may seem culturally inappropriate in some teaching contexts. Teachers do, however, develop teaching personalities that recognise teaching as a kind of performance (Tobin *et al.*, 1990).

This teacher attempted to draw on analogies and metaphors that could anchor the abstract concept to be learnt in terms of things and experiences that student would be familiar with. Most students know that a sequence of stills can be obtained by filming a moving object. Students know something (something that is denser than air, that is) thrown into the air will follow a trajectory that has a maximum, even if the object never stops at that point. Students will know about, and many will have ridden on, fair rides that have maxima, and this can be linked to concrete material model of a reaction profile that acts as a marble track. Many students will have experience of marble tracks, but – if not – have probably seen similar tracks used with toy cars. Students will appreciate it is virtually impossible to get a marble to stop at the top of the track (and if not, this will be demonstrated). There is such little friction that it is incredibly difficult to give the marble enough energy to reach the peak without it arriving with enough momentum to continue moving. This is theoretically achievable (if very unlikely in practice), but as a marble resting at the very peak would be in a state of unstable equilibrium if this was somehow achieved, the slightest breeze or vibration of the support would set it rolling again.

Lecturing was caricatured above as presenting information clearly, but not offering support for the interpretive processes by which that information is made meaningful. The teaching episode was quite different (and used much more teaching time) as the teacher was constantly making teaching moves (eliciting prior learning, introducing comparisons, showing a model, mooting a position for critique) designed to frame the information being represented, and to link it to the interpretive resources it was assumed the students could access, to facilitate sense-making for the students. The teaching was also more interactive: the learners were invited to undertake activities, and to give their opinions, to discuss points among themselves, and to contribute to demonstrations.

In principle, a good lecturer does teach, and uses many of the same techniques and devices. Teachers in school teach many lessons by using laboratory activities or group-work where students engage with various learning resources whilst the teacher moves around the groups. In a large, packed, lecture theatre it may be difficult to include that kind of activity (although lectures can be complemented by laboratory classes, and supplemented with examples classes and the like). However, lecturers can use dialogic teaching, even if the set-up of the room, or the cultural norms of the institutional context, make it difficult for students to move around or to do much of the speaking to the class.

The lecturer can intersperse the presentation of information with diagnostic questions to test prior knowledge; and advance organiser activities (Ausubel, 1978), such as activities ('scaffolding planks') that elicit relevant prerequisite learning and ask students to process it in a way that prepares them for new information to be presented in the lecture (Taber, 2018a); and students can be asked to consider and discuss questions in pairs or small groups without leaving their seats in the theatre. The lecturer can use the narrative device of what a previous (real or hypothetical) student said to present views that may, or many not, be canonical for critique. The lecturer can use models, diagrams, graphs, analogies, metaphor, simile, personification, narrative, and so forth. Perhaps the lecturer can offer a historically dubious story about the Napoleonic wars to engage students' imaginations and make a teaching point. (I was taught that the French army had tin buttons that changed allotropic form and crumbled away in the cold of the Russian Winter so the soldiers had to fight with one hand keeping their trousers up.)

In other words, the difference between lecturing and class teaching is a matter of degree. School chemistry teachers just entering the profession are often amazed at how little new material they can introduce in a single lesson. The amount of 'content' that might be dealt with in a paragraph in a university lecture can become the basis of a whole school lesson (as we have seen with the introduction of the {transition state} concept above). University lecturers are unlikely to have the luxury of time to develop ideas in the way our fictional school teacher did above. But they can use the same toolbox of teaching techniques to help engage students' interpretive resources, even if they need to use these devices more sparingly – which is justified when the students are more mature, are high-achievers, are highly motivated in the subject, have strong metacognitive knowledge, and strong study skills and habits, *etcetera*. The more students differ from such an ideal, the more the lecturer needs to work in a similar mode to the school teacher to help students acquire, retain, and develop authoritative chemical concepts. Even when students do come close to this ideal, the process of teaching still requires the lecturer to not only have good subject knowledge, but also a well-developed model of the interpretive resources the students bring to class. Otherwise the lecturer will be presenting information, but not sharing chemical concepts.

References

Ausubel D. P., (1978), In Defense of Advance Organizers: A Reply to the Critics, *Rev. Educ. Res.*, **48**(2), 251–257.

Herron J. D., Cantu L., Ward R. and Srinivasan V., (1977), Problems associated with concept analysis, *Sci. Educ.*, **61**(2), 185–199.

Kind V., (2009), Pedagogical content knowledge in science education: perspectives and potential for progress, *Stud. Sci. Educ.*, **45**(2), 169–204.

Kuhn T. S., (1970), *The Structure of Scientific Revolutions*, 2nd edn, Chicago: University of Chicago.

Leach J. and Scott P., (2002), Designing and evaluating science teaching sequences: an approach drawing upon the concept of learning demand and a social constructivist perspective on learning, *Stud. Sci. Educ.*, **38**, 115–142.

Mercer N., (1995), *The Guided Construction of Knowledge: Talk Amongst Teachers and Learners*, Clevedon, Somerset: Multilingual Matters.

Scott P. H., (1998), Teacher talk and meaning making in science classrooms: a review of studies from a Vygotskian perspective, *Stud. Sci. Educ.*, **32**, 45–80.

Scott P. H., (2007), Challenging gifted learners through classroom dialogue, in Taber K. S. (ed.), *Science Education for Gifted Learners*, London: Routledge, pp. 100–111.

Taber K. S., (1997), Student understanding of ionic bonding: molecular versus electrostatic thinking? *Sch. Sci. Rev.*, **78**(285), 85–95.

Taber K. S., (2002), *Chemical Misconceptions – Prevention, Diagnosis and Cure: Theoretical Background* (vol. 1). London: Royal Society of Chemistry.

Taber K. S., (2005), Developing Teachers as Learning Doctors, *Teacher Development*, **9**(2), 219–235.

Taber K. S., (2009), Learning at the symbolic level, in Gilbert J. K. and Treagust D. F. (ed.), *Multiple Representations in Chemical Education*, Dordrecht: Springer, pp. 75–108.

Taber K. S., (2010), Straw men and false dichotomies: Overcoming philosophical confusion in chemical education, *J. Chem. Educ.*, **87**(5), 552–558.

Taber K. S., (2014), *Student Thinking and Learning in Science: Perspectives on the Nature and Development of Learners' Ideas*, New York: Routledge.

Taber K. S., (2018a), *Masterclass in Science Education: Transforming Teaching and Learning*, London: Bloomsbury.

Taber K. S., (2018b), Scaffolding learning: principles for effective teaching and the design of classroom resources, in Abend M. (ed.), *Effective Teaching and Learning: Perspectives, Strategies and Implementation*, New York: Nova Science Publishers, pp. 1–43.

Taber K. S., (Forthcoming), Experimental research into teaching innovations: responding to methodological and ethical challenges.

Tobin, K., Kahle, J. B., and Fraser, B. J. (ed.), (1990), *Windows into Science Classrooms: Problems Associated with Higher-level Cognitive Learning*, Basingstoke, Hampshire: Falmer Press.

Vygotsky L. S., (1978), *Mind in Society: The Development of Higher Psychological Processes*, Cambridge, Massachusetts: Harvard University Press.

CHAPTER 13

How Do Students Acquire Concepts?

This chapter offers a brief account of the formation of concepts. It reports some current thinking on this topic, including some of the key theory. It begins, however, with the beginning of a short vignette based on the author's own experience in writing this book.

13.1 A Personal Conception of Positive-ray Spectrum: Implicit Concept Formation

It was reported in Chapter 5 how Thomson discovered that neon had isotopes when investigating the positive-ray spectrum of neon (Harkins and Liggett, 1923). The authors of that account use a concept {positive-ray spectrum}, which they seem to assume is meaningful to readers. On coming across that reference, I did not have an active concept {positive-ray spectrum}, but did not initially notice a lack of understanding. The term was incidental to the information I was seeking in reading the paper, and I vaguely made sense of {positive-ray spectrum} as referring to the energies of radioactive emissions from neon. It seemed viable that neon could have radioactive isotopes and so the term 'made sense'.

Before yesterday, as far as I am currently aware, I had no concept {positive-ray spectrum}. Yesterday when I read about the positive-ray spectrum in the paper by Harkins and Ligget (written a decade before Frédéric and Irène Joliot-Curie reported positron emission as a form of induced 'artificial' radioactivity), I interpreted the term 'positive ray' as likely suggesting positrons, and so referring to nuclear emission. In effect, I formed a concept {positive-ray spectrum} from my reading, by drawing upon my existing

---

Advances in Chemistry Education Series No. 3
The Nature of the Chemical Concept: Re-constructing Chemical Knowledge in Teaching and Learning
By Keith S. Taber
© Keith S. Taber 2022
Published by the Royal Society of Chemistry, www.rsc.org

knowledge and, in particular, associations for the term 'positive ray'. This effectively happened tacitly, as my reading was focused elsewhere, and although I was aware this was not a familiar term, I did not stop to reflect upon what it meant, as I immediately formed a vague idea of what it *could* mean that satisfied my immediate purposes – and so I carried on reading. (To be continued...)

13.2 Three Levels of Description

There are different levels at which conceptual learning can be described. Three of these levels can be labelled as behavioural, processing system, and physiological–anatomical. Describing learning at these levels can provide complementary accounts – that is accounts at these three different levels describe learning from different viewpoints and need not be seen as competing. A more detailed discussion of the issues raised in this section can be found in an earlier book (Taber, 2013).

13.2.1 The Behavioural Level of Description

The sense in which we might usually feel we can observe learning concerns behaviour. A student taking an examination (a school test, a PhD *viva voce* examination, *etc.*) or being evaluated in the classroom is judged in terms of certain behaviours. The student: makes a verbal presentation, builds a model, calculates the rate of a reaction, infers the metallic cation present in an unknown sample, draws a labelled diagram, and so forth.

If a question in a test asks 'what shape is a molecule of ammonia' and one student writes 'pyramidal' and another student writes 'tetrahedral' then an examiner will likely draw different conclusions about their knowledge from how they behaved in response to the question. The behaviour (here, what was written) is indicative, but may not lead to valid inferences. We can imagine a student who has not developed an appropriate {molecule} concept, nor anything more than a vague {ammonia} concept but has learnt the association 'ammonia-pyramidal' in rote fashion. The behaviour (giving the correct answer) may indicate rather limited understanding. It is even possible that a student is simply using rather limited cues – that they had studied past papers and learnt that often when there was a question with the word 'shape', the right answer was one of 'linear', 'angular', 'tetrahedral', 'pyramidical' and 'square planar', and so they picked one of these options and wrote it down.

It is also possible that the student who wrote 'tetrahedral' knows that ammonia is said to be a pyramidal shape, but was very nervous in the test, and read methane for ammonia; or saw the question was about ammonia, thought about the electronic configuration, pictured the arrangement of electrons pairs around the nitrogen, imagined what shape the four atoms would make up, realised this was pyramidal shape, went to write 'pyramidal' but somehow wrote 'tetrahedral' instead. We all occasionally experience 'slips of the tongue' and of the pen or typing fingers.

The behavioural description treats learning as a black box. The behaviourist school in psychology (Watson, 1924/1998) went as far as to suggest there was no value in accounts of learning in terms of mental events and processes, as these could never be observed, and a scientific account should limit itself to what can be observed and independently corroborated. (Consider the discussion of objectivity and subjectivity in Chapter 11.) Even if we reject such a perspective, we all, necessarily, use behavioural evidence in evaluating the knowledge and understanding of others.

Behaviour can be misleading, but this is not to dismiss this level. If these imaginary students were asked to explain the basis for their answers then there would be a more revealing set of behaviours on which to make an evaluation. The PhD student who has prepared a research thesis, including results that have been published in several peer-reviewed research journals of high repute, and has given answers to probing examiners' questions about his work in a two-hour in-depth conversation, has produced sufficient evidence through her behaviours such that expert examiners are unlikely to make a gross error in inferring the level of conceptual understanding developed.

As behavioural evidence may lead to false inferences, there is another problem evaluating learning from teaching through observing such behaviours. Learning is a change in the potential to behave in some way (which may simply mean the potential to report that ammonia has a pyramidal molecule, or may mean something more elaborate and substantial), so to evaluate learning we have to show that a learner can now behave in a way they could not previously (Taber, 2009). However, it is only possible to observe behaviours, not *the potential* for behaviour, so failure to observe a behaviour before teaching is not definite evidence that there was not already potential for that behaviour. Our nervous student may not have learnt anything new from teaching, but perhaps mistakenly wrote that ammonia was tetrahedral in the pre-test, but then did not make the same slip in the post-test, so learning appears to have occurred. Alternatively, she could have made the slip in the post-test and not the pre-test – which, if taken at face value, might suggest that teaching has undermined earlier effective prior learning.

We need to use observed behaviours as evidence of students' conceptual understanding, and of their conceptual development (see Chapter 15), but we also need to keep in mind that the process of drawing such inferences is fallible. We can reduce incidences of errors of inference by observing more of a learner's behaviour before reaching conclusions, and by framing the behaviour we observe. If we are interested in finding out about students' understanding of how electronic structures of molecules impact on molecular shape (rather than if they know the shapes of some common molecules) then we should elicit behaviour by a question requiring an explanation rather than simply asking for the assignment of shapes.

13.2.2 The Physiological–Anatomical Level of Description

Learning produces a change in potential through a physical change in the structure of the learner. This seems clear, and moreover it seems that these changes are synaptic changes. The brain processes information through neural nets. The brain contains a vast array of highly connected neurones. When one nerve cell is activated, it sends an electrical signal to other neurones. Depending on the specifics of the connections to those other neutrons, that is the synapses, the signal may change the activity of those other neurones: if the incoming signal is strong enough they may be triggered to fire themselves or their activity may actually be reduced – that is, their firing may be inhibited.

Although the details of how neurones and their connections function is understood in increasing detail, how this links to actual conceptual understanding is still quite vague. The subjective experience of understanding something (and the objective behaviours this might facilitate – such as giving an oral explanation for why the ammonia molecule is pyramidal) is considered to be supported by the way learning is represented in the synaptic connections in the cortex. Learning the valence-shell electron-pair repulsion theory involves changes to some of the synaptic connections in the brain – but we are very far from being able to find direct correlates. We cannot show that some particular synaptic connections were changed on learning VSEPRT, and, indeed, we do not really have a clear idea of how many synapses would be involved, nor how localised or distributed the changes in the brain would be.

There is an intriguing analogy between how the concepts themselves are embedded in networks linked with various direct and indirect associations of different strengths (see Chapter 2), and the idea of a neural net of brain cells with a (superficially, at least) similar structure: but we should not be tempted to map concept nodes (*e.g.* {potassium}) onto neutrons and associations linking concepts (*e.g.* 'potassium is a metal') onto individual synapses. The scale of the neural apparatus that correlates with a concept or a specific conception is likely to be considerably greater than single cells and their connections.

Somehow, synaptic changes, which are in effect the anatomical representation of past mental experiences in memories, facilitate changes in thinking that change the potential for behaviour. A living human's brain is undergoing a continuous modification of its structure at the synaptic level, partly due to new experiences, and partly due to ongoing maintenance processes (some to repair damage, some to consolidate learning).

For a natural scientist, there is something reassuring to think that we understand something of what is going on in terms of material changes that correlate with learning. Yet, despite much attention to neuroscience and how it might inform education (Goswami, 2006), our models of how knowledge is represented in neural networks currently offers limited support for teachers of chemistry (or other subjects) – there is a vast chasm (rather than a tiny synaptic cleft) between our models of the synaptic change underpinning learning and the actual behaviours that teachers are observing as evidence for student knowledge and understanding. Perhaps the relationship between the synaptic changes and the observed behaviours is analogous to the relationship between reaction mechanisms at the molecular level and the reactions observed at the bench: but, so far, we have much stronger theories to connect levels in the latter (*i.e.*, reactions) than the former (*i.e.*, learning) case.

In Chapter 2, I introduced the notion of the 'congenst' (the {congenst} concept) – where the congenst is a concept-generating structure that resources a person's thinking. The congenst is part of the representation of past experience in the physical structure of the brain. The congenst is not a concept, but provides the substrate for conceptual thinking. However, that thinking will also depend upon the general state of activation of the brain (both in terms such as mood, level of fatigue, and also in terms of what the person has recently been thinking about and so the relative activation of different resources) and the information being processed due to external stimulation. When a person seems to apply a concept in a different way to previously, this could be the unchanged congenst being applied in a different (internal cognitive, and/or wider external) context, or could be due to the underlying congenst having itself been modified.

We might ask someone "what do you consider the most characteristic property of a metal?", and they might reply "it will conduct electricity". On another occasion, we might ask the same person the same question and instead be told "it will tend to form cations". This might reflect conceptual change, and some substantive modification of the underlying congenst – or could instead just reflect the activation of basically the same underlying structure under different conditions. Perhaps these questions were asked of a student in a high-school chemistry class, and on the first occasion the student had just come from a physics class where electrical conductivity was being discussed and the conductivities of different materials calculated, but on the second occasion the student had just come from a biology class looking at the absorption of micronutrients from the diet.

The teacher (or researcher) asking the question will not necessarily be aware of the wider context that might modify precisely what is brought to mind when a person is asked a question: in particular, a person's recent subjective experiences. If we ask the same question at different times, and get different answers, we cannot immediately discriminate between whether this is activation of a *modified* underlying anatomical structure, or differential activation of essentially the *same* structure. Congensts are constantly being modified – it seems that each time they are activated this leads to slight shifts in synaptic strengths throughout the network. The conditions in which some thinking (such as considering a response to a question) takes place are never completely reproducible: if, as the proverb suggests, we can never step in the same river twice, nor perhaps can we ever think (exactly) the

How Do Students Acquire Concepts?

same thought again. This clearly complicates research into conceptual change – and indeed the educational tasks of teachers and their students.

13.2.3 The Information Processing Level of Description

The third level to be considered examines cognition as a process that occurs in an information-processing system with specialised components. Those components are not on the scale of individual neurones and synapses, but relate to functional aspects of how information is processed during cognition. It is assumed that these functional components are correlated with actual anatomic structures (which may already be known or assumed), but that is not the focus at this level.

Data enters the system, and is moved about, acted upon, perhaps in some sense stored, and may lead to output. This level of description is closer to our usual way of discussing mental activity than the other two levels considered above, but – as discussed below – should not be equated with it. Consider a description at this level when a teacher asks a question of a student, in terms of the cognitive system of the student. An input transducer (*i.e.*, ears) converts information from sounds into information that can be handled in the system (electrical pulses). That information is processed such that the original sound becomes interpreted as speech (and those with computers that are meant to respond to our voice commands know this is not a trivial matter), and then moved on to other parts of the system where it can be processed into a form suitable for conscious attention. The outcome of that processing will be sent to working memory that acts as a kind of executive module. This will trigger the search for and accessing of particular representations in long-term memory; and then working memory will process the information from these two sources (perception + memory) to produce some kind of output – this may be formulated in language, and instructions sent to an output transducer (in effect the larynx) that will generate sounds, encoding information in language.

Generally, within the system, information is interpreted, compared, represented (remembered), accessed, integrated, modified, *etcetera*. Some of that processing is initiated by new information received from outside the system. Some of the processing is motivated internally, and is part of on-going processes that act to organise and refine the system to make it more effective at processing new information. This may include setting up new associations between stored representations, and making edits in stored information – changes that have no obvious immediate trigger in terms of current sensory input (and perhaps take place during sleep).

13.2.4 The Mental Register (Revisited)

Generally, we tend to discuss cognition in terms of a mental register (See Chapter 1) that is used in everyday life as well as by professionals in areas such as education (Taber, 2013). So, we think of a student hearing the teacher's question, and *understanding* it, and *thinking* about it, in terms of prior *learning* represented in *memory*, and suggesting an answer. People *think, imagine, dream, remember, consider, believe, etc.* (But to do this they rely on a system that can process information.)

So, conceptualisation can be treated as a black box that generates behaviours, or it can be understood at one level in term of information processing in a system, and at another level as reflecting synaptic connection strengths and electrical activity in neural circuits – even if these are not the most helpful ways for teachers to think about cognition (Taber, 2013). In education, it is usually more productive to think less about brain activity, than in terms of the mind. The scientific perspective on mind is that it is an emergent property of complex brains, such that mind relies upon the anatomic structures of the brain, but it may be studied in its own right (thus, the existence of psychology).

When Kekulé had his vision of the snake biting its tail, and the insight of the benzene molecule having a cyclic structure, this was his subjective experience – but correlated with information being processed in this brain. When a student looks at an image showing some lines and marks on a flat page, and 'sees' this as representing a pyramidal ammonia molecule, this is a subjective correlate of processing of information in the central nervous system: processing that occurs at various levels involving the retina as well as various structures in the brain itself (*n.b.*, the retina is sometimes considered to be functionally part of the brain). When a student pours some solution from a conical flask to a boiling tube, this behaviour is correlated with (and it might be suggested, caused by) some information processing in the brain. When a student is sleeping, and learning is being consolidated, then this is a matter of information processing in the brain (which leads to changes in neural circuits through modifications of synaptic connections – these descriptions are complementary).

So, we have different ways of describing cognitive processes. In everyday life we talk of learning, forgetting, understanding, and so forth: but in analysing these events in education it can be useful to use complementary ways of describing what is going on – and often the information processing level is helpful (Johnstone, 1989). However, there is a major impasse. Despite the great attention that has been given to consciousness both by philosophers, and more recently by neuroscientists, we still do not really understand how to bridge between the information description and the subjective experience. We can use an information processing perspective to explain a behavioural description by mooting models of what must be going on in terms of functions (data input, information retrieval, *etc.*), and we can reasonably assume that *in principle* a valid information processing description could eventually be matched to a description in physiological-anatomical terms. We are very aware that most of the information processing that goes on in thinking occurs at a preconscious level, and only a fraction reaches conscious awareness. But quite how to best explain the subjective experiences of thinking, understanding, being confused, being convinced, *etcetera*, remains a mystery at this time. This should be borne in mind (for example, in reading the rest of this chapter, and indeed Chapter 15) as this means our knowledge of learning and other cognition remains deficient in a very important sense.

13.3 Concept Formation

We form concepts largely without reflecting on the process. In Chapter 8, it was suggested that a chemist can *see* salt dissolve in water to form a solution where a novice will see salt disappear. This is not simply suggesting that they *describe* the process differently (perhaps because of a difference in vocabulary) but that it is actually perceived and conceptualised differently. We are not able to by-pass our conceptual structures and observe with naivety – as a result of the experiences that led to our expertise, the structures of our brains have changed so that sensory information is interpreted differently. A chemist and a novice learner may look at the same symmetrical hexagon pattern, but the expert will *instantly* 'see' a benzene ring structure: not a diagram that on reflection and after some deliberation, they can make sense of as representing such a structure. The representation is being interpreted, but, in the expert, this occurs preconsciously – it has become automatic, perhaps as automatic as seeing a tree.

13.3.1 Implicit Learning in Everyday Life

Supposedly, when explorers first met with members of indigenous populations that had no connection with modern technology the local people were unable to make sense of photographs they were shown by the visitors – they had no experience that allowed them to interpret the images. We learn to decode representations. There is a story that Picasso was approached at a showing of his work by a man who commented that Picasso's representations of women did not look anything like real women. The man took a photograph of his wife from his pocket to show Picasso how a woman really looked. Picasso asked if the woman was really like the photograph, and, when this was confirmed, commented on how small and flat the man's wife was. (This may well be as apocryphal as the disintegration of Napoleonic uniform buttons referred to in Chapter 12, but does useful pedagogic work as a parable.)

We *learn* to see what is represented in photographs, and to recognise objects and events represented on movie or television screens. A baby may be transfixed by the patterns on the television screen, but has to learn that they are representations of something else. In the language of film, cameras pan and move to close up, in a way that would surely be disconcerting, if not disorientating, to someone who had never before seen such a representation. Films originally shot in monochrome, now usually only used for making an artistic statement, or to suggest a different level of reality for some scenes, may seem to lack realism to modern audiences, but the author remembers when television programmes were only be transmitted in 'black and white' and when this was the norm, it did not stop viewers engaging with what was represented. Indeed, discovering that some films familiar from being seen on television as a child were actually made in colour (or seeing modern digitally coloured versions of familiar classic monochrome films), may initially be slightly disconcerting. Perhaps in a future where all such entertainments are based on an immersive multimedia experience, current state-of-the-art high-definition television shows with hi-fi surround soundtracks will lack realism to those not used to an image being presented within a limited 'window' within a familiar room.

We (that is, we who are inducted into the cultures that use these forms of representation) all learn these conventions and automatically appreciate, for example, that the camera is moving on our behalf, but if this was a new experience it could present problems in interpretation as visual data indicates we are moving through space, when proprioceptive data tells us we are stationary. We have learnt to conceptualise this in a different framework, such that we sense no contradiction or confusion – just as people who wear special glasses that invert their vision (so everything is upside down) adapt, and come to function, not by coping with the world appearing to be upside down, but by (over a few days) learning to see the world the right way up regardless (Kornheiser, 1976).

For many viewers today, the works of painters of the French Impressionist school clearly and obviously represent – this is a pond with water lilies, say – and it is difficult to appreciate how revolutionary their style was when their works were first exhibited to the public. Beethoven is perhaps one of the most recognised of Classical composers – but again his work must have initially been very difficult to engage with at the time it offered the shock of the new. This is something that often causes inter-generational conflict when parents cannot understand why their children play music that seems little more than 'noise'.

Any new style that one is not familiar with initially *is* noise (literally, that is, not just metaphorically) until one becomes familiar with its patterns. When I first heard the record of the Concert for Bangla Desh (by George Harrison and friends) I could make little of the opening piece in classical Indian style, by Ravi Shankar, and was relieved once this was 'out of the way' and the more familiar pop/rock tracks played. Many years later (having in the interim listened to music by groups such as Shaki, that incorporated Indian styles) I saw a televised BBC 'Prom' concert where Anoushka Shankar (one of Ravi's daughters) played the suite 'Passages', co-written by her father and the American ('minimalist') composer Philip Glass, with other Indian musicians and a chamber orchestra. The Indian styles and instrumentation were by this time much more familiar. I was so impressed that the next day I bought a copy of the original recording. Readers can probably bring to mind their own non-chemical examples of how they learnt to perceive meaning and order in initially unfamiliar and seemingly chaotic contexts.

How Do Students Acquire Concepts? 283

Professional chemists, including chemistry teachers, are likely so familiar with the conceptual basis of our subject that we may find it difficult to remember how complex it must appear, with its various conceptual nuances, to novice learners. If so, perhaps we can instead call upon how we have learnt to perceive differently in other areas of life – whether that be the first time of viewing a Picasso painting, or the first taste of curry, or the first time hearing the music of Stockhausen – some domain where we can recall moving from the 'shock of the new' to a level of familiarity that allows us to experience the 'same' phenomenon in an entirely different way.

13.3.2 Studying Concept Formation

Psychologists have studied concept formation in laboratory studies (Howard, 1987). Often, in these studies, an artificial concept is created for the research to avoid the experimental subjects (as participants may be considered in such studies) having prior experience of the concept. For example: the target concept could be of (any) two-dimensional shape with three corners that is not drawn in red. This could be called a balvoid (or some other name intended to give no obvious clue). Subjects may be asked to suggest which of a set of offered figures represents a balvoid, and be given feedback on their guesses. With time the subjects will get better at discriminating balvoids form non-balvoids – to the point where their discriminations of previously unseen figures have high accuracy, and this may happen before they can offer any verbal description of the characteristics of balvoids. That is, they may be operating with an effective balvoid concept even before they are consciously aware of the formal criteria used to make discriminations. As the brain operates as an extensive neural network this should not be surprising: electronic models of neural nets can be 'trained' to undertake such tasks without using any formal algorithms reflecting explicit criteria by providing feedback on performance.

Psychologists can undertake studies that are microgenetic (Brock and Taber, 2017; Siegler, 2006), which means they are taking observations at a fast-enough rate to observe the changes in conceptualisation. This is possible in an artificial concept acquisition task, such as forming a balvoid concept. Whilst such work is of interest, normal concept formation occurs in a much less controlled social environment.

13.3.3 Inherent Pattern Recognition

Humans start learning concepts from a very young age, even before they have natural language to enter into conversation with others. Studies with babies very soon after their birth show that even new-borns are not completely 'blank slates' (Goswami, 2008). One example is that there seems to be an innate bias in the cognitive system to recognise faces (something clearly of an advantage to a baby relying for survival on forming a bond with its mother). Young babies have also been found to have a reluctance to crawl across glass shelves. There seems to be an innate sense that one should not venture where there is no visible source of support. Of course, young children are not thinking in these terms, that is, thinking of 'no visible source of support', using verbal descriptions. This is more like an instinct such as the instinct that leads to parent birds feeding chicks when they see a particular visual cue (that we would describe as the open beaks of chicks – but the behaviour can be initiated with pretty crude imitations of the natural stimulus).

Babies are also thought to have expectations about some aspects of the physical world. Studies suggest that babies show surprise if an apparently solid object disappears – say, if it is clearly placed under a cushion, but is not there when the cushion is lifted. The surprise is not communicated as such, and is inferred from indirect evidence (time spent staring exceeding that spent gazing at unsurprising events). This can be interpreted as the baby having expectations about how the world behaves – but again we should be aware that these 'expectations' are the external observers' inferences about the internal subjective experience. No-one knows how a baby feels when a surprising event happens or how they experience their expectations when nothing surprising is noticed (but see below, about p-prims). Here, we are inferring cognition from behaviour (see above), over which is laid an interpretation in terms of our own subjective experiences – what it is to have expectations about how things work, what it is like to be surprised when something unexpected happens.

Regardless of the extent to which the human cognitive system is truly inherently biased – for example to learn to see faces, or expect constancy in aspects of the physical world – people seem to start to abstract from experiences at a very young age. The cognitive system (the brain, with its sensory 'transducers') appears to have a strong capacity to notice similarity in experiences, and build implicit knowledge elements from these abstractions.

13.3.4 Phenomenological Primitives (P-prims)

These implicit knowledge elements are sometimes known as phenomenological primitives, or p-prims for short (diSessa, 1993). It is important to emphasise that although referred to as knowledge elements, p-prims operate at a preconscious level (Taber, 2014). A person is not aware they are abstracting experience to form p-prims, and is not aware that they have particular p-prims. P-prims are considered to be significant in cognition, but are not open to direct introspection. In this sense, they are different to formal concepts, or what might be termed 'explicit' concepts.

A reader of this book can reflect on their personal {atomic nucleus} concept, say, which is an explicit concept. At least part of the representation of that concept is accessible to introspection. Perhaps the reader can bring to mind an image of an atomic nucleus, or a definition. Certainly, the reader can produce a verbal account of aspects of their concept – a set of statements linking an atomic nucleus (in abstract) with mass, charge, their {chemical element} concept and their {isotope} concept, and so forth. Indeed the reader could, if they wished, produce a concept map (Novak, 1990) demonstrating manifold associations they have for the concept.

We are not aware of our p-prims, so cannot tell anyone else about them (as we might talk about our concept of the atomic nucleus), although it is possible to infer they exist from indirect evidence. Indeed, the physics educator Andreas diSessa (1993) undertook extensive interviews with students from which he inferred the presence of a wide range of p-prims that seemed to be common among his study participants.

As they are abstractions of commonalities about past experiences, p-prims act as expectations about the nature of the world. If whenever a baby drops an object, it falls down, then the baby comes to 'expect' this to happen. If they 'dropped' a helium filled balloon, or dropped a toy boat in the bath, they may be 'surprised' at this event as it did not fit with the expectation based on past experience. It is important to emphasise that expectation need not mean a prediction made consciously, but refers to something about the state of part of the cognitive system. The young child would not usually 'notice' things falling after a while, but if something slipped out of their hand whilst distracted would look for the object on the floor. We can say the child would 'know' to look on the floor, but again this is not explicit knowledge in the young child. The child looks on the floor (behaviour) but does not deliberate on the situation that has arisen and make a conscious decision to look on the floor.

If something happens in a way that seems contrary to the expectations provided by past experience then the (information processing) system is likely to bring this to conscious attention as something worthy of notice. Although we cannot fully put ourselves in the place of a young, pre-verbal, child, we might recognise experiences we have had that might be comparable. I certainly recall occasions when I have dropped something, and having crawled around the floor for a while find myself with some 'cognitive dissonance' when I believe I have thoroughly inspected all places where the object could feasible have fallen or rolled, but without finding it. The internal subjective feelings I experience (as I mutter to myself 'it *must* be here, somewhere') are perhaps not so different to those of an infant surprised when a psychologist has tricked them by making a toy seem to vanish into thin air.

P-prims cannot be considered as *chemical* concepts here, but are something quite different. But if educational theorists are correct, there is not an absolute distinction between p-prims and formal concepts of the kind taught and learned in chemistry – I discuss some examples below. P-prims are described as 'implicit' knowledge elements due to their inaccessibility to introspection. It is not surprising that the cognitive system includes such elements. Information from the visual field is processed to identify edges of objects and the like, and the action of these system components is not available to our conscious thinking – indeed perception of the world is usually experienced in terms of things and events, not patches of colour and movement that we have to consciously decide is a bus or a conical flask.

The processing, the interpretation of data, has (nearly always) occurred by the time we are aware of seeing something. The final outcome of a series of processing stages is available to us consciously, but the processing itself is all preconscious.

If we are in the street and pass strangers talking in a language we are not familiar with, we hear sounds and probably realise it is language; but when we hear conversations in our own language we automatically hear the words as the preconscious processing has identified them. For a new born baby, there would just be noise – William James's (1890) blooming buzzing confusion – and all talk would initially just be undifferentiated sound. However, the brain is able to build up circuits to recognise words in English (or Spanish, *etc.*) that discriminate sounds into words, and can become very successful in dealing with different voices, accents, and listening conditions. The brain abstracts patterns from the environment and constructs system components to scan aural input to detect speech and then interpret it as streams of words.

In early life, then, our hearing becomes tuned to discriminate the sounds common in the language or languages we hear – the system uses experience of hearing talk to abstract common patterns and use these as the basis for expectations about future talk. This usually works well, at least until we decide to learn a language that uses a different set of sounds (phonemes), at which point we may have great trouble hearing (what to us are) subtle differences that may sound obvious to someone exposed to that language since early in life. The construction of interpretive apparatus tuned to detect words in one language enables and privileges a particular way of discriminating sounds, and this actually changes the way sound is heard. In effect, the signals arriving in the system are tidied up to better fit the set of expected sounds of our native language (and so become treated differently in speakers of different languages).

This is analogous to computer software that can sharpen up low-resolution images: doing so is not magic, it involves applying an algorithm to the data actually available, to in effect predict other data that has not been collected. The picture may look sharper, but this is an artificial process and detail missing from the original data set that cannot be predicted from the data actually collected will still be missing. For example, we might imagine a poor-quality spectrum that has very noisy lines that are only able to show the most prominent peaks. It would be possible to tidy the spectrum up to give a clean smooth line where the peaks stand out well from the noise. This might look better in a student's lab notebook. However, such a tidied spectrum would lack small details (of chemical significance) that a trained observer would expect to see at the apparent resolution of the smoothed spectrum. In this situation, the cleaner image could actually be misleading – as instead of a judgement (i) that there might be a small peak at such-and-such wavenumber, but the noisiness of the data makes it difficult to be sure, an observer may, when looking at the smoothed line, conclude (ii) there is no such peak. In a similar way, the brain of a person who has learnt one language

How Do Students Acquire Concepts? 287

hears an unfamiliar, very different, language in a way shaped by the data-tidying apparatus built upon expectations deriving from their own familiar language.

The brain then has systems to recognise commonalities in data, which are used to build expectations into the system to allow the more effective processing of information – perhaps like identifying some people arriving at a country who can be allowed to enter freely (perhaps their passports show they are citizens of that country) without going through detailed checks that may be applied to others. This process does not just apply to early life when we first learn languages, but seems to continue to operate at different levels. I am aware I have to be careful about the spelling of certain words (for example I often seem to hear 'Istanbul' as 'Instanbul'). Presumably I misheard some words when I first came across them, or they were poorly pronounced at my first introduction, but now this is how I tend to hear them. This may explain why a student I taught talked about how, in atoms, electrons were arranged in shields – I suspect he was actually hearing 'electron shield' when others referred to 'electron shells'. I only recognised this when I replayed recordings of interviews with him, as I tended to hear his 'electron shields' as 'electron shells' until I noticed the discrepancy – after which it became easier to appreciate he was consistently using the alternative term. One could imagine that without this having been identified through the specific focus on the precise language used when analysing research interviews, this could have continued unnoticed. One can imagine a hypothetical discussion between two students each hearing what they expect to hear:

"So, the sodium atom has three shields?"
"Well, it has three shells with electrons, yeah."
"And the third shield just has one electron?"
"Yeah, and the electron in that shell is missing in the ion."
"So, is the third shield still there when the atom is ionised?"
"What, like an empty shell. I'm not sure."
"Yeah, can you have a shield just there in case it is needed for an electron? Or is there only a shield when there is at least one electron in it."
"Dunno. If you can have empty shells, then I guess all atoms have them."
"Yeah, so then sodium would have a fourth shield, even if it was never used."
"So, we need to know if the shells are shells of electrons of shells for electrons to go into."
"Yeah, is a shield made of the electrons, or is a shield part of an atom that does not need an electron to exist?"
...

The (bootstrapping) process by which the brain uses information from the environment as the basis for developing its own apparatus for more efficiently processing future information does not just apply to perception. P-prims develop that reflect commonalities in various experiences. One should be careful in describing processing that is preconscious using language that we normally use to refer to deliberate thought, but the p-prims can be considered to represent ways in which we expect future experience to unfold, and so comprise a set of resources for making sense of that experience.

13.4 Two Systems of Knowledge

There is a contrast being made here between two modes of cognition. One mode is conscious and deliberative, and the other is preconscious and automatic (Claxton, 1997; Evans, 2008; Kahneman, 2011). We are aware of using the former mode when we solve problems, and plan, and reflect on our experiences. We become aware of the latter mode when we have a sudden moment of insight that seems to arrive in consciousness as if from nowhere – a 'lightbulb moment' (*i.e.*, as if someone had turned a light on in a very dark room) or when we have a strong feeling about something that we cannot rationally justify.

We might think of the deliberative mode as a 'higher' form of cognition that is presumably not available to many non-human animals (presumably earth worms do not deliberate on things, and only have an automatic mode of cognition). However, we should not dismiss the importance of preconscious processing in human cognition. As is suggested above, in perception, preconscious stages of processing are necessary to allow conscious deliberation: as by the time we are aware of something, it has nearly always been meaningfully interpreted. Indeed, most sensory input is processed preconsciously and never reaches consciousness – the system, in effect, filters out most input as not needing to be addressed by the executive!

Without this, we could not function, as (a) we would be dealing consciously with raw data that did not make any sense, and (b) we would be continuously distracted by a vast amount of sensory (including from proprioceptors) data – relating to the light levels, objects in the visual field, the position of the limbs, ambient noise in the environment, *etcetera*. Instead, most 'input' is processed and judged as not worthy of attention (we notice a distinct change in light level, a sudden noise, an unexpected movement – not a constant update on the *status quo*), and what is presented is already interpreted. The system develops biases to process information preconsciously, and is selective about what is brought to conscious mind. These biases do not always work perfectly for us – for example when sitting concentrating on work, or a good book, in a room that is very slowly getting colder, we may not notice we are feeling cold until after it would have been wise to take action. In general, though, these biases make our mental lives possible.

It is also important to appreciate that deliberative thinking is not always superior and 'higher level'. The automatic model of thinking works far quicker, and allows us to respond to threats or opportunities that do not offer the leisure for reflection and planning. But even in scientific work, the deliberative system needs to be complemented by the automatic system. Polanyi (1962) emphasised how in scientific work the scientist develops expertise that is tacit, and cannot be fully described or justified. Scientists make use of insight (Brock, 2015). That is, experts in a field develop a level of insight not available to novices (Cianciolo et al., 2009).

Scientists deliberately involve the automatic system in problem solving. When faced with particularly challenging problems, it is useful to explore the issue consciously, but when no immediate solution is apparent it may be sensible to move on to concentrate on something else – and allow the preconscious system to work on the problem in the background (Keller, 1983). It is not unusual for the breakthrough (that is, the insight in consciousness of a solution arrived at preconsciously) to appear whilst a scientist is doing something quite different, and indeed often relaxing (walking, taking a bath, etc.). In reporting their work, scientists have to present a logical argument for their findings – but the route to a solution often arrives as an insight, a creative moment (Taber, 2011), which then has to be checked and supported by logical analysis of data.

13.4.1 Spontaneous Concepts

This raises an issue of how we understand 'thinking' and indeed concepts, if, indeed, much human cognition goes on outside of the field of awareness of our consciousness and without deliberate oversight. Knowledge elements such as p-prims are not deliberately engaged by a thinker, and they are not open to reflection – they are considered 'encapsulated' units in cognition – like little black boxes whose operation is not visible and that cannot be deliberately modified. Yet, they are involved in making discriminations, in making sense of experience as fitting patterns of past experience.

The term 'spontaneous concepts' has long been used to refer to concepts that it is considered are developed by a person from their direct experience of the environment without mediation by a teacher. Famously, the developmental psychologist Jean Piaget (1970/1972) studied how children below a certain age failed in particular conservation tasks, that older children seemed to naturally understand. If several similar sticks were aligned on the table the young child will agree they are the same length, but moving one out of alignment (even though observed by the child) changes this judgement. A perceptual cue suggests one is longer, and the child does not seem able to appreciate that the sticks will not have changed length, such that if they were the same length before the translation then they must still be the same length. Parallel results are found across a range of tasks: two matched lines of checkers or coins will be seen as having the same number of items, but if the interviewer spreads out the objects in one of the lines so they are spaced further apart (again watched by the child), the child then judges there are more objects in the more diffuse set: again the argument that the number of items has not changed, so if the two sets matched before then they must still have the same number of items after the change, does not seem to be available to a young child. If the child is shown two similar vessels containing a coloured liquid, which are adjusted till the child confirms there is the same amount in both, then pouring the liquid in one vessel into a different shape container (so there is a different height of liquid) leads to a judgement that now there is now more liquid in one container than the other.

These probes are not set up as tricks – the aim is not sleight of hand. The interviewer seeks to ensure that the child is carefully watching the transformations and sees what is going on – so the aim is to test their conservation concepts. Some of these probes have been criticised (Donaldson, 1978) suggesting that the process confuses children who misunderstand what is wanted (*e.g.*, perhaps thinking along the lines that 'obviously the amount of water has not changed, so the adult must be asking me about something else – perhaps height'), but the effects are highly reproducible and there are clear shifts towards conservation as children mature. To view a child who has decided the chocolate has not been fairly shared as the interviewer has two slabs, and the child only one, then agree that the sharing becomes fair when the interviewer breaks the child's single slab into two smaller slabs, seems incredible (and funny) from our adult perspective. The voice and face of the child appear to reflect true satisfaction that he now has his fair share, even though the amount of chocolate he has been given has not changed in the slightest.

Piaget was primarily interested in how the human organism seems to be able to develop concepts, such as {conservation of number}, {conservation of volume} and so forth, based directly on their experience of the world, and without needing explicit teaching. He developed a 'constructivist' model that saw the infant's brain as having the potential to learn from action in the environment, and in doing so building cognitive structures that could support more sophisticated learning. Although much of the specifics of the theory he developed has been questioned since, this central insight is reflected in current thinking about how cognition develops.

13.4.2 Learning Scientific Concepts

This particular focus was complemented by the work of Piaget's contemporary Lev Vygotsky (1934/1986, 1978), who was particularly interested in learning that was mediated through culture – that is, through people being introduced to practices and ideas by others who had already mastered them. So, whereas the child will in time spontaneously develop the ability to recognise that two rows of the same number of counters retain the same number despite changes in spacing of one of them, the child will not spontaneously realise that ammonia is NH_3, or that metal surfaces can catalyse some gas-phase reactions, or that benzene has delocalised bonding. In the former case, Piaget tells us that sufficient experience of handling objects will lead to the development of the conservation concept. Arguably, scientific discoveries have come about in a similar way – in principle. Sufficient experiences with the phenomena have supported the development of scientific concepts. Yet this does not happen unmediated.

The reader might wish to consider whether the following terms are meaningful to them, and if so how familiar they were with the terms *before* they started reading this book:

- quanticle;
- congenst;
- accredited concept.

If one or more of these terms is/was unfamiliar, but is/are now associated with some particular meaning, then how did you acquire that meaning given that you were not shown and given the opportunity to handle quanticles, congensts or accredited concepts? Did you notice the use of the word mentipulate in Chapter 2, and if so, was this word familiar? If not, did you need to look up its meaning, or did you infer the intended meaning? As a final example, when reading Chapter 5, did you already know that the positive-ray spectrum was, or did you like the author (see the opening of this chapter) have to form a new concept of what the term meant?

We can consider chemical concepts such as 'potassium' and 'acid' (as these were discussed in some detail in Chapters 5 and 6). These concepts were developed over an extended period of time, by people with intense experience in the area of work, each of whom had a specialised education. Humphry Davy did not have the kind of scientific education a career scientist today would seek (as it was not available when he was young) – but the young Davy's learning about the chemistry that was known in his time was mediated by others more advanced in their learning. Davy may not have had a high opinion of his schooling, but experiences such as working in the apothecary when apprenticed to a surgeon, and reading the work of Lavoisier, meant he was not developing all the concepts (and skills) he would need in his scientific work *ab initio*. He later worked in institutions that were intended to support scientific work.

Although some concepts may be spontaneously developed by individuals without mediation, canonical scientific concepts have been developed by communities over extended periods and become available to other individuals when they are inducted into the culture through joint participation in activity, and by others representing the concepts using symbolic systems such as natural language. Vygotsky (1934/1994) referred to concepts as 'scientific' or 'academic' (he was writing in Russian and his works have been translated) when they were only acquired through formal systems of education. People also have 'everyday' concepts that are also mediated through interaction with others in the culture in more informal ways.

In formal education, a scientific concept, such as 'acid', would be presented in teaching through demonstration and example, using the shared symbolic resources of language. What Vygotsky was very aware of, however, is that although abstract concepts can be represented symbolically in language, this is not sufficient for them to be understood. This links to Ausubel's notion of meaningful learning (where what is taught is understood in terms of prior learning) and the suggestion above (see Chapter 11) that communication of information is not enough for teaching, as that information has to be interpreted with the resources the learner has available.

13.4.3 Melded Concepts

Vygotsky considered that the concepts a person developed spontaneously (which are of the nature of p-prims, where a p-prim is something that operates automatically in cognition, and is not the subject of deliberation) were related to the concepts provided by the culture through formal teaching. Vygotsky wrote of the two types of concepts growing towards each other: spontaneous concepts provide the grounding for understanding abstract concepts, and scientific concepts provide the basis for deliberating on, and talking to others about, spontaneous concepts abstracted from direct experience.

It may seem this is often not necessary, as taught concepts are usually understood in terms of other formal abstract concepts that have previously been learned. However, this is only deeply meaningful learning if the concepts we are using to make sense of new learning are themselves meaningful – so ultimately (even if often indirectly) our concepts need to be ground in experience. It has been pointed out that human language betrays a tendency to ground abstractions on experience of the physical world (Lakoff and Johnson, 1980a, 1980b), as when we talk of period 1 being at the 'top' of the periodical table (even when the table is horizontal) – and in everyday language we describe all sorts of abstractions as heavy, deep, high, and in terms of being inside/outside or adjacent to/on top of/beneath/behind, *etcetera*, other abstractions. I expect readers of this book spontaneously appreciated what I meant above by 'deeply' meaningful learning and concepts being 'ground' in experience.

Our concept system is ground in our experience of the world. When we learn new abstract concepts, which we understand in terms of existing concepts, we are indirectly drawing on the experiential underpinnings of those concepts. So, in practice, we may consider the personal concepts we operate with as melded concepts – concepts that have evolved from (or formed by hybridisation of) both our spontaneous concepts and the formal representations of abstract concepts that are presented to us in language and other symbolic means (gesture, diagrams, graphs, *etc.*) that we can interpret.

13.4.4 Engaging Implicit Knowledge in Building Deliberative Concepts

Vygotsky was writing almost a century ago. More recent theorists have considered how the implicit knowledge elements supporting spontaneous concepts may be recruited in deliberative thinking. One model recognises that the brain seems to have the ability to not only represent experience, but to make new representations of its own representations. Piaget considered natural conceptual development to provide 'formal' operations, which means that the person could internally work with mental entities and mentipulate them. Science involves a lot of working with abstractions that are compared, combined, modified, *etc.* The kinds of concepts I have referred to as meta-concepts and discussed in Chapter 8 (concepts of models, laws, theories) involve these kinds of meta-abstractions, where abstract concepts are formed by reflecting on and relating the concepts (*i.e.*, abstractions) that reflect categories of object or event.

Karmiloff-Smith (1996) considered how human brains seem to have the ability to work with different types of representation that vary in terms of their plasticity and their accessibility to introspection. P-prims work outside of couscous awareness, and cannot be modified as they work automatically. However, Karmiloff-Smith suggested that the brain can in effect clone (well, re-represent) such elements at a different 'level' (using a metaphor of a physical dimension moving from direct perceptual data to abstract conceptualisation) in the brain where they were no longer unmodifiable. Through several stages of re-representation the brain produces representations that are available to introspection, and allows us to produce images or verbal descriptions, and indeed to be able to modify these mental representations. The original p-rims are not themselves modified (and continue to process information automatically in cognition) but provide a kind of starting point, a template perhaps, for forming a different kind of processing unit with different properties supporting a different kind of cognition (moving from the 'fast, automatic' to the 'deliberate, reflective' system).

A somewhat different description is provided by those who suggest that higher-level conceptual structures take the form of what are called 'coordination classes'. Coordination classes draw directly on the implicit knowledge elements (at the level of p-prims) in various combinations (diSessa and Sherin, 1998). In what is sometimes referred to as a 'knowledge-in-pieces' approach (diSessa, 1988), it is the p-prims that are seen as stable, whereas the concepts used by students in consciously thinking about the chemistry are built from these units. One might suggest a light-hearted analogy for concepts and conceptual change here. P-prims are cognitive atoms that are built into conceptual molecules with different properties. Different molecules can be built from the same repertoire of atoms, and under appropriate conditions the conceptual molecules may be decomposed and new ones will appear. The conceptual atoms themselves, however, are like those of Democritus – indivisible and unchanging.

13.4.5 Teaching Informed by the Knowledge-in-pieces Model

The pedagogic significance of this model is that it assumes that p-prims develop from experience and that by the time the chemistry teacher meets a student he or she will have developed an extensive set of p-prims that the student is not actually aware of, and the teacher can do little about. However, as the student is taught chemistry they will construct explicit concepts that will be resourced by their repertoire of p-prims. The student will not be aware they are doing this, as it is an automatic process, but precisely which p-prims are engaged in learning particular concepts depends upon how teaching is intuitively understood by the student. In principle then, different teaching approaches – different metaphors and analogies for example – can lead to a student constructing a concept from a different set of resources (a different combination of p-prims).

This is a very powerful idea. In practice, however, if neither teachers nor students are aware what the available p-prims are, then it not possible for teachers to plan teaching accordingly. The seminal work of diSessa (1993) suggested a good many p-prims that seemed to be drawn upon by college physics students, but to date there is limited work testing these ideas in chemistry to see if the same p-prims seem to be operating (and if so, how), or if perhaps there are others that diSessa did not report.

Although these ideas have been discussed in the context of chemistry learning (Taber, 2014), there has as yet been little direct research based on this approach. However, where work has been framed in this way, some evidence has been found that supports the idea that some of the p-prims that seem to be operating in physics learning may also be channeling understanding of chemistry (as would be expected, as p-prims are considered domain-independent abstractions) and that other p-prims not identified in the context of physics learning may also be influential (Taber and García Franco, 2010). It also seems that at least some of the research identifying explicit alternative conceptions might be productively reframed from a 'knowledge in pieces' perspective (Taber and Tan, 2007). These ideas about how concepts are formed will be followed up in the next two chapters that consider the characteristics of student conceptions, and how conceptual change takes place.

References

Brock R., (2015), Intuition and insight: two concepts that illuminate the tacit in science education, *Stud. Sci. Educ.*, **51**(2), 127–167.

Brock R. and Taber K. S., (2017), The application of the microgenetic method to studies of learning in science education: characteristics of published studies, methodological issues and recommendations for future research, *Stud. Sci. Educ.*, **53**(1), 45–73.

Cianciolo A. T., Grigorenko E. L., Jarvin L., Gil G., Drebot M. E., and Sternberg R. J., (2009), Practical intelligence and tacit knowledge: advancements in the measurement of developing expertise, in Kaufman J. C. and Grigorenko E. L. (ed.), *The Essential Sternberg: Essays on Intelligence, Psychology and Education*, New York: Springer.

Claxton G., (1997), *Hare Brain, Tortoise Mind: How Intelligence Increases When You Think Less*, London: Fourth Estate.

diSessa A. A., (1988), Knowledge in pieces, in Forman G. and Pufall P. (ed.), *Constructivism in the Computer Age*, New Jersey: Lawrence Erlbaum Publishers.

diSessa A. A., (1993), Towards an epistemology of physics, *Cogn. Instr.*, **10**(2&3), 105–225.

diSessa A. A. and Sherin B. L., (1998), What changes in conceptual change?, *Int. J. Sci. Educ.*, **20**(10), 1155–1191.

Donaldson M., (1978), *Children's Minds*, London: Fontana.

Evans J. S. B. T., (2008), Dual-Processing Accounts of Reasoning, Judgment, and Social Cognition, *Annu. Rev. Psychol.*, **59**(1), 255–278.

Goswami U., (2006), Neuroscience and education: from research to practice?, *Nat. Rev. Neurosci.*, 7, 406–413.

Goswami U., (2008), *Cognitive Development: The Learning Brain*, Hove, East Sussex: Psychology Press.

Harkins W. D. and Liggett T. H., (1923), The discovery and separation of the isotopes of chlorine, and the whole number rule, *J. Phys. Chem.*, **28**(1), 74–82.

Howard R. W., (1987), *Concepts and Schemata: An Introduction*, London: Cassell Educational.

James W., (1890), *The Principles of Psychology*. Retrieved from http://psychclassics.yorku.ca/James/Principles/index.htm 29th December 2018.

Johnstone A. H., (1989), Some messages for teachers and examiners: an information processing model. *Assessment of Chemistry in Schools (Vol. Research in Assessment VII)*, London: Royal Society of Chemistry Education Division, pp. 23–39.

Kahneman D., (2011), *Thinking, Fast and Slow*, New York: Farrar, Straus and Giroux.

Karmiloff-Smith A., (1996), *Beyond Modularity: A Developmental Perspective on Cognitive Science*, Cambridge, Massachusetts: MIT Press.

Keller E. F., (1983), *A Feeling for the Organism: The Life and Work of Barbara McClintock*, New York: W H Freeman and Company.

Kornheiser A. S., (1976), Adaptation to laterally displaced vision: A review, *Psychol. Bull.*, **83**(5), 783–816.

Lakoff G. and Johnson M., (1980a), Conceptual metaphor in everyday language, *J. Philos.*, **77**(8), 453–486.

Lakoff G. and Johnson M., (1980b), The metaphorical structure of the human conceptual system, *Cogn. Sci.*, **4**(2), 195–208.

Novak J. D., (1990), Concept mapping: a useful tool for science education, *J. Res. Sci. Teach.*, **27**(10), 937–949.

Piaget J., (1970/1972), *The Principles of Genetic Epistemology*, Mays W. (Trans.), London: Routledge & Kegan Paul.

Polanyi M., (1962), *Personal Knowledge: Towards a Post-critical Philosophy*, Corrected version edn, Chicago: University of Chicago Press.

Siegler R. S., (2006), Microgenetic analyses of learning, *Handb. Child Psychol.*, **2**, 464–510.

Taber K. S., (2009), *Progressing Science Education: Constructing the Scientific Research Programme into the Contingent Nature of Learning Science*, Dordrecht: Springer.

Taber K. S., (2011), The natures of scientific thinking: creativity as the handmaiden to logic in the development of public and personal knowledge, in Khine M. S. (ed.), *Advances in the Nature of Science Research – Concepts and Methodologies*, Dordrecht: Springer, pp. 51–74.

Taber K. S., (2013), *Modelling Learners and Learning in Science Education: Developing Representations of Concepts, Conceptual Structure and Conceptual Change to Inform Teaching and Research*, Dordrecht: Springer.

Taber K. S., (2014), The significance of implicit knowledge in teaching and learning chemistry, *Chem. Educ. Res. Pract.*, **15**, 447–461.

Taber K. S. and García Franco A., (2010), Learning processes in chemistry: Drawing upon cognitive resources to learn about the particulate structure of matter, *J. Learn. Sci.*, **19**(1), 99–142.

Taber K. S. and Tan K.-C. D., (2007), Exploring learners' conceptual resources: Singapore A level students' explanations in the topic of ionisation energy, *Int. J. Sci. Math. Educ.*, **5**, 375–392.

Vygotsky L. S., (1934/1986), *Thought and Language*, London: MIT Press.

Vygotsky L. S., (1934/1994), The development of academic concepts in school aged children, in van der Veer R. and Valsiner J. (ed.), *The Vygotsky Reader*, Oxford: Blackwell, pp. 355–370.

Vygotsky L. S., (1978), *Mind in Society: The Development of Higher Psychological Processes*, Cambridge, Massachusetts: Harvard University Press.

Watson J. B., (1924/1998), *Behaviorism*, New Brunswick, New Jersey: Transaction Publishers.

CHAPTER 14

What Is the Nature of Students' Conceptions?

This chapter sets out a range of characteristics of student conceptions, in terms of a set of dimensions that concepts appear to vary along (Taber, 2014). Some of this diversity is clearly linked to ideas introduced in earlier chapters of the book. A key theme is that not only do students not all share the same conceptions, but that even within a single individual, a person's concepts will not all be of the same kind.

14.1 The Person on a Learning Journey

It should not be surprising that any one person's conceptions are not all of the same kind in the sense of sharing the same characteristics or qualities. The brain seems to have inherent mechanisms to modify the representations it makes of past experience over time in response to new experiences. So, the notion that we are all lifelong learners is not just a slogan or aspiration, but part of our biological make-up.

Any person then is always 'unfinished' in terms of their learning: the brain is always primed to modify understanding on the basis of new evidence. In that sense, at least, our learning journeys can be compared with science itself. People have differential levels of experience in different areas, and so their conceptual understanding of different domains can be at quite different levels. It is considered that developing real expertise in any domain requires many years of close engagement (Gardner, 1998). So, an expert in chemistry – or chess or cricket statistics or collecting baseball cards – may have quite limited knowledge and understanding relating to medieval theology or antique clocks or non-avian theropod dinosaurs or Beatle song lyrics.

14.2 Dimensions of Student Conceptions in Chemistry

There are a range of dimensions that students' concepts relating to their chemistry learning can vary across. These can be labelled as:

- canonicity;
- commonality;
- explicitness;
- commitment;
- multiplicity;
- connectivity.

As the following treatment suggests, although these dimensions relate to distinct qualities that conceptions have (one could imagine any particular conception placed on a multidimensional graph, having a position on each dimension), they are not entirely independent.

14.3 Canonicity: More or Less Alternative Conceptions

A big issue in teaching students is that they often either arrive in class with alternative conceptions that are inconsistent with canonical concepts, or they interpret teaching so as to develop alternative conceptions. Yet, we also know many students do manage to acquire concepts that sufficiently match the target knowledge in the curriculum. A student may have some concepts that are creditable, and some others that encompass alternative conceptions and are not consistent with what is being taught.

Moreover, within any particular class, different students may have different areas where they have more or less canonical concepts. When teaching a particular topic, new learning may depend upon canonical understanding of other concepts that would be considered 'prerequisite' (Taber, 2015), as prior learning (see Chapter 10). The extent to which this prior learning is in place will often vary across a class. Holding alternative conceptions is certainly not restricted to the lower achieving students.

In most classes, it is usually sensible to assume that some students will hold versions of the identified prerequisite concepts that are at odds with the canonical versions; and that the students holding alternative conceptions will be different for different topics. It is also important for teachers to bear in mind that holding alternative conceptions does not necessarily make it difficult for the learners to understand teaching (although it can sometimes) – rather, they may well understand it quite *differently* from intended (Gilbert et al., 1982).

Teachers will be more aware of students who do not understand teaching (as they themselves are aware they are not making sense of information presented, and may be frustrated or disengaged) than of students who are making sense of teaching in non-canonical ways. The author taught a student in an advanced chemistry class (*i.e.*, an elective level post-compulsory but pre-university course only available to those who performed well on school-leaving examinations) who thought that the charge signs on atoms denoted deviations from an octet or full shell (Taber, 1995). So, for Annie (as she was called in the reported study), for example, Na^+ was a neutral atom: the species with a sodium nucleus and one electron more than a full shell. This only became clear at the end of her course after a series of research interviews. Annie had managed to spend two college years in classes with peers, listening to teachers, reading textbooks, practicing past examination questions, without her, or her teachers (myself included), becoming aware she had an alternative conception. Annie made sufficient sense of teaching and classroom dialogue not to appreciate she did not understand material in the way that was intended by her teachers or textbook authors. Perhaps this is an extreme case, but as it only came to light because Annie volunteered to be interviewed in depth about her thinking, it seems likely there are many students like Annie who pass through chemistry courses making sense of different aspects of teaching in alternative and sometimes idiosyncratic ways without this being recognised.

Annie had a conception of charges on ions that worked for her, and she understood teaching that included references to Ca^{2+}, or Cl^-, accordingly: that is, she (subjectively) understood, but she did not understand the chemistry canonically. (In terms of the discussion in Chapter 11, the objective information 'Ca^{2+}' was communicated, but the subjective meaning given to it was not the intended meaning.) Learning is an iterative process of constructing new knowledge on existing conceptual foundations. Learning a new concept, when drawing upon an alternative version of a concept that is needed for understanding the new learning (*i.e.*, having an alternative conception of some prerequisite knowledge), will often mean the new concept is not understood as intended either, and so a new alternative concept is constructed upon the existing one.

14.4 Commonality: 'Popular' and Idiosyncratic Alternative Conceptions

People who grow up in the same physical environment (and we all live in a universe with the same basic laws and principles), and the same social setting, will have many similar experiences of their natural and cultural worlds. To *some extent*, then, there are good reasons why we often share similar concepts (even before we consider formal teaching that deliberately seeks to align people's knowledge and understanding).

However, everybody's actual experience is somewhat different, and as learning is an iterative process, as we understand experience as it is interpreted through the resources we have already developed by abstracting commonalities from past experiences, there is also a tendency for something akin to 'conceptual momentum' to operate such that once a person understands things in a certain way, that influences their thinking, and their further learning. So, differences in people's conceptions tend to channel further conceptual learning in various ways. Left to our own devices, each of our conceptual schemes would continue to diverge, but the social interactions between us moderate this effect and support a degree of sharing our ways of thinking (see Chapter 9). Conceptual development continues then under these different pressures.

It is not surprising, then, that research suggests that there are some common alternative conceptions, where a good many students come to understand particular chemical ideas in similar but non-canonical ways. Many studies into alternative conceptions in various topics (Kind, 2004), report non-canonical ideas that seem to be shared by substantive proportions of those surveyed. However, nor is it surprising that individuals will also develop some idiosyncratic conceptions that are quite different from how most of their classmates think.

I have never come across another student who shared Annie's alternative conception of the meaning of charge symbols. I taught a student (mentioned in the previous chapter) who referred to what everyone else in the class were calling 'electron shells' as 'electron shields' but again in my experience this was a one-off. As the term electrons shells is based on metaphor (see the discussion in Chapter 13 on how we understand abstract concepts through more familiar ideas), it might have been just as appropriate if a different metaphor had come to be used canonically: electron shields would probably have worked just as well as electron shells. After all, 'shielding' is used in the context of the {effective nuclear charge} concept and explaining patterns of ionisation energies. However, electron shells became canonical terminology, and electron shields did not.

These particular conceptions seem rare. However, often different students hold similar enough alternative conceptions for those conceptions to be considered (to a first approximation, at least) as shared. One of the most common, perhaps the most common, alternative conception in learning chemistry relates to the significance of full shells or octets of electrons (Taber, 1998). Learners will commonly suggest that:

- bonds form so that atoms can obtain full outer shells/octets of electrons;
- chemical reactions occur to allow atoms to obtain full outer shells/octets of electrons.

This thinking has been found among many students studying chemistry at high-school level and beyond. The former proposition might be viewed somewhat sympathetically, given the octet-rule heuristic that suggests most common stable ions and molecules do have these configurations. However, this is usually understood by learners in the context of what has been called an atomic ontology (Taber, 2003a): the idea that when we think about chemistry we start with sets of discrete atoms (when actual chemistry has probably never, certainly seldom, started from that point: not in the laboratory, nor in nature; neither on earth, nor, indeed, elsewhere in the universe). The second statement, suggesting that this is a driving force for chemical reactions, is, however, not only non-canonical, but contradicted by just about all the chemical reactions that students learn in school chemistry (which nearly always concern reactants, as well as products, which 'satisfy' the octet rule). These alternative conceptions, and a range of related ones (discussed below as being part of a conceptual framework), have been reported in a range of studies from different parts of the world (Taber, 2013), such that teachers at secondary-school level or college level might expect that students in their classes are likely to come to lessons thinking this way.

14.5 Explicitness: Learners Are Aware of Some, but Not All, of Their Thinking about Chemical Topics

The importance of preconscious processing in cognition has been discussed in Chapter 13, where the idea of 'p-prims', implicit knowledge elements that are active in cognition and can shape student thinking, was introduced. Implicit knowledge is not directly accessible by introspection to the person who applies it. This perhaps explains the truism that it is easier to spot bias in others, than our own biases.

Biases (unlike prejudices) are not inherently bad, and may be well founded. As one example, a common experience we all have as youngsters is that intensity tends to be greater nearer a source. That is framed in verbal terms, as we might make the point as adults, but a young (pre-verbal) child can abstract a pattern from different experiences that biases their cognition to expect future experience to follow a similar pattern. We come to expect a lamp to be brighter as we get closer, a fire to feel hotter, a loudspeaker to sound louder. If this pattern leads to expectations that are not fulfilled we are likely to notice (*e.g.*, perhaps a child wearing headphones walks towards the amplifier expecting the music to get louder). These expectations may be productive in teaching. A learner has no direct experience of the attraction between an atomic nucleus and an electron, but will readily accept that there is a greater attraction when the separation is less. (However, the same p-prim might channel thinking in terms of this intensity–separation pattern, but in a less helpful way, when a learner is asked why it is warmer in summer than in winter.)

Biases lead to intuitions that can be useful in learning, but may be misleading: so, if a student measures a potential difference of little more than one volt between a copper strip and a zinc trip when they are dipped in an electrolyte solution, then common sense might lead to them expecting they could scale up the voltage by using pieces of metal twice as large. The knowledge-in-pieces perspective (see Chapter 13) suggests that learners have a reservoir, or toolkit, of different implicit knowledge elements, feeding different intuitions, and teaching can in principle find better matches by shaping teaching presentations to engage the most pertinent intuitions.

Although p-prims are inaccessible to direct introspection, they can be inferred to some extent from the deliberative thinking that is built upon them. One study explored students' explanations of natural phenomena, and noted that although students might offer scientific forms of explanations (if not always canonical explanations), when they were subjected to a sequence of 'why' questions (*i.e.*, each explanation was itself interrogated) they tended to soon reach a point where they could only reply along the lines 'that's just the way things are' or 'it's just natural' (Watts and Taber, 1996). It is interesting that this was often not just a matter of 'I do not know', but rather that no further explanation was to be expected or needed. This seems to reflect how the patterns we abstract from experience in the world when very young seem to negate epistemic hunger, so we are more likely to seek explanations when these expectations are confounded (*e.g.*, why doubling the size of the metal plates *did not* double the voltage between them, rather than why adding more magnesium ribbon to a beaker of acid *did* lead to more bubbles). When our expectations are met, there is no cognitive dissonance (Cooper, 2007) to give us a feeling that we do not understand.

This is not meant to be critical of learners. Although science is always seeking deeper, more fundamental, explanations, even science must eventually settle on 'that's just the way it is'. (In a sense science is seeking to find just that.) Even as teachers, encouraging curiosity and enquiry, we might not relish the student who continuously and persistently asks questions such as "well, why is the electronic charge -1.6×10^{-19} C?" and would eventually revert to reporting that this is just what has been observed to be the case.

Implicit knowledge elements such as p-prims are not domain specific – they seem to operate at a level in the cognitive system that activates prior to any sense of which realm of activity a phenomenon is most related to. One of the longest-standing elements of this kind is known by the label of the experiential gestalt of causation (Lakoff and Johnson, 1980; Andersson, 1986). The term gestalt refers to something that is perceived as a coherent whole, as when we automatically recognise a bus as a bus rather than as an array of distinct colours and parts. It was suggested that this 'gestalt', abstracted from experience in the world, has widespread relevance in learning science. The experiential gestalt of causation leads to us perceiving many events as happening to something (referred to as the patient), acted upon by some agent, using some form of instrument. This simple pattern is widely applicable, as when the teacher uses a lighted splint to ignite the gas mixture leaving the Bunsen burner. Something is passive and acted upon. Something else is active and causes the action, using some kind of tool or device. In science, this can lead to oversimplified understandings of systems.

In physics, students may commonly think that the earth is attracting the moon (but not the other way around) using gravity. In circuits, students tend to look for simple linear cause–effect relationships rather than appreciate how there is a system of inter-related parts.

In chemistry, we find that it is common for students to think of chemical reactions in ways framed by this expectation that causes are often due to one agent acting on a patient (Taber and García Franco, 2010). For example, reaction mixtures may be stirred to increase mixing, but this may be seen as a process of someone initiating the reaction (which is actually spontaneous) through the instrument of the stirring rod. The reaction between two substances is often understood as if one substance was the aggressive agent, and the other was a passive patient acted upon (so perhaps the acid forces the metal to react). Readers will appreciate that this tendency is hardly countered by some of the metaphorical language canonically used in our subject: such as oxidising 'agent'; nucleophilic 'attack'; the 'substrate' on which an enzyme 'acts'. Such ways of thinking are not necessarily problematic, but because they are based on implicit cognition, intuitions about how the world is, they tend to be applied automatically without reflection and awareness, and so uncritically, whether they are helpful, or not.

If one plots the pattern of ionisation energies across period 2 (Taber, 2018) or period 3 (Taber and Tan, 2011) one gets the familiar shape of a general rising trend and two dips. It is possible to construct a trend line by 'allowing for' the two dips relating to the shift between electrons being removed from (i), the s and p orbitals, and (ii), from a singly occupied to a doubly occupied p orbital, to 'correct' for these shifts and give a single rising line. The trend in each case is quite close to a straight line, with the final point (Ne or Ar) being pretty close to where extrapolation from the other data points would suggest it might be. There is a trend moving across the period for increasingly more energy to be needed to remove an electron. We can explain this in terms of the increasing core charge, and the decreasing atomic radii such that the outer-shell electrons become closer to the nuclear charge.

Students will commonly suggest that there is a particular stability associated with a full shell of electrons (or an octet – Ar does not have a full outer shell, although not all students appreciate that). This does not reflect the evidence – although Ne is more stable than F (and Ar than Cl) – this difference can be explained without seeking anything 'special' about the octet of electrons, as it follows the pattern found across the period as a whole. It seems likely that intuitive ideas about the special nature of things that are complete/full/whole are channeling many students towards the idea that the full shell has some inherent stability, a bit like a security perimeter that could easily be breached if it was incomplete. There is something intuitively appealing about seeing the full shell as special, even if the evidence is that its properties can be explained by other factors that apply to all configurations. Students will often apply this intuition to the extent of judging such chemically dubious species such as C^{4+}, or Na^{7-}, as stable because they have these special configurations (Taber, 2009). If that seems incredible to any teachers, they can check the thinking of their own students using diagnostic materials (Taber, 2002) available for free download from the Royal Society of Chemistry website (http://www.rsc.org/learn-chemistry/resource/res00001102/chemical-stability?cmpid=CMP00002081).

The student conceptions that learners themselves are aware of, are, by definition, those that are explicitly represented in cognition, and that students can readily represent in their talk and writing/drawing. However, these explicit concepts are often channelled by the insidious activity of implicit concepts. Such aspects of cognition cannot be directly accessed, but can sometimes be inferred indirectly by the way people make discriminations intuitively (Kelly, 1963).

14.6 Commitment: Belief or Suspicion?

Another important dimension along which our conceptions vary is the degree of commitment we hold to them. This is related to the grounds on which our thinking is based (see Chapter 4), and is also relevant to our openness to conceptual change (discussed further in Chapter 15).

Early research on students' alternative conceptions tended to emphasise how some notions students acquired seemed to be retained despite teaching, such that learners sometimes held on to their ideas tenaciously in the face of various evidence and arguments (Gilbert and Watts, 1983). Indeed, what sometimes seems to happen is that even teaching that appears to have been convincing about the merits of a more canonical account over an alternative conception at the time of the lesson (so learners seem persuaded), may not be sufficient to lead to long-lasting shifts in the conceptions students use (Gauld, 1989).

However, by no means all of the concepts that students develop, and indeed draw upon in chemistry classes, are so strongly committed to. Piaget (1929/1973) had noticed in his interviews with children (see Chapter 13), that sometimes when a question was asked that a child did not have a ready answer to, they would 'romance' a response. This is neither surprising, nor necessarily unfortunate. In some circumstances where a person is not able to provide an answer to a question, it is better to simply respond that one does not know (if asked in a court of law who committed a crime, it is inappropriate to speculate or guess if we have no strong grounds for believing we know who the criminal was). In teaching, however, teachers often ask questions to encourage students to think things through. Imagining possibilities is an essential part of the scientific process, and an authentic science education will sometimes encourage learners to undertake creative acts of imagination as well as to apply logic to test out ideas. Without the imaginative step, there would be no ideas for scientists to test in the laboratory (Taber, 2011).

Student conceptions then are generally mooting viable possibilities – notions of how things might be that are under active consideration. Some may be rather speculative – where students are not committed to them and

What Is the Nature of Students' Conceptions? 305

are quite open to having them challenged. Some have become strongly committed to. We might say that some learners' concepts approach the status of beliefs – although genuinely scientific conceptions should always fall somewhat short of belief, as scientific ideas are always potentially open to re-examination in the light of new evidence, or a new argument for how evidence should be understood.

Of course, there are some conceptions we may hold that, in practice, we strongly doubt science will ever require us to revise:

- water is a compound of oxygen and hydrogen;
- ammonia has the formula NH_3;
- carbon has atomic number 6.

...

We are perfectly happy for students to hold such conceptions and strongly commit to them, but we would not wish them to commit to concepts that include such propositions as:

- an atom can never have more than eight electrons in its outer shell;
- there are two types of bonding – ionic and covalent;
- a sodium atom will spontaneously donate its outer electron to a chlorine atom;
- at equilibrium, a reaction mixture contains equal amounts of reactants and products;

....

There is no reason a teacher should be concerned if a student *entertains* such ideas, and indeed it may be useful for a student to moot some such notions as the starting point for an exploration of how they fit with canonical chemistry. Chemistry learning is, however, impacted (and may be 'derailed' or misdirected) when students become strongly committed to alternative conceptions.

14.7 Multiplicity: Unitary and Manifold Conceptions

One of the great strengths of the human cognitive apparatus is that it allows 'contrary imaginations'. That phrase borrows from a classic, if now somewhat dated (Hudson, 1967), study that explored the thinking of post-compulsory school students (actually, all boys in this study) who were divided into those who had elected to either take art/humanities subjects or science subjects. One of the themes was of convergent and divergent thinking. A caricature of an extreme convergent thinker would be someone who when asked to list possible uses of a shoe box could only suggest it could be used to keep shoes in – whereas a divergent thinker might suggest using it as a model space station, a makeshift burglar alarm placed beneath an accessible window, to write on as a diary, somewhere to keep some air, a seat for a teddy bear to sit on, something to carry around everywhere to arouse interest in others, a makeshift set-square, something to keep the rain off one's head, the cause of a security alert when left in a public place, a target for a pretend drop of relief aid from a toy plane, a gallery room for displaying very small artworks. . . .

It seemed that often those choosing science were measured to have high IQ (intelligent quotient) scores, but they were less likely to be divergent thinkers. One would hope this is no longer so, as science depends upon highly creative people being prepared to think things that are outside the (shoe) box: Einstein imagining the invariance of the speed of light for all observers; Kekulé dreaming up molecular structures without terminal atoms; Rutherford considering that much of the atom might be in effect empty space, and Meitner imagining the atomic nucleus could be thought of for some purposes as if a liquid drop; Lavoisier being prepared to link a whole class of reactions to one substance (oxygen). . . . All chemical discoveries require someone to imagine some new possibility that can then be tested.

For science to progress, it is important that established ways of thinking (*e.g.*, phlogiston, inert nature of noble gases, intransmutability of elements) can be supplemented by considering other possibilities. Despite the suggestion that scientific revolutions succeed because the adherents of the old ideas retire or die, at least some of those recognised as part of the community (and so inducted into the existing norms) need to be able to see things in a different way (Kuhn, 1970). Once there was a community of chemists who worked with the phlogiston theory to make sense of reactions, it was virtually impossible that a completely rival community could come into existence completely independently and suggest the oxygen theory of combustion: rather some of the people using the earlier ideas had to be open to also considering a new possibility.

We should not be surprised then that it is not always the case that a person has one single way of thinking about a topic: a unitary conception. A student might have a unitary {acid} concept based around the idea that an acid is a substance that releases hydrogen ions when in solution. They will discriminate between acids and non-acids on the basis of that criterion (see Chapter 6 for a discussion of the development of the chemical {acid} concept). It is possible that, in time, this student's concept will change to admit substances as acids in terms of being electron acceptors. It is also possible a learner could hold two somewhat discrete conceptions of an acid – that is, the student may 'know' an acid gives rise to hydrogen ions in solution, but also know there are some acids that act as electron acceptors and do not release hydrogen ions. Such a student has a multiple or manifold conception. Possibly the holding of distinct manifold conceptions may be an intermediate stage in developing a more nuanced unitary concept that encompasses different facets.

What Is the Nature of Students' Conceptions? 307

I worked with a student over an extended period and explored his ideas about chemical bonding in a wide range of contexts (Taber, 2000). Tajinder (the assumed name used in the study) held manifold conceptions. According to Tajinder:

- bonds form to allow atoms to fill their outer electron shells;
- bonds form because of electrical interactions;
- bonds form to minimise the energy of a system.

Although Tajinder saw these as alternative explanations, he also saw them as complementary (rather than, 'here are three possibilities, I think one of them is right'). He thought these were three different ways of thinking and talking about chemical bond formation that all had merit. (Something of how Tajinder's ideas shifted over time will be discussed in the next chapter.)

Holding manifold conceptions may give way to adopting one of the options as the preferred way of thinking; or it may be an intermediate stage in constructing a more inclusive, sophisticated (but unitary) understanding (where different descriptions reflect different facets of the same concept, not alternatives); or the discrete alternatives may be retained over time if there are no good grounds for either merging them or choosing between them. The language here may imply a deliberative process, but as discussed earlier, the development of a more coherent set of ideas may result from automatic features of the cognitive system that occur outside of conscious awareness – and, quite possibly, mostly when we sleep (Walker and Stickgold, 2004).

14.8 Connectivity: Discrete Conceptions and Conceptual Frameworks

The final dimension considered here relates to how well integrated a person's conceptions are. In Chapter 2, concepts were characterised as embedded within networks. So, to some extent, all concepts are connected to other concepts rather that existing in isolation. Meaningful learning (rather than rote learning, see Chapter 11) requires the learner to understand teaching in terms of existing conceptual resources, so links new information with prior learning. However, there can still be different degrees of integration. Some material is learnt as relatively isolated ideas, whereas other learning is tightly linked to extensive frameworks of ideas.

Some conceptual links are, to a first approximation, fairly discrete, and so can be modified without more widely disrupting conceptual structures. As one example, potassium salts give rise to a lilac-coloured flame. Of course, it is conventional to describe this flame as lilac, but a student seeing the flame test without being told this description might spontaneously think of the flame they observe as purple or pink. A student who learns this may expand their {potassium} concept to include this property (an additional proposition) – and clearly this may be linked with a flame test {concept}, the general concept {flame}, and so forth. If a student saw the flame as pink, and was only told later that officially (for purposes of examination questions, perhaps) this colour was lilac (perhaps a first introduction to the concept {lilac} as a colour), the substitution of the descriptor lilac for pink – or indeed the additional proposition that this version of pink is called lilac in chemistry, would not be a major disruption of a conceptual network. It would not, for example, substantially impact upon understanding the nature of potassium compounds, or the meaning of other associations for potassium (*e.g.*, 'potassium is an alkali metal' does not mean something substantially different within a conceptual network when 'potassium salts give a pink colour to a flame' is changed to 'potassium salts give a lilac colour to a flame').

There are, though, other situations where the degree of integration of an idea into conceptual structures is such that changing a conception would be quite disruptive of the wider network. An obvious historical example would be the shift from understanding combustion in terms of the release of phlogiston to understanding combustion as a reaction with oxygen. Not only is this a completely different way of thinking, but the phlogiston theory had been used to interpret and describe a wide range of chemical phenomena to provide a somewhat coherent account of much of the discipline (Thagard, 1992).

In student learning it is often found, as discussed above, that many students in most classes acquire an explanatory principle based on the desirability of full shells or octets (Taber, 1998): that is, that atoms 'need' or 'want' octets/full outer shells. Such anthropomorphic language is often used by students without any sense it may not be appropriate in a scientific account (Taber and Watts, 1996). The core ideas are:

- bonds form so that atoms can obtain full outer shells/octets of electrons;
- chemical reactions occur to allow atoms to obtain full outer shells/ octets of electrons.

As chemical reactions and chemical bonds are core ideas in chemistry, these alternative conceptions become widely applied. Moreover, much teaching is subsequently understood by students in terms of these conceptions. Often in school chemistry, students initially learn about two types of chemical bond: covalent and ionic. Covalent bonding is often understood using the 'sharing' metaphor, and it is understood that by sharing electrons atoms can achieve octets/full outer shells. Ionic bonding is often understood – rather than being the attraction between ions, such as those that exist in solution when reagent solutions are mixed – as a means of atoms obtaining desired electron configurations whereby a metal atom donates its electron to a non-metal ion (see Chapter 10).

Ionic bonding is often then understood in terms of a molecular conceptual framework of several linked notions (Taber, 1997). First, a bond only exists where an electron has been transferred between the metal atom that gave rise to a metal cation and the non-metal atom that gave rise to the anion (a 'history conjecture'). So, in sodium chloride formed by neutralisation and subsequent evaporation of the solvent, students will consider each Na^+ ion is ionically bonded to the one Cl^- ion that it gave its electron to (even though according to canonical chemistry these ions already existed in the distinct acid and alkali solutions mixed together in the procedure).

Given the electronic configurations of the sodium and chlorine atoms and ions, only one ionic bond can be formed by each ion (*i.e.*, with the counter ion it has donated an electron to, or accepted one from) – a 'valency' conjecture. As the crystal of NaCl clearly has structural integrity, something must be holding together the ions that are not (in this model) ionically bonded – and this is explained in terms of the attractions between the ions – seen as something other than a bond (a 'just forces' conjecture). This is referred to as molecular framework as, even when the term molecule is not used, discrete (molecule-like) bonded NaCl ion pairs are considered to exist in the structure. Indeed, students shown an image of NaCl structure will often quite happily suggest which of the neighbouring anions is actually bonded with a particular cation, and which are just attracted to it. Watching students confidently make such apparently arbitrary assignments is an interesting experience.

Students who have developed these ideas may later find it difficult to understand polar bonds as intermediates on a dimension from pure ionic to pure covalent bonds, as the covalent–ionic distinction is considered dichotomous (as bonds are formed by sharing or electron transfer – understood as quite different mechanisms). Metallic bonding may be understood as a variation on ionic or covalent bonding (Taber, 2003b); hydrogen bonding may be understood as a type of covalent bond (or, alternatively, just not a proper bond); and dispersion forces/van der Waals' forces and salvation interactions are considered 'just forces', and so not really bonding. Understanding of ionisation energy may be distorted by the notion that there is something inherently stable about a full shell (see above) that links to this being the driving for force for chemical change.

The interconnectedness of different concepts acting as nodes of an extensive network allows alternative conceptions to act as the basis of extended conceptual frameworks (as in this example). It is possible for a teacher to miss the connections students are making because of their different, unique, conceptual networks.

If a student suggests that the sodium cation is more stable than the sodium atom then this can be readily understood by a teacher as a sensible comment in the context of a metallic lattice, or a salt crystal, on in solution – all situations where the Na^+ ion is stabilised by interactions with other species. However, students will often also mean that an isolated sodium atom is unstable compared to the cation separated from its outer electron. Whilst this is a chemically unlikely context, it is the abstract context within which ionisation energy is understood – where students are expected to learn that energy is needed to pull the negative electron away from the positive residue of the atom.

There is a conceptual parallel to the inductive effect (see Chapter 2). Just as a bond between two atoms of the same element cannot be assumed to have totally symmetrical electron distribution without taking into account what else those atoms are bonded to, so the meaning of any chemical concept (even when represented in language suggesting canonical understanding) is subtly modified by the concepts it is closely linked to. A student who suggests that normal rain water is acid may seem to have acquired a canonical proposition: but if for that student an acid is always a dangerous, corrosive liquid then the seemingly canonical statement does not reflect a canonical understanding. So, although, by their nature, all concepts are linked to some extent, the degree of integration, and the strength of the connections, can vary considerably.

14.9 Conclusion

If students' conceptions can vary so much, then although it is important for teachers to identify learners' alternative conceptions, this is just a starting point in appreciating how such conceptions may influence student thinking and further learning. Conceptual change then, the theme of the next chapter, is not only influenced by the conceptions learners already have, but the nature and status of those conceptions.

References

Andersson B., (1986), The experiential gestalt of causation: a common core to pupils' preconceptions in science, *Eur. J. Sci. Educ.*, **8**(2), 155–171.

Cooper J., (2007), *Cognitive Dissonance: Fifty Years of a Classic Theory*, London: Sage.

Gardner H., (1998), *Extraordinary Minds*, London: Phoenix.

Gauld C., (1989), A study of pupils' responses to empirical evidence, in R. Millar (ed.), *Doing Science: Images of Science in Science Education*, London: The Falmer Press, pp. 62–82.

Gilbert J. K., Osborne R. J. and Fensham P. J., (1982), Children's science and its consequences for teaching, *Sci. Educ.*, **66**(4), 623–633.

Gilbert J. K. and Watts D. M., (1983), Concepts, misconceptions and alternative conceptions: changing perspectives in science education, *Stud. Sci. Educ.*, **10**(1), 61–98.

Hudson L., (1967), *Contrary Imaginations: A Psychological Study of the English Schoolboy*, Harmondsworth: Penguin.

Kelly G., (1963), *A Theory of Personality: The Psychology of Personal Constructs*, New York: W W Norton & Company.

Kind V., (2004), *Beyond Appearances: Students' Misconceptions about Basic Chemical Ideas*, 2nd edn, London: Royal Society of Chemistry.

Kuhn T. S. (1970). *The Structure of Scientific Revolutions*, 2nd edn, Chicago: University of Chicago.

Lakoff G. and Johnson M., (1980), *Metaphors We Live By*, Chicago: University of Chicago Press.

Piaget J., (1929/1973), *The Child's Conception of The World*, J. Tomlinson and A. Tomlinson, Trans., St. Albans: Granada.

Taber K. S., (1995), Development of Student Understanding: A Case Study of Stability and Lability in Cognitive Structure, *Res. Sci. Technol. Educ.*, **13**(1), 87–97.

Taber K. S., (1997), Student understanding of ionic bonding: molecular versus electrostatic thinking? *Sch. Sci. Rev.*, **78**(285), 85–95.

Taber K. S., (1998), An alternative conceptual framework from chemistry education, *Int. J. Sci. Educ.*, **20**(5), 597–608.

Taber K. S., (2000), Multiple frameworks?: Evidence of manifold conceptions in individual cognitive structure, *Int. J. Sci. Educ.*, **22**(4), 399–417.

Taber K. S. (2002). *Chemical Misconceptions – Prevention, Diagnosis and Cure: Classroom Resources* (vol. 2), London: Royal Society of Chemistry.

Taber K. S., (2003a), The atom in the chemistry curriculum: fundamental concept, teaching model or epistemological obstacle? *Found. Chem.*, **5**(1), 43–84.

Taber K. S., (2003b), Mediating mental models of metals: acknowledging the priority of the learner's prior learning, *Sci. Educ.*, **87**, 732–758.

Taber K. S., (2009), College students' conceptions of chemical stability: The widespread adoption of a heuristic rule out of context and beyond its range of application, *Int. J. Sci. Educ.*, **31**(10), 1333–1358.

Taber K. S. (2011). The natures of scientific thinking: creativity as the handmaiden to logic in the development of public and personal knowledge, in M. S. Khine (ed.), *Advances in the Nature of Science Research – Concepts and Methodologies*, Dordrecht: Springer, pp. 51–74.

Taber K. S. (2013). A common core to chemical conceptions: learners' conceptions of chemical stability, change and bonding, in G. Tsaparlis and H. Sevian (ed.), *Concepts of Matter in Science Education*, Dordrecht: Springer, pp. 391–418.

Taber K. S., (2014), *Student Thinking and Learning in Science: Perspectives on the Nature and Development of Learners' Ideas*, New York: Routledge.

Taber K. S., (2015), Prior Knowledge, in R. Gunstone (ed.), *Encyclopedia of Science Education*, Berlin-Heidelberg: Springer-Verlag, pp. 785–786.

Taber K. S., (2018), Alternative Conceptions and the Learning of Chemistry, *Isr. J. Chem.*, DOI: 10.1002/ijch.201800046.

Taber K. S. and García Franco A., (2010), Learning processes in chemistry: Drawing upon cognitive resources to learn about the particulate structure of matter, *J. Learn. Sci.*, **19**(1), 99–142.

Taber K. S. and Tan K. C. D., (2011), The insidious nature of 'hard core' alternative conceptions: Implications for the constructivist research programme of patterns in high school students' and pre-service teachers' thinking about ionisation energy, *Int. J. Sci. Educ.*, **33**(2), 259-297.

Taber K. S. and Watts M., (1996), The secret life of the chemical bond: students' anthropomorphic and animistic references to bonding, *Int. J. Sci. Educ.*, **18**(5), 557-568.

Thagard P., (1992), *Conceptual Revolutions*, Oxford: Princeton University Press.

Walker M. P. and Stickgold R., (2004), Sleep-Dependent Learning and Memory Consolidation, *Neuron*, **44**(1), 121-133.

Watts M. and Taber K. S., (1996), An explanatory gestalt of essence: students' conceptions of the 'natural' in physical phenomena, *Int. J. Sci. Educ.*, **18**(8), 939-954.

CHAPTER 15

How Do Students' Concepts Develop?

This chapter considers how conceptual change occurs, in the light of what has been discussed about chemical concepts in previous chapters. It begins, however, with a continuation of the personal vignette that began Chapter 13.

15.1 A Personal Conception of Positive-ray Spectrum: Reflective Concept Development

In Chapter 13, I described how I was not familiar with the term 'positive-ray spectrum', a term that I came across when doing research for writing this book, and I automatically formed a concept {positive-ray spectrum} to make sense of the source as referring to the energies of radioactive emissions from neon. It was only when returning to the text the next day, that I paused to deliberate on what the term 'positive-ray spectrum' actually meant, and questioned whether radioactive neon was known at that time, or could have been investigated ahead of identifying the presence of isotopes of the element. It is now known that neon has some very short-lived isotopes that decay with positron emission, but at the time J. J. Thomson was working neon would not have been considered radioactive.

So, a positive-ray spectrum was something else. Thomson developed a method of investigating what he called positive ray spectra to identify constituents of gases through their atomic weights. For example, using this technique, he identified spectral 'lines' of mass:charge ratio of 163 ± 5 and 260 ± 10 that he suggested might be due to molecules of inert gases:

> These lines, might, as far as their atomic weights go, originate from molecules of krypton and xenon respectively, and I am inclined to think

---

Advances in Chemistry Education Series No. 3
The Nature of the Chemical Concept: Re-constructing Chemical Knowledge in Teaching and Learning
By Keith S. Taber
© Keith S. Taber 2022
Published by the Royal Society of Chemistry, www.rsc.org

that this is the correct interpretation, there are, however, many properties of these lines which seem at first sight to be incompatible with this explanation (Thomson, 1922, p. 292).

What Thomson was investigating were rays produced in vacuum tubes, where current was allowed to flow through a partially evacuated tube containing a gas at low pressure. If the cathode contains holes, some charged particles pass through these holes and can be detected (by their effect on the gas in the tube, or a photographic plate) behind the cathode. Strong magnetic fields would cause the rays to follow parabolic paths, with a curvature depending upon their mass:charge ratio. The term cathode ray is today largely associated with electrons: some of the electrons making up the current in the vacuum tube pass through the holes in the cathode and continue to travel beyond. Thomson was interested in rays with very different mass:charge ratios:

> ...particles which produce the parabolas on the photographic plates may be divided into the following classes:
>
> 1. Positively electrified atoms with one charge.
> 2. Positively electrified molecules with one charge.
> 3. Positively electrified atoms with multiple charges.
> 4. Negatively electrified atoms.
> 5. Negatively electrified molecules. (Thomson, 1913, p. 7)

Thomson pointed out that the parabolas indicated the mass:charge ratio, and he suggested means of determining the charges involved, so allowing the apparatus to be used to identify "the masses of all the particles in the tube, and thus identify the contents of the tube as far as this can be done by a knowledge of the atomic and molecular weights of all its constituents" (Thomson, 1913, p. 11). In effect, Thomson had developed the prototype of the mass spectrometer. The positive-ray spectrum of neon (Harkins and Liggett, 1923) referred to the mass spectrum obtained for neon using this kind of method.

Before yesterday, I had no concept {positive-ray spectrum} although I did have a concept {mass spectrum}. Yesterday, I formed a concept {positive-ray spectrum} from my reading, by drawing upon my existing knowledge and, in particular, associations for the term 'positive ray'. In effect, I formed an alternative conception, a misconception, somewhat different from the author's intended meaning. Although (in terms of the discussion in Chapter 11) the *information* communicated ('positive-ray spectrum') was clear enough, the sense I made of it did not match the author's intended meaning. This morning, I returned to check what I had drafted yesterday, and was uneasy about referring to positive ray spectra without checking what this meant: my initial conception seemed unconvincing. So, I did some more reading, which led to my realising that a positive-ray spectrum would fit under my existing concept of {mass spectrum}.

My recently formed concept {positive-ray spectrum} changed considerably, and become subsumed under the existing concept of {mass spectrum}. Moreover, my existing concept {mass spectrum} also shifted as I learnt a little more about how mass spectroscopy developed, and how the first (what we might now call mass spectra) were obtained, and how they were conceptualised at the time.

This episode could be described in the language Piaget (1970/1972) used in discussing his theories of cognitive development (see Table 15.1). First, I came across an unfamiliar term ('positive-ray spectrum') that I *assimilated* by acquiring a new concept {positive-ray spectrum} understood in terms of my existing concepts (*e.g.*, {positrons}, {radioactivity}). However, there was some *disequilibration* as my initial understanding of the term seemed non-viable in terms of my existing understanding of the contemporary state of the knowledge and the technical developments I was reading about. Finally, I resolved this and *accommodated* the new (for me) concept within my existing conceptual framework by understanding positive-ray spectrum as a form of mass spectrum. Another learning theorist contemporaneous with Piaget, Vygotsky (1934/1986), would likely highlight how this was a dialectical process: not only did I form a new meaning for 'positive-ray spectrum' but in doing so modified my existing concept {mass spectrum}.

15.2 Meaningful Learning Revisited

Learning can be considered to be more or less 'meaningful' in nature (Ausubel, 2000). Learning 'by rote' means learning without understanding. In the distinction drawn in Chapter 11, it means committing information to memory, without being able to interpret that information in terms of the available prior learning. A young child could be taught to commit to memory

Table 15.1 Applying Piaget's scheme of conceptual learning to a personal instance of acquiring a new chemical concept.

Step	Description
Assimilation ↓	The term 'positive-ray spectrum' (encountered in external environment) assimilated into conceptual structure, by being understood in terms of exiting concepts {spectra}, {positrons}, {cathode rays}, {nuclear instability}, {radioactivity}, *etcetera*.
Disequilibration ↓	The modified conceptual structure includes the concept of 'positive-ray spectrum' as a spectrum of energies of positrons emitted from radioactive neon – but being measured using naturally stable isotopes of neon; before the phenomenon of positron emission was discovered.
Accommodation ↓	'Positive ray spectrum' is re-conceptualised in terms of the deflection of accelerated neon ions being related to their mass:charge – so linked to existing concept {mass spectrum}.
Equilibration	A new concept of {positive-ray spectrum} that is consistent with prior concepts has been added to, and incorporated in a coherent way into, conceptual structure.

the phrase "alumina is an amphoteric oxide" without having any basis for making sense of what the phrase was intended to mean.

It can require considerable effort to learn meaningless material by rote, and it tends to rely on a good deal of rehearsal, whereas meaningful learning may more readily occur (sometimes without deliberate effort). In practice, chemistry learning in classrooms is never likely to be entirely by rote, entirely meaningless learning, but the degree to which information presented in teaching can be readily related to prior learning can vary. It is also important to note that where a students' prior learning consists of alternative conceptions (see Chapter 14), meaningful learning may well occur, but the information presented in teaching may come to be understood in quite different ways from those intended (Gilbert *et al.*, 1982). It is also possible for students to form alternative conceptions through meaningful learning if they make links with prior learning that seems relevant to them, but which is not canonically connected (Taber, 2005). For example, it seems unlikely the substantive proportions of secondary-age learners who suggest that the electrons in an atom are bound to the nucleus by gravity (Nakiboglu and Taber, 2013) have been explicitly taught that is the case.

15.3 The Metaphor of the Conceptual Ecology

One metaphor that has been applied to the mental context of learning is that of an ecology – a conceptual ecology. A person has a diverse pool of resources represented in his or her cognitive system that can potentially be useful in making sense of teaching, and so this provides a kind of mental environment in which some teaching is understood well, other teaching less well. Moreover, conceptions or conceptual frameworks may be considered to sometimes compete within this environment. Some conceptions may grow and thrive, and others may wither away, or different conceptions may co-exist. Clearly such a metaphor should not be treated as a formal model (there tend to be ways that all analogies, metaphors, similes, *etc.*, are not like the target idea), but it offers a useful way of thinking about conceptual development.

The conceptual ecology is not just about formal concepts. The cognitive *skills* a person has available clearly influence how concepts may be learnt. Piaget (1970/1972) suggested that younger learners could not readily make sense of highly abstract ideas as their thinking was linked to concrete situations and examples. Common basic habits of thinking may seem to be shared aspects of human cognition, but to some extent these may need to be learnt through modelling and mediation by others demonstrating those ways of thinking to the uninitiated (Vygotsky, 1978). As one example, the logic of syllogism (*e.g.*, given that the empirical sciences seek to explain natural phenomena; and given that chemistry is an empirical science; therefore, it logically follows that chemistry seeks to explain natural phenomena) may only tend to develop in cultural contexts where such thinking is publicly practised (Luria, 1976). The scientific way of classifying natural entities and events by developing typologies based on ontological similarities (as discussed in Chapter 7) may not be consistent with a more natural attitude to associate things functionally (*e.g.*, a round-bottomed flask might be more naturally categorised as belonging with a reflux condenser with which it might be used, rather than with a conical flask just because they are both kinds of flask).

15.3.1 Epistemological Sophistication

An important aspect of the cognitive ecology would seem to be the epistemological assumptions a learner has. A student who thinks science produces absolute, proven, factual knowledge may understand teaching differently from a student who considers that science develops models and theories that make up conjectural, theoretical knowledge. Learning about an orbital model of the atom, after having previously learned a planetary orbit model, is quite a different task within these contexts. Similarly, for someone with a more sophisticated epistemological stance, oxidation states can more readily be understood as reflecting answers to a *hypothetical* question about how molecules may be most readily split up using heterolytic bond fission (even when those molecules would tend to undergo homolytic bond fission).

Oxidation state is a useful idea, even if it is based on imagining a fictional process – for example, the answer to the question 'if a methane molecule were to be decomposed into a set of simple ions, which ions would be produced?' (the basis of assigning oxidation numbers) is seen as informative, even if the process is not feasible in any likely chemical conditions. This is potentially a context that may be understood differently by different groups of students. The most sophisticated thinkers will have no difficulty with the conjectural nature of the imagined process, whereas other students may get stuck on 'but that would not happen'. However, research also suggests that many students at high-school level may simply accept this decomposition process as realistic and unproblematic, and would consider an ion such as C^{4-}, with its outer shell octet, as stable and perfectly viable (Taber, 2009).

Research into cognitive development suggests that people pass through a number of key shifts in the sophistication of their thinking during development (Kuhn, 1989). The first step is the realisation that accounts of the world may not match to reality (people can lie, or simply be mistaken – parents are not the oracle). However, at this point, accounts tend to be taken as either true or false, right or wrong. We might imagine a child at this stage seeing a periodic table that showed hydrogen placed at the head of the halogens wanting to know if this was indeed where hydrogen *should* be placed, *or not* (rather that considering there might be a credit-worthy argument for placing hydrogen there that falls short of being completely convincing). A response to realising that this approach is insufficient, and is inadequate for dealing with the complexity of the world, is often a shift to relativism: that all statements are just a matter of opinion. Some people prefer to put hydrogen above the alkali metals; others prefer to put hydrogen above the halogens – it is really just a matter of personal choice.

The final stage of development, involves an acceptance that often in life we cannot rely on developing definitive knowledge to establish an absolute truth, but we can still seek to hold a coherent set of values and seek to develop strong positions that we can defend with rational argument. That is, the intellectually mature person does not assume they are right, but seeks to develop a position that has strong grounding (Krathwohl *et al.*, 1968). This mature position clearly has strong links with a modern understanding of the nature of science as a means of generating knowledge (Popper, 1961/2011).

Readers working in universities may suspect that such findings are of less significance for them than for school teachers who are dealing with younger, less mature learners. This is a fair point, but it is worth bearing in mind that work undertaken in the mid-twentieth century with undergraduates at elite Colleges in the United States (Harvard, Radcliffe) found many students were still in the process of moving through and beyond a form of relativism (Perry, 1970).

15.3.2 Rich Ecologies

We saw in Chapter 12 that in order to help *make the unfamiliar familiar* (which might be taken as a motto or slogan for teaching) teachers may employ not only telling and demonstrating, but models, narrative, gesture, diagrams, metaphor, analogy, animism, and so forth. Interpretive resources that may usefully populate a learner's conceptual ecology are not limited to their formal concepts, but also include the images and experiences of various kinds that they have represented and can call upon to make sense of teaching,

Learning is therefore supported by a learner having a richly populated set of 'resources' for interpreting and making sense of teaching. Teaching involves – at least implicitly, and preferably deliberately – applying a model of how learners will make sense of the information being presented (see Chapter 12), and teaching will be more effective when the teacher's predictions about how material can be understood are accurate. Yet, given the complexity of the process (especially in large class contexts) some incidents of mismatch between the assumptions made by the teacher, and how learners actually make sense of teaching, are inevitable (Taber, 2001a). A richer set of interpretive resources available to a learner makes it more likely teaching can be understood, but also more likely that teaching may be construed in unintended (and, therefore, sometimes unhelpful) ways.

The 'knowledge-in-pieces' perspective (diSessa, 1988), introduced in Chapter 13, sees the teacher's role to be, at least in part, guiding students in constructing formal concepts by selecting the most appropriate resources they have available. A student who is being introduced to the notion of chemical bonding will have available within their conceptual ecology representations of objects that hold things together (*e.g.*, cuff-links, handcuffs, screws, bolts, rivets) but if we want them to think of bonding as a process rather than bonds as objects, then it may be useful to steer them to thinking of a familiar process such as holding hands. However, that image may bring unhelpful associations of agency and sentience when we know students too readily accept and use anthropomorphic explanations in chemistry (Taber and Watts, 1996). So, in some ways, bonding is like atoms being held together by being handcuffed, and, in some ways, it is like them holding hands, but although both images offer a useful way of making the unfamiliar familiar, the new concept must be developed by discriminating it from (as well as comparing it to) what is familiar.

A pair of repelling ring magnets mounted horizontally on a central vertical spoke offers another comparison where a neutral equilibrium is maintained by the balance of (gravitational and magnetic) forces – but again the differences are important as well as the similarities. This phenomenon may not be familiar to students, but can be readily demonstrated in class, and can be manipulated by the students so they can 'feel' the force acting – so it becomes represented within the conceptual ecology. Exploring comparisons of a new concept with *a range* of referents expected to be familiar to students, each with different useful (but also less helpful) features, is likely to be most productive. However, simply guessing which resources particular students have available, and how they understand them, will not always be effective. Teaching needs to have the flavour of dialogue, then, where the teacher or lecturer is not just communicating to learners, but is constantly seeking feedback on how material is being understood (such as in the vignette illustrating a teacher introducing transition states in Chapter 12).

15.4 Two Types of Conceptual Change?

The term 'conceptual change' can be used to mean any kind of change in a person's conceptions, but is sometimes reserved to refer to major shifts in understanding, such as those from an alternative conception to more canonical conception. In this chapter, the term is meant in an inclusive sense, to cover all kinds of changes in the concepts that a person has available to resource their thinking.

It is certainly a simplification, but nonetheless an approach often mooted, to consider that conceptual change can fall into two major categories. One category involves learning that is relatively straightforward, and the other category learning that is inherently more difficult to achieve. Various terminology has been used to describe these two classes – including assimilation *versus* accommodation (although these terms are also used in Piaget's model as complementary parts of an overall process of conceptual change, see Table 15.1); or 'weak' and 'strong' restructuring; or restricting the term conceptual change to only the latter, more 'radical', type of change, and labelling more straightforward modifications as something other than conceptual change. The idea is that some conceptual learning does not disrupt existing conceptual structure, whereas other changes rely on substantive modifications.

The treatment of concepts set up in the early chapters of this book suggest that any kind of conceptual learning must involve some modification in conceptual structures (as concepts are embedded in networks), even if this can be to very different degrees. This treatment also suggests that the kind of conceptual change where a person begins with a particular well-established way of understanding some class of phenomena, perhaps an alternative concept; and then shifts to a completely different way of thinking, such as applying a more canonical concept; is not likely to happen as a simple sudden substitution of one conceptual structure for another as a result of a single teaching event. The historical evidence, as well as the evidence from studies of conceptual change in learners, suggests such 'radical' conceptual change is more complex than this. Consider the following scenario:

Once upon a time, a chemistry teacher addressed a question to a student.

Teacher: Why does hydrogen react with fluorine?

Student: Because the hydrogen atom only has one electron and wants two, and the fluorine atom is one electron short of a full outer shell, and it needs a share in another electron as it wants to complete its octet. So they react together, and form a molecule where they are both happy with their electronic arrangements.

Teacher: That's an interesting suggestion, but it is not right. Atoms are not sentient beings with desires, and, in any case, hydrogen and fluorine are molecular and so the reaction does not occur between separate atoms.

Student: Yes, of course, how silly of me. I see now that the reactants already have the kinds of electronic configurations that stable molecules and ions tend to have. Why does the reaction occur then?

Teacher: It is better if you think in term of the potential electromagnetic interactions between the charges in the two molecules if they come together or 'collide' in the gas state, and whether a reconfiguration is possible that leads to an energetically more stable state overall.

Student: Oh, I see: the products have a lower energy state than the reactants, so it would have been more sensible if I had been considering the bond energies associated with the H–H, F–F and H–F molecules, and the enthalpy change in a possible reaction. Yes, in future, in this and any other examples, I will always bear in mind the actual structures of the reactants, which are not likely to be atomic, and I will always think in terms of the physical interactions in the system, and will never again try to explain chemistry in terms of the desires of inanimate entities like atoms.

And the student never again applied alternative conceptions based on atoms wanting or needing to obtain full electron shells, and lived happily, explaining chemistry canonically, ever after.

How Do Students' Concepts Develop? 321

I trust that readers with experience of working closely with learners will consider such a 'fairy tale' scenario unfamiliar: it is not very feasible given the way human cognition and learning work.

I suggest that dividing conceptual change into two types, whilst useful as a general principle, creates a dichotomy when it is likely there is more of a continuum between conceptual changes that are modest and relatively local within a person's conceptual structure, and those changes that involve greater modification of a conceptual networks. The model offered in Chapter 14, of how concepts vary along such dimensions as explicitness, commitment, multiplicity, and connectivity, certainly implies that conceptual changes will 'disrupt' existing concepts to varying degrees.

Although there is much research that discusses conceptual change, actual detailed empirical studies of conceptual change are somewhat rarer, given the methodological issues involved in detailing and monitoring a person's slowly shifting concepts. This is one area where more research is certainly needed (see Chapter 16). It is, however, possible to offer an analysis of some kinds of conceptual change.

In this chapter I will discuss some hypothetical examples of types of conceptual change, hoping to show that even what might at first sight seem to be a simple discrete change due to the addition or modification of a single conception may entail complications in how concepts are represented in the brain. I will, here, discuss three types of conceptual change (but intended as reflecting points on a continuum): adding new material; making a discrete and distinct modification to knowledge; and shifting to a new way of thinking about come topic. The treatment will reflect how these should not be treated as entirely distinct types of change. I will then briefly discuss some findings from actual research studies and relate them to the theoretical understanding of conceptual change.

15.5 Accretion – Conceptual Addition

One category of conceptual change that may seem simple to understand is adding to existing conceptual repertoires. This might mean adding a new concept, or 'just' adding a new characteristic or example to an existing concept.

We might consider a student who had a {metallic element} concept, and knew a range of metals – perhaps iron, copper, lead, *etc.* The student would therefore have an {iron} concept, and a {copper} concept, and so forth, associated with the more abstract {metallic element} concept (*e.g.*, iron is a metallic element, *etc.*). Perhaps (consider this a thought experiment) the student has heard of tungsten in the context of lamp filaments, but has no conception of what it is beyond being a material, and has never heard of niobium.

The student might be taught that tungsten is a metallic element. Learning this involves two changes in conceptual structure. Adding the conception 'tungsten is a metallic element' modifies both the {tungsten} concept and the {metallic element} concept. The {tungsten} concept is developed by having a link to the {metallic element} concept such that there is a more specific conception of tungsten. The {metallic element} concept is not changed as fundamentally, but now has an additional connection, providing a further example of a metallic element. After this conceptual change, activation of the {tungsten} concept is likely to lead to activation of the {metallic element} concept and activation of the {metallic element} concept is more likely (than before, when it was unlikely) to lead to activation of the {tungsten} concept.

The language here ('activation') reflects the physiological/anatomical level of description (see Chapter 13), but in everyday terms, thinking about metals may now lead to tungsten coming to mind, where that would have been much less likely before this new learning. It was suggested earlier in the book that learning is a change in the potential for behaviour, and this new conceptual linkage changes the potential to behave in particular ways (for example in response to questions such as 'give examples of metals' or 'what kind of material is tungsten?') There is now a subsuming relationship between the two concepts such that the {tungsten} concept is now understood as not only subsumed under the {material} concept, but more specifically under the {metallic element} concept. Indeed, a similar point can be made about the link between {tungsten} and the {element} concept.

If the learner was taught that (previously unfamiliar) niobium is a metallic element then they would create a new concept, {niobium}, to be linked to their {metallic element} concept. Whereas the learner already has some notions about {tungsten} they would, without further information about niobium, only initially form a {niobium} concept linked to existing concepts such as {element} and {metallic element} and drawing upon these concepts. However, as suggested in Chapter 2, concepts are located in nets or web of associations so linking a new concept to an existing concept relates it to existing associations of that concept.

The learner could deduce, *inter alia*, that niobium is probably a good conductor of electrical current and heat, may be sonorous; is likely to be malleable and ductile; that clean samples are likely to have a lustre; that all niobium atoms have nuclei with the same proton number; that niobium can be located somewhere on the periodic table, and probably not in the p-block; that niobium is more likely to form cations than anions.... In other words, syllogistic logic applies: metals are good electrical conductors; niobium is a metal; therefore, niobium is a good electrical conductor; *etcetera*.

After these two learning episodes, activation of the learner's existing {cation} concept (that is, when thinking about cations) is more likely to lead to the activation of the revised {tungsten} concept and may lead to activation of the new {niobium} concept – that is, the learner may think of the possibility of tungsten and niobium cations: even if during the new learning there had not been any explicit references made to these elements forming cations. There are now associations (indirectly though the conceptual network) making it more likely the learner will bring tungsten and/or niobium cations to mind, without being formally introduced to the specific concepts {tungsten cation} and {niobium cation}. So, formation of a {niobium} concept, within a particular conceptual ecology, may be sufficient to also lead to subsequent formation of a {niobium cation} concept through spontaneous, internal mental processes.

If a learner saw and handled some niobium, and was able to observe some of its characteristics then this would support learning. However, as was discussed in Chapter 13, many concepts are learned without direct experience, through teaching in terms of symbolic communication (such as verbal language: 'niobium is a metal'). A student who had never seen any niobium or any picture of any niobium, would still be able to form a mental image of some niobium – being told 'niobium is a metal' is interpreted in terms of prior learning, and a learner would likely picture niobium (if perhaps vaguely) drawing upon their experience of metals such as iron, copper, zinc, lead, *etcetera*.

That is not to say that the learner would have some magical telepathic power that enabled them to know what niobium actually looks like, but they could readily imagine a feasible (if not actually accurate) image based on background knowledge. Almost certainly, niobium would be imagined as a solid – even by students who were aware that mercury is a metal that is liquid at room temperature. This reflects the creative aspect of science, where scientists imagine possibilities beyond current knowledge, forming conjectural knowledge that can in principle be tested empirically (Taber, 2011).

So, as concepts are nodes embedded in conceptual networks, even being taught an isolated item of information – 'niobium is a metallic element' – may have indirect as well as direct consequences for what is learned (*cf.* the conceptual inductive effect, see Chapter 2), and cannot be considered as just the addition of an isolated 'fact'. In some ways, this hypothetical account of an individual student learning has parallels in the history of chemistry. When potassium and sodium were first discovered (see Chapter 5) they had somewhat different properties to other known metals (softer, less dense, likely to spontaneously react with water....) Even though the {potassium} and {sodium} concepts could be subsumed under a {metallic element} concept, this cannot be seen as simply adding examples. The addition of concepts such as this does not only provide additional links to a person's {metallic element} concept, but potentially starts to shift their understanding of that concept, modifying the concept by changing its total 'content'.

15.6 Conceptual Realignment – Correcting Discrete Conceptions

Another type of conceptual change might be considered as correcting the conceptual network. Consider that a learner held the conception that copper was a magnetic material: that they had a {copper} concept that included this association, and (considering the link in the other direction) a concept {magnetic materials} that included copper as an example.

This learner may later come to have some experience which they interpret as contrary to this conception. This may be direct laboratory experience of samples of copper not being attracted to magnets. At face value this suggests that the association 'copper is a magnetic material' needs to be modified to 'copper is not a magnetic material': a modification in the {copper} concept (changing a property) and the {magnetic materials} concept (deleting an example).

If the experience that copper did not seem to be in the least attracted to magnets was very surprising, and was the focus of checking and perhaps discussion with lab. partners, followed by a strongly motivated search of sources (the web, textbooks, adult experts), then it is quite possible conceptual change will occur – new experience will have been interpreted in a way that leads to a modification of the existing representation of past experience. Yet, it may be useful to remember how researchers understand this process of changing such representations to occur in the brain.

Perhaps the learner has a chemistry revision notes book (or a personal study wiki), in which she had written 'copper is a magnetic material', or which contains a table of magnetic materials (including copper), or a section on properties of copper (including it being magnetic), or a concept map with a link between 'copper' and 'magnetic materials' labelled 'example'. It would be easy to change the representation of the conception here – deleting, moving, adding the word 'not' at an appropriate place. However, as discussed in Chapter 13, the learner's actual conception that 'copper is magnetic' is understood to be represented in terms of the synaptic strengths of connections between neurones in neural networks in the brain. The learner can deliberately update their revision notes, but revising the brain's representation of experience (*i.e.*, memory) is not a deliberate act, but relies on automatic processes that evolved long before human beings acquired formal language or developed the practice of studying in classrooms. Of course, there are techniques to encourage such modifications – making representations such as notes and diagrams, using the new way of thinking in problem solving, reimagining the experience, deliberately contextualising the new information in relation to related ideas, revising the new information – but if the brain processes for revising the stored representations are not triggered then the new information may be forgotten (or, more strictly, weakly represented in memory, and so later not readily brought to mind).

The human brain does not maintain a totally consistent representation of experience, so it is possible to 'remember' or 'know' contrary things. The brain does, however, appear to have evolved processes of 'memory enhancement' (Walker and Stickgold, 2004), which can modify representations towards consistency: but this occurs over an extended period of time. This seems sensible from an evolutionary perspective, given that our experiences may be misperceived or misinterpreted (so it would usually be inappropriate to overthrow an established way of thinking based on one apparent counter-example). Ultimately, contradiction suggests incomplete understanding – but if new information that was inconsistent with prior experience was never admitted there would be no possibility of correcting false representations of the world. Brains seem to work towards eliminating contradictions in our representations of experience, but usually only slowly. (As our new experiences are understood through the interpretive resources we have available, many potential inconsistencies are never admitted in the first place as we are channelled to experience the world in consistent ways.)

In this hypothetical example, I have suggested that a new experience, actually handling samples of copper and magnets, might be inconsistent with a prior conception (that copper was magnetic). As discussed in this book, much that is learned in chemistry classes cannot be based on such direct experience as a good deal that is taught concerns abstract theoretical ideas that are imposed on, and may be several layers of interpretation away from, what can be directly observed in the laboratory (see Figure 8.3). So, for example, the teacher may have learned to interpret an observation as the presence of double bonds, but, in making the same observation, the student simply sees the colour of a mixture in a test tube fade somewhat.

In general, we might expect the actual handling of materials in this hypothetical case – direct empirical experience – to be more likely to bring about conceptual change than learning from a teacher's comment – 'no that's not right, copper is not a magnetic material', or from a statement spotted on a website or in a textbook. However, research has shown that even when learners are faced with laboratory evidence that their conceptions are false, *and* notice the significance of the evidence, *and* seem to 'change their minds' at the time – that is verbally report no longer holding their prior alternative conceptions – this may not lead to a change that survives in the medium term. Some weeks later students may remember the laboratory demonstration, but with an outcome that is consistent with, and reinforces, their prior alternative conception (Gauld, 1989). Learning has taken place – some conceptual change has occurred – but the brain has adjusted memory to allow a more coherent representation of experience (so what should have acted as a counter-example is now recalled as additional support for an established alternative conception).

Even at the time of meeting counter-evidence, there are many ways a student may decline to be convinced by laboratory experience if it seems inconsistent with their expectations. Kuhn (1970) has argued that major conceptual revelations in science depend upon a scientist recognising the significance of an anomaly that other scientists tolerate. A naive model of science based on simple falsifications would suggest that scientists should pay great attention to findings that contradict hypotheses, but in practice many can genuinely be explained away as human error, measurement error, instrument failure, *etcetera*. It has been argued that when working within a generally productive research programme, it often makes sense to 'quarantine' anomalies (Lakatos, 1970), that is to notice that a result seems, at face value, to contradict current understanding, but then to treat this as a puzzle to be put aside and addressed at another time.

In our hypothetical example of the student failing to attract copper samples towards magnets, the student might suspect that the magnets are not actually magnets, perhaps having become demagnetised; or that the supposed specimens of copper are not actually copper, perhaps being a 'fool's copper'; or that they are composed of copper that has been treated in some way to make it seem non-magnetic, or are perhaps an unusual allotrope/phase of copper that is atypical in this respect. Chemists and other scientists habitually make such moves in their professional scientific work, so we should not be critical of our students for doing the same. The student is using their creativity to imagine hypothetical possibilities (which in principle can be explored further by seeking new evidence). Indeed, it is clearly very sensible, and consistent with the scientific attitude of always being critical and seeking alternative explanations, to consider various potential explanations for what seems an anomalous result.

In terms of conceptual change, the shift from 'copper is a magnetic material' to 'copper is generally/usually a magnetic material' may be a more manageable shift (and perhaps, if the original concept seemed well motivated, even wiser – being based on one particular episode of finding counter-evidence) than the conclusion 'copper is a non-magnetic material'. It could also be a metaphorical weigh station, an intermediate conception (Driver, 1989), on a trajectory towards the more substantive shift: 'copper is a magnetic material' does not admit examples of non-magnetic copper, but 'copper is generally/usually a magnetic material' invites exploration of questions such as 'to what extent?', 'under what conditions?', that can motivate and admit further shifts.

Whether, in this hypothetical case, conceptual change is likely to be triggered by the anomalous experience – such that some months later the student will when probed demonstrate evidence of holding the conception 'copper is a non-magnetic material' and not of holding the conception 'copper is a magnetic material' – will then depend on the particular properties of this prior conception. These relevant dimensions were discussed in Chapter 14. If the student's conception 'copper is a magnetic material' derives from an implicit concept that is activated automatically, rather than an explicit concept deliberately accessed, it will be more difficult to extinguish the tendency to think this way (*i.e.*, this intuition). If this is explicit knowledge, usually accessed in verbal form, it may be easier to change.

There is also the issue of how well committed the learner is to 'copper is a magnetic material'. If this is an incidentally acquired notion of no great significance, it is more likely to be changed than if the student feels this is some fundamental principle (consider a scientist finding evidence suggesting that energy is not conserved in some chemical reactions, or discovering new elements with nuclei containing non-integral proton numbers – such as an element between nitrogen and oxygen).

There is also the issue of embeddedness. In Chapter 2, it was suggested that concepts are nodes embedded in an interconnecting network. Some concepts are more highly connected than others. A conception that is part of a more extensive network may be more difficult to shift. In our hypothetical example, the conception 'copper is a magnetic material' may not be an isolated 'fact' but may follow from how the student's {copper} concept is linked to their {metals} concept. That is, if copper is understood as a fairly prototypical metal and for this student there is a strong association between the {metals} and {magnetic materials} concepts such that being magnetic is considered one of the core characteristics of metals, then this may be a tenacious conception. In this case, questioning the magnetic nature of copper raises questions about the magnetic nature of other metals: iron, lead, zinc, *etcetera*. This issue is discussed further in the other examples later in the chapter.

Finally, there is the issue of manifold conceptions. Even if a learner has the conception 'copper is a magnetic material' represented in their conceptual structure, this may not exclude an alternative conception 'copper may not always be a magnetic material'. This would be logically excluded if there was total commitment to the conception 'copper is a magnetic material', but if that conception had a status of being probable, feasible, likely, but was not definitively committed to, then the individual could have this conception represented as well as an alternative conception inconsistent with it.

Perhaps as a young child the person developed the idea that all metals were magnetic, but since has read and heard things that suggest otherwise. So, perhaps a conception along the lines 'metals are magnetic' is part of the {metals} concept, but no longer treated as highly committed, and not part of the 'hard core' (Lakatos, 1970) of that concept, and so a conception along the lines 'metals may not always be magnetic' is also represented. In this case the new experience is not just contrary to one well-established conception, but also potentially supportive of a potentially competing conception that is also available. If this is the case, (a) there are available interpretive resources to make sense of the laboratory experience without questioning the obvious interpretation (this copper is not magnetic); (b) that alternative (*i.e.*, here *more* canonical) conception will be activated in terms of the new experience, and the activation of a conception tends to strengthen it, making it even more readily activated and accessed in future; (c) a representation of the laboratory experience in memory can potentially be linked to a prior conception (without distortion) such that the memory of the actual experience is more likely to be recalled later, and if so, is less likely to be distorted (as in remembering the copper being attracted to the magnet, even though that was not what was observed) when recalled.

Strengthening the 'metals may not always be magnetic' conception in this way by making it more easily activated (so 'brought to mind') and by anchoring a mental representation of new experience that can be interpreted as evidence to support it, changes the relative strength (and therefore ease of activation) of the two contrary conceptions within the conceptual ecology, and may be part of a shift that makes up conceptual change – albeit not instantaneous conceptual change, as in the sense of suddenly no longer ever thinking X and now always thinking Y.

Probably, the prior conception would not be eliminated in the sense of being edited out of the brain by deliberate synaptic changes. It seems likely there is no active erasing of memory. The electrical stimulation of brain areas of conscious patients during surgery – a common process used to avoid mistakes in neurosurgery – has been known to activate clear memories that the patient claims not to have accessed for decades. So, it seems likely that natural processes gradually degrade brain circuits, and regular use strengthens them and supports their maintenance. Conceptions tend to fall into disuse, and consequently become more difficult to activate (as students who do not revise tend to find out), rather than being actively eliminated from the system.

15.7 Restructuring Conceptual Frameworks

The previous example concerned the magnetic properties of copper, and what might be involved in a student learning that, contrary to their conception of metals as magnetic, copper is not a magnetic material. The metallic nature of iron and steels, and perhaps nickel and cobalt, is met in school. That metals are not generally magnetic is usually also learned during compulsory education.

15.7.1 But Perhaps Copper Was a Magnetic Material after All?

More sophisticated notions of magnetism are not usually met until higher education. So more advanced students may learn about antiferromagnetism, paramagnetism, and diamagnetism, as different kinds of magnetic behaviour or properties that materials can have. So, generally, when students learn about these classes of magnetic properties this is against the background of a well-established conceptual framework of ideas about magnetism: its link with ferrous metals, and its giving rise to a strong enough effect to be observable without special equipment – the magnetic catch that keeps the fridge door closed; the magnet used to pick up pins or other small ferrous objects from the floor; the needle in a compass that points in a reference direction. They probably know that a sample of magnetic material needs to be magnetised in some way before it acts as a magnet. A learner probably knows something about electromagnetism – which may be more or less well integrated into their understanding of magnetic materials.

Studying in an advanced course, where the student has already learned background material to support learning about different magnetic behaviour, and so has acquired prerequisite concepts (perhaps {electronic configurations}, {electron spin}, {the magnetic quantum number}) therefore introduces new notions of magnetism that are relevant to, but do not initially fit well with, an existing conceptual framework for understanding magnetism.

Within the perspective of the advanced course, and the new magnetic concepts, copper (metal) is diamagnetic. Learning that copper exhibits diamagnetism can develop a learners' {copper} concept as they can add a new 'copper is diamagnetic' conception associating their well-established {copper} concept with the newly acquired {diamagnetism} concept. A more interesting question is how their well-established {magnetic materials} concepts is related to new learning.

If a student's concepts had been represented in a set of (paper-based) concept maps that they maintained and updated, we can appreciate the changes that might be indicated. In effect, the existing concept of {magnetic materials} needs to be relabelled, changing the concept name to {ferromagnetic materials}, and subsumed under a new, more encompassing, overarching {magnetic materials} concept (and the {magnetism} concept relabelled to {ferromagnetism} and subsumed under a new broader {magnetism} concept). It may be possible to do this with some corrections and additions – the corrections might involve crossing-out, erasing with a rubber, or using correction fluid or sticky labels to cover-up old representations that needed to be deleted or changed. Modifying a computer-based concept map might involve a little typing and a few mouse clicks.

Of course, even if a student does externally represent their knowledge in this form, and can carry out such changes, the internal mental representation is not open to deliberate interventions to cross-out or white-out material that needs to be updated. In addition, perhaps on looking at their revision notes, the student might decide that changing a concept map to fit the new learning might be impractical due to the amount of crossing out and additional material needed: it may be better to start again on a new page, and draw a new concept map not constrained by the structural legacy of the previous representation. That certainly does not have a ready mental analogue – conceptual representation is not kept on a series of internal memory boards that we can pull out to slot in replacements when appropriate, but rather is based on an extensively connected set of neural circuits.

What seems more likely is that the existing conceptual framework, including the individual's {magnetic materials} and {magnetism} concepts, which is probably perfectly adequate for making sense of most aspects of magnetism met in everyday life outside an advanced chemistry course, is maintained and continues to be applied in what seem pertinent contexts, and does not undergo any dramatic modification.

However, a new conceptual framework is developed due to the linking of various concepts that are being closely associated in the material presented in teaching – concepts that may be relatively well established, having initially been acquired some years before (*e.g.* {electronic configuration}, {quantum numbers}, {(fermionic) spin}, *etc.*) and new concepts relating to classes of magnetic behaviour: {ferromagnetism}, {antiferromagnetism}, {diamagnetism}, {paramagnetism}; each with an associated concept relating to a class of material: {ferromagnetic materials}, *etc.* Those new concepts will be subsumed under a new concept, likely understood as {magnetism}.

If so, at this point the learner has two discrete {magnetism} concepts:

- One is long established, and widely applicable in most contexts, but deficient when answering certain types of university examination questions or as the basis for going into chemistry research or practice in some specialised areas.
- The other is newly formed (and so, at this point, less robust), and is perceived as relevant in some particular technical contexts; but is seldom especially pertinent to understanding how the idea of magnetism is used in most everyday situations.

If the new framework is reinforced over time through regular explicit references in classes and use in problem-solving, *etcetera*, it will become consolidated (and so more easily accessed), and will come to be readily brought to mind in contexts where it is recognised as relevant; but these will mostly be specialised rather than everyday uses.

15.7.2 Integrating Conceptual Frameworks

Clearly there is potential for linking these two frameworks. A good student can form a strong association between the new {ferromagnetism} concept (that is subsumed within the new {magnetism} concept) and the pre-existing well-established {magnetism} concept, and similarly with the new {ferromagnetic material} concept and the pre-existing {magnetic material} concept – which I will for clarity denote {magnetic materials$_{established}$} rather than the new {magnetic materials$_{subsuming}$} concept. So, for example, there may be many relevant associations that are part of the pre-existing {magnetic materials$_{established}$} concept that are never raised in chemistry lectures when ferromagnetism is presented, as they are not of particular relevance to the chemistry curriculum. A student with a strong association between {magnetic materials$_{established}$} and {ferromagnetic materials} concepts along the lines of the conception that 'ferromagnetic materials are what are usually just called magnetic materials' will be able to apply associations from the {magnetic materials$_{established}$} concept to discussion of ferromagnetic materials without having been explicitly taught these associations apply (just as how in the example discussed above, a student learning niobium is a metallic element may think of niobium cations without anyone suggesting this possibility to them).

Will these two distinct frameworks coalesce? Unlike a representation on paper, the integrity of distinct representations of concepts in the human brain are not based on physical separation but on the strength of synaptic

How Do Students' Concepts Develop?

connections – so even if two representations are physically separated in the cortex, they could become so well associated that they are strongly interlinked and, in principle, could become subsumed into what is functionally a single entity. The representations of our fictional student's {magnetic materials$_{established}$} and {ferromagnetic materials} concepts within conceptual structure should not involve any contradictory elements, so presumably with sufficient activation of links between them they *could* effectively coalesce in terms of how a representation in memory is activated and used, even if distributed across brain areas. Indeed, there is evidence that representations can be well integrated and coherent despite being distributed physically in the cortex (Alvarez and Squire, 1994).

However, such integration would require sufficient co-ordinated activation of the two different concepts to strengthen the links between them. More likely, many chemistry undergraduates largely activate their {ferromagnetic materials} concept only in particular study contexts, with little need to relate this to other associations of their {magnetic materials$_{established}$} concept, and after graduating (there will be exceptions of course) seldom have reason to call upon their {magnetic materials$_{subsuming}$} concept; and so their {magnetic materials$_{established}$} concept would be preferentially activated in contexts where magnetism is relevant. Likely, they would seldom, then, activate their {ferromagnetic material} concept.

The reader may reasonably question this hypothetical narrative, which is not based on any direct study of students learning advanced magnetism concepts. However, the account is informed by a range of research and scholarship that suggests that complex conceptual change is usually more a matter of gradual shifts than simply learning a new way of thinking that effortlessly replaces conceptions now recognised to be deficient.

15.8 Studies of Conceptual Change

There have been various studies on conceptual change relevant to chemistry learning. Thagard analysed historical cases, using a notion of explanatory coherence (Thagard, 1989), and has suggested that the resistance of chemists such as Priestley to abandoning the phlogiston theory and adopting the ideas of the Lavoisiers can be seen as entirely rational (Thagard, 1992). If Priestley had spent many years making sense of his laboratory experiences in terms of the phlogiston theory, and had become adept at conceptualising chemical phenomena in this way, then he would have developed extensive, well-integrated, explanatory frameworks, whereas the less familiar oxygen theory would have subjectively, but quite reasonably, seemed to have offered much less explanatory power. Although Lavoisier may have initiated a revolution in chemistry, it is much less clear that when his ideas were first proposed they offered a clearly superior way of thinking that justified abandoning phlogiston (Chang, 2012).

Chi and her associates have looked in particular at conceptual change in terms of shifting ontologies (Chi, 1992). Many aspects of learning chemistry can be usefully analysed in terms of how learners understand the basic categories of things that exist in the world. For example, how categories such as metals, acids and oxidising agents are understood can be seen as an ontological issue (see Chapter 7). So, for example, learning chemistry may involve shifting from a metal *vs.* non-metal dichotomy, to an electronegativity dimension. In parallel, students may need to abandon a dichotomous ontology of covalent *vs.* ionic bonding (where polar bonds tend to be seen as a subcategory of covalent bond) to a continuous dimension (Taber, 2013), or, better, several dimensions with 'pure' forms of bonding representing ideal apices that may not have any real representatives (*e.g.*, no substance has 'pure' ionic bonding nor 'pure' metallic bonding). Individual students may have their own idiosyncratic conceptions that, from a canonical perspective, require some modification. I taught one student who heard about ionic and covalent bonding in some of her chemistry classes, but only about single and double bonds from another teacher (who was teaching introductory topics in organic chemistry and had little need to refer to ionic substances). She had reached a point of questioning whether these were actually referring to quite different things, and whether she needed to modify her understanding to integrate these two set of ideas.

Chi has, in particular, explored how people categorise 'things' on fundamentally different ontological trees. So, material objects (conical flasks) are different from processes (oxidation) and mental experiences (imagining a methane molecule). Chi argues it may not be viable to shift something from one basic ontological tree to another. This has much relevance to learning science where a child may understand heat and electricity and chemical bonds as 'things' rather than (as is implied in the teaching they meet) 'processes'. After all, early chemistry used phlogiston and caloric as key explanatory ideas, and these would be seen as substantial – as materials. Moving beyond these ideas meant forming an ontologically distinct way of thinking. A modern understanding of combustion posits it as a chemical reaction *between* oxygen (or perhaps chlorine or fluorine) and another substance: not simply the addition or removal of oxygen rather than the addition or removal of phlogiston. From this perspective, definitions of redox processes that are framed in terms of addition/removal or oxygen/hydrogen/electrons may be unhelpful to learners in focusing on an overall process.

Joan Solomon (1983) found that learners can often hold multiple frameworks of the same concept – thinking and talking about energy one way in the science class, but quite differently among peers in everyday contexts. The science teacher and philosopher Bachelard (2002) analysed the nature of modern scientific concepts, and argued that the way these are learnt and used betrayed evidence of how they have evolved over different historical stages of development of scientific thought. He suggested that the way scientists think about fundamental ideas (such as energy, mass, *etc.*) reflects not only the current canonical understandings but also fossilised impressions of earlier notions. Scientists' own concepts have this multi-facetted nature, which he pictured as a kind of conceptual profile. Mortimer (Mortimer and El-Hani, 2014) has developed this idea of conceptual profiles and has undertaken extensive work suggesting that people often have a manifold set of ways of thinking about some notions, with different versions being more or less likely to be activated overall, and with contextual cues tending to trigger different 'versions'.

Although conceptual change is such an important topic, there have not been a great many studies that explore it in chemistry learning in any detail. One study that is very relevant (although framed as physics learning) looked at high-school students developing understanding of a quantum model of the atom. Petri and Niedderer (1998) described how one student (referred to as Carl) began with an 'initial conception' of the atom in terms of a planetary model. They discuss how this remained part of the developing {atom} concept, as it became complemented by other models: the probability-orbit model; the state-electron model; and the electron-cloud model. They discuss a 'learning pathway' where "Carl developed new conceptions about the atom and integrated them into a new, fairly complex and stable cognitive system, which ultimately contained several parallel conceptions" (p. 1079).

As this study employed a range of data-collection methods, and had a longitudinal element, following students through a teaching sequence, Petri and Niedderer were able to follow Carl's conceptual development during the study, and characterise "the final state of Carl's cognitive system is an association of co-existing conceptions" (p. 1083):

> several conceptions co-exist and are connected to form different layers of the cognitive system, with a metacognitive layer on top. Carl was able to reflect on differences, problems and advantages of each conception. The most powerful element in this configuration was the planetary conception; it is the first to be considered, but the status of the other elements is higher. (Petri and Niedderer, 1998, p. 1083)

So, in this case, Carl operated with, and was aware he operated with, a concept that consisted of several distinct facets. Carl knew that the new models he had learnt had more scientific worth than this initial planetary model (he could deliberate on that): however, that model was well established and was more readily activated and so tended to be brought to mind first.

There are parallels here with the case of Tajinder, introduced in Chapter 14. Tajinder started his college chemistry course with an initial conception of bonds as being the means by which atoms were able to obtain octets or full shells (Taber, 2001b). He retained this as part of his {chemical bond} concept throughout his two-year (pre-university) college course, supplementing it with explanatory principles based upon Coulombic interactions and minimising energy. As with Carl, Tajinder showed metacognitive insight into his thinking – he considered the three principles as complementary narratives that he could choose between, and that might all be valid at the same time. Slowly, the tendency to activate ideas about needing full shells was attenuated, and explanations based on electrical forces came to be used more. The idea of a conceptual ecology, introduced earlier in the chapter, seemed helpful in thinking about this case:

> The notion of a conceptual ecology is metaphorical, but fertile. This in-depth case study of science learning has provided evidence to show how a learner may simultaneously hold manifold explanatory conceptual schemes for a particular topic area .... evidence has been presented to show why one of Tajinder's explanatory principles (minimum energy) failed to put down deep roots, and how, over time, the topology of Tajinder's cognitive structure came to favour the growth of a new alternative scheme (Coulombic forces) over the 'native species' (octet rule). This on-going conceptual revolution did not involve a gestalt-shift between the competing conceptions, but rather a gradual process of succession within the shifting sands of the conceptual ecology. (Taber, 2001b, p. 750)

So, Tajinder showed conceptual development, but it was not a matter of replacing one conception with another, but (as with Petri and Niedderer's Carl) a case of adding new conceptions as complementary layers of a concept. As teaching reinforced the use of more canonical ideas (like Coulombic forces) these tended to be activated more readily, but without the initial facet (atoms wanting octets) being extinguished. Tajinder went on to study a science-based course in university, but found this did not give him occasion to apply his chemistry learning. Several years after completing his college chemistry course Tajinder was interviewed again (Taber, 2003), at which time he was found to readily activate ideas about bond formation minimising energy, but as an adjunct to ideas about octets. Without regular reinforcement in the interim, his learning about Coulombic forces giving rise to bonds was not readily activated, but the prior conception in terms of atoms wanting octets of electrons (which had slowly become called upon less frequently for explanations when he was actively learning chemistry) was once again most readily brought to mind. Ecologies shift under applied pressures, and if we do not work to maintain our gardens they can evolve in ways quite different from how we might prefer.

That particular study reminds us that, as some of the early work on alternative conceptions suggested, some conceptual frameworks that students develop are resistant to being challenged by teaching once they are established in a learner's imagination. Teachers can seek to guide students' conceptual development, but sometimes there are considerable biases in students' starting points that are difficult to overcome:

> Students seem to find the Octet framework persuasive, and it has been found to be tenacious ... so that students commonly interpret chemistry

through aspects of this conceptual framework even after they have been taught about more scientifically valid alternatives. Over his course Tajinder learnt to rely less on this way of thinking about chemistry... [Later, however] Tajinder's thinking ... showed a number of features of octet thinking: applying more restricted categories of bonding; defining covalent and ionic bonding in terms of electron sharing and transfer; considering bonding electrons in molecules to 'belong' to specific atoms; considering polar bonds to be covalent or ionic; ignoring bonding between molecules; and explaining reactions as if the reactant species were discrete atoms. In particular Tajinder's ... highly anthropomorphic explanations (of atoms thinking, aiming, liking, trying, striving and feeling) is very typical of the alternative framework...

Of the three explanatory principles for bonding that Tajinder had commonly used during his college course, his Coulombic Forces [principle] should have been the highest status...: it was coherent, applied both to examples that could and could not previously be explained, it had a wide range of application and was consistent with explanations used in the other parts of the course (about atomic structure, patterns in ionisation energy, the shapes of molecules, *etc.*). ... Yet [later] Tajinder seemed to have 'regressed' to largely using his pre-college level ideas as a starting point for answering interview questions. (Taber, 2003, p. 273)

Conceptual change occurs within a complex 'ecological' context. Teachers cannot be expected to engineer conceptual change simply by designing well-considered presentations of the concepts of chemistry – even when those presentations are informed by the literature on learners' alternative chemical conceptions. Rather, teaching is more like gardening or managing a natural space like a wood or a park. The teacher has to constantly venture into the environment and see what is growing and where – and then proceed to actively interact with that environment. Teaching is an ongoing process of encouraging shifts within learners' diverse and actively developing conceptual ecologies.

References

Alvarez P. and Squire L. R., (1994), Memory consolidation and the medial temporal lobe: A simple network model, *Proc. Natl. Acad. Sci. U. S. A.*, **91**, 7041–7045.

Ausubel D. P., (2000), *The Acquisition and Retention of Knowledge: A Cognitive View*, Dordrecht: Kluwer Academic Publishers.

Bachelard G., (2002), *The Formation of the Scientific Mind: A Contribution to a Psychoanalysis of Objective Knowledge*, McAllester Jones M. (Trans.), Manchester, UK: Clinamen.

Chang H., (2012), *Is Water H_2O? Evidence, Realism and Pluralism*, Dordrecht: Springer.

Chi M. T. H., (1992), Conceptual change within and across ontological categories: examples from learning and discovery in science, in Giere R. N. (ed.), *Cognitive Models in Science* (vol. XV), Minneapolis: University of Minnesota Press, pp. 129–186.

diSessa A. A., (1988), Knowledge in peices, in Forman G. and Pufall P. (ed.), *Constructivism in the Computer Age*, New Jersey: Lawrence Erlbaum Publishers.

Driver R., (1989), Students' conceptions and the learning of science, *Int. J. Sci. Educ.*, **11**(special issue), 481–490.

Gauld C., (1989), A study of pupils' responses to empirical evidence, in Millar R. (ed.), *Doing Science: Images of Science in Science Education*, London: The Falmer Press, pp. 62–82.

Gilbert J. K., Osborne R. J. and Fensham P. J., (1982), Children's science and its consequences for teaching, *Sci. Educ.*, **66**(4), 623–633.

Harkins W. D. and Liggett T. H., (1923), The discovery and separation of the isotopes of chlorine, and the whole number rule, *J. Phys. Chem.*, **28**(1), 74–82.

Krathwohl D. R., Bloom B. S., and Masia B. B., (1968), The affective domain, in Clark L. H. (ed.), *Strategies and Tactics in Secondary School Teaching: A Book of Readings*, New York: The Macmillan Company, pp. 41–49.

Kuhn D., (1989), Children and adults as intuitive scientists, *Psychol. Rev.*, **96**(4), 674–689.

Kuhn T. S., (1970), *The Structure of Scientific Revolutions*, 2nd edn, Chicago: University of Chicago.

Lakatos I., (1970), Falsification and the methodology of scientific research programmes, in Lakatos I. and Musgrove A. (ed.), *Criticism and the Growth of Knowledge*, Cambridge: Cambridge University Press, pp. 91–196.

Luria A. R., (1976), *Cognitive Development: Its Cultural and Social Foundations*, Cambridge, Massachusetts: Harvard University Press.

Mortimer E. F. and El-Hani C. N., (2014), *Conceptual Profiles: A Theory of Teaching and Learning Scientific Concepts*, Dordrecht: Springer Science & Business Media.

Nakiboglu C. and Taber K. S., (2013), The atom as a tiny solar system: Turkish high school students' understanding of the atom in relation to a common teaching analogy, in Tsaparlis G. and Sevian H. (ed.), *Concepts of Matter in Science Education*, Dordrecht: Springer, pp. 169–198.

Perry W. G., (1970), *Forms of Intellectual and Ethical Development in the College Years: A Scheme*, New York: Holt, Rinehart & Winston.

Petri J. and Niedderer H., (1998), A learning pathway in high-school level quantum atomic physics, *Int. J. Sci. Educ.*, **20**(9), 1075–1088.

Piaget J., (1970/1972), *The Principles of Genetic Epistemology*, Mays W. (Trans.), London: Routledge & Kegan Paul.

Popper K. R., (1961/2011), Facts, standards, and truth: a further criticism of relativism, in Popper K. R. (ed.), *The Open Society and Its Enemies* (vol. Addenda), London: Routledge, pp. 485–511.

Solomon J., (1983), Learning about energy: how pupils think in two domains, *Eur. J. Sci. Educ.*, **5**(1), 49–59.

Taber K. S., (2001a), The mismatch between assumed prior knowledge and the learner's conceptions: a typology of learning impediments, *Educ. Stud.*, **27**(2), 159–171.

Taber K. S., (2001b), Shifting sands: a case study of conceptual development as competition between alternative conceptions, *Int. J. Sci. Educ.*, **23**(7), 731–753.

Taber K. S., (2003), Lost without trace or not brought to mind? – a case study of remembering and forgetting of college science, *Chem. Educ. Res. Pract.*, **4**(3), 249–277.

Taber K. S., (2005), Learning quanta: barriers to stimulating transitions in student understanding of orbital ideas, *Sci. Educ.*, **89**(1), 94–116.

Taber K. S., (2009), College students' conceptions of chemical stability: The widespread adoption of a heuristic rule out of context and beyond its range of application, *Int. J. Sci. Educ.*, **31**(10), 1333–1358.

Taber K. S., (2011), The natures of scientific thinking: creativity as the handmaiden to logic in the development of public and personal knowledge, in Khine M. S. (ed.), *Advances in the Nature of Science Research – Concepts and Methodologies*, Dordrecht: Springer, pp. 51–74.

Taber K. S., (2013), A common core to chemical conceptions: learners' conceptions of chemical stability, change and bonding, in Tsaparlis G. and Sevian H. (ed.), *Concepts of Matter in Science Education*, Dordrecht: Springer, pp. 391–418.

Taber K. S. and Watts M., (1996), The secret life of the chemical bond: students' anthropomorphic and animistic references to bonding, *Int. J. Sci. Educ.*, **18**(5), 557–568.

Thagard P., (1989), Explanatory coherence, *Behav. Brain Sci.*, **12**(03), 435–467.

Thagard P., (1992), *Conceptual Revolutions*, Oxford: Princeton University Press.

Thomson J. J., (1913), Bakerian Lecture: Rays of Positive Electricity, *Proc. R. Soc. London*, **89**(607), 1–20.

Thomson J. J., (1922), On the Analysis by Positive Rays of the Heavier Constituents of the Atmosphere; Of the Gases in a Vessel in which Radium Chloride Had been Stored for 13 Years, and of the Gases Given off by Deflagrated Metals, *Proc. R. Soc. London*, **101**(711), 290–299.

Vygotsky L. S., (1934/1986), *Thought and Language*, London: MIT Press.

Vygotsky L. S., (1978), *Mind in Society: The Development of Higher Psychological Processes*, Cambridge, Massachusetts: Harvard University Press.

Walker M. P. and Stickgold R., (2004), Sleep-Dependent Learning and Memory Consolidation, *Neuron*, **44**(1), 121–133.

CONCLUSION

CHAPTER 16

Lessons for Chemistry Education

This chapter considers what the discussion and analyses in the earlier chapters might suggest in terms of implications (i), for teaching; and (ii), for fertile directions for further research.

16.1   Lessons for Chemistry Teachers

Careful readers of the preceding chapters of this book will likely have drawn their own conclusions about the implications of the account offered here for teaching. Perhaps what we know about conceptual learning may even be reassuring to some teachers. There is an important message that teaching that is based on strong subject knowledge, carefully considered and sequenced, and presented clearly, is not enough to ensure learners understand as intended – and certainly not enough to make sure students develop canonical concepts. These are necessary, but not sufficient, conditions for teaching the abstract and often nuanced concepts of the discipline. At one level, that explains and justifies how many students may so commonly perform so badly in tests and examinations (and why so many do not feel competent they could continue their study of chemistry further).

Teaching involves applying a model of how the learner will understand teaching, that is, a model of how learners will understand teaching presentations in turn of existing prior learning. Teaching goes wrong because learners do not make the expected links, either because they do not make links at all – the expected prerequisite learning is absent; or the relevance of prior learning is not recognised, so it is not brought to mind – or they make different links – their prior learning of a relevant concept involves an alternative conception; they make a creative non-canonical link to prior learning that seems relevant, but is unhelpful and distorts the intended meaning of teaching (Taber, 2001, 2005).

16.2 Responsibilities of the Chemistry Teacher

Perhaps some might suggest it does not really matter if only a minority of learners master the conceptual content of chemistry instruction. If enough students do well enough in school chemistry to feed into chemistry and related courses at degree level; and if teaching on degree courses is good enough to enable sufficient numbers of students – perhaps mostly those with strong and canonical prior learning, perhaps often those with very high levels of motivation, and perhaps usually those with more sophisticated epistemological commitments – to do well enough on their degree programmes to feed into research and industry, then that is good enough.

It might even be suggested, mischievously, that if we *could* teach all the students in chemistry classes well enough that they all develop a strong understanding of all the concepts in the curriculum, and therefore that they all acquire well-established personal concepts sufficiently consistent with canonical chemical knowledge to enable them to successfully answer all our test questions, then we *would* have a problem. Certainly, if we see education as being largely about sorting the students, and identifying the best candidates for progression.

Such a stance is not acceptable (as surely to make the study of chemistry worthwhile for students, all learners should be able to access the ideas being taught, and learn them sufficiently to acquire creditable concepts they can apply in canonical ways), and invites the criticism that a chemistry curriculum is not suitable for many learners if they are expected to fail to make sense of, and meaningfully learn, much of it. In compulsory school courses (or indeed in subsidiary degree courses that are mandatory for students taking, say, health-science degrees), there is a responsibility to teach at a level matched to the students; and in elective courses, such as chemistry degrees, that are only open to well-qualified candidates, there is a responsibility to select students suitably prepared, or provide them with appropriate bridging support (be that foundation years, extra tutorial support, mentoring, or whatever can be shown to suffice).

In any case, it seems unlikely we would ever be able to teach that well that we created the problem of all our students scoring maximum marks on summative tests. In addition, if we could, we should not find reasons to complain! That would simply require us to design more challenging test items to differentiate within classes of students who all held well-taught, well-learnt, and canonical, chemical concepts.

16.3 Understanding Differently

We probably do not have to worry about such a scenario – as the earlier chapters in this book have suggested, teaching a subject such as chemistry is inherently challenging. Most students in most chemistry classes have some missing prerequisite knowledge for what is being taught, and alternative conceptions of some key concepts – including alternative conceptions that they consider to be just what they were previously taught. Students often do not fully understand the material being taught in class. Students often misinterpret the information being presented: they come to interpret it differently from that intended. That much will have been familiar to many readers before setting out on this volume – although I hope readers will now have a better feel for why this is so often so, and appreciate how it derives from inevitable complications that go beyond simplistic notions of poor teaching or weak students.

Even when learners understand teaching as intended, they may later – and sometimes not that much later – appear to have forgotten what they were taught, or remember it quite differently. That is probably also somewhat familiar to most teachers, even if it seems a little more mysterious – especially to teachers who were lucky enough to find most chemistry they studied as largely fitting their intuitions and so being readily learned (Taber, 2004). Being a very strong learner of chemistry may not always be an ideal qualification for being an insightful teacher of many of the learners in chemistry classes. Strong subject knowledge is certainly important for a teacher – but having acquired strong subject knowledge without having to sometimes struggle may actually make it more difficult to appreciate the experience of many students who, in some topics, may, even when committed to studying, find it very difficult to ground what they can take from teaching in anything they already understand. I hope the accounts offered in this book at least explain why studious, motivated, hard-working students do not automatically learn what was taught.

16.4 Checking for Shared Understandings

At one level the challenge is immense – especially when teaching large classes. However, there are some simple principles to support effective teaching. These are not panaceas, but they will at least help in taking a scientific approach that attempts to shape teaching so to better support learning. These points may potentially appear trite enough to be little more than proverbs – but the groundwork in the earlier chapters suggests they are not trivial when deliberately and consistently built into a teaching approach:

- being explicit – about what is assumed in terms of excepted prior learning, and how teaching is meant to be understood;
- seek a dialogue to check student understanding at all points during teaching (not only through the examination).

Teaching is an activity that is intended to bring about specific learning, and so is essentially about how the student understands the information presented, and not just how the teacher understands it. Teaching that focuses on the latter without considering the former is destined to often be of limited effectiveness. Teachers, therefore, need to be concerned with the range of mental resources that students bring to class, and how these are used to make sense of teaching. It is fine to give students some responsibilities, to monitor their learning, and check they understand material, and organise how to revise their classes, but most cannot do that effectively without support. In particular, although a student should be aware whether they feel they understand something or not, that does not give any assurance that when they do understand, they understand canonically. The whole extensive literature on student alternative conceptions (Taber, 2009) can be seen to be just as much as a celebration of how the human imagination is able to find alternative ways – sometimes very creative and idiosyncratic ways – to understand, as it is as a catalogue of how so often learning 'goes wrong' from the perspective of what is set out in a curriculum.

16.5 Always Imagine, but Never Assume

As students may show diverse conceptualisations in learning chemistry, the effective chemistry teacher needs to be creative in imagining the possibilities for how students think. Teaching is a good deal like doing science, in that there is a creative step when possibilities are imagined (for example, about a potentially productive analogy or simile: 'students in the class likely know about X, and so can appreciate how Y can be somewhat like X in some ways'), but this conjecture must not be assumed, but must rather be seen as a hypothesis to be tested in the laboratories of the classroom and the students' minds. 'Always imagine, but never assume' could be taken as the credo of both the scientist and the teacher.

Teaching involves making the unfamiliar familiar, and often in chemistry we cannot simply do this by showing examples of the actual entities we refer to (what can we actually point at on the laboratory bench when suggesting 'look, here is a double bond/transition state/nucleophile/inductive effect...'). Even when we can in principle point at an exemplar ('look here is a redox reaction'), it may be difficult for learners to abstract the critical characteristics (*e.g.*, what is it that various redox reactions have in common that make them examples of redox – and how does this relate to what can be directly observed in those reactions? (See Figure 16.1, *cf.* Figure 8.4)). So, a good deal of chemistry teaching involves making the unfamiliar familiar by pointing out how the unfamiliar is *somewhat like* something that is already familiar. That may involve models, diagrams, analogies, metaphors, and so forth, as well as material previously taught on a course.

This can be the basis of effective teaching, but – as described in earlier chapters – the nature of concepts and conceptual learning presents challenges. The teacher operates (implicitly, if not deliberately) with a mental model of how teaching will be understood. An analogy will be offered because it relates to something assumed to be familiar to learners, and so

Lessons for Chemistry Education

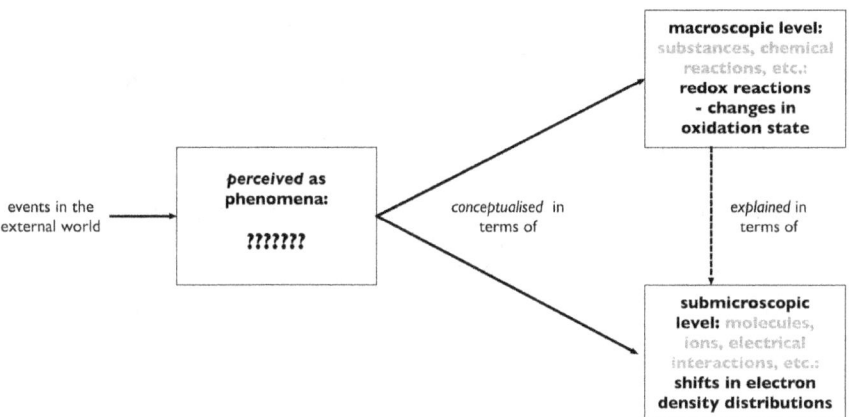

Figure 16.1 What is the common phenomenon that chemists abstract as being 'redox'?
Adapted from Taber, 2013a with permission from the Royal Society of Chemistry.

where it is assumed students will be able to recognise the positive analogy (the features of the analogue structure that map onto the target structure) and also appreciate which other features of the analogue they should ignore. The teacher imagines this is a possibility, and one that can be employed pedagogically. But, the cardinal rule is that the teacher should never assume. So, a teacher should:

- *never assume* that students have the prerequisite learning expected at this stage of their course (prior learning may be missing, or they may have alternative conceptions);
- *never assume* that students have the particular interpretive resources needed to appropriately interpret models, analogies, metaphors, *etcetera*;
- *never assume* that students will automatically appreciate which features of an analogy, metaphor, model, *etcetera* are intended to be understood as relevant to the material to be learned, and which should be discounted.

This is certainly not to suggest such devices should not be used: they may be essential for helping learners to appreciate the target knowledge, especially when abstract ideas are being taught. Moreover, teaching is usually less dry for learners when a wide range of such devices are included in teacher presentations. So, the mantra should be: never assume, always check:

- *check* that students have the prerequisite learning expected at this stage of their course;

- *check* that students have the particular interpretive resources to interpret analogies, metaphors, *etcetera*;
- *check* that students appreciate which features of an analogy, metaphor, model, *etcetera*, are intended to be understood as relevant to the material to be learned; and which should be discounted.

One can imagine manifold points in teaching episodes where the sensible next teaching move is to pause and 'check'. The teacher needs to keep in mind that the descriptor 'obvious' is relative: what is obvious from the expert perspective, may be obscure (or, at the very least, may benefit from reiteration) to the novice. So, for example, even if it seems obvious, the teacher may usefully check:

- (Check) do students appreciate that this two-dimensional figure is meant to represent a three-dimensional structure?
- (Check) do students appreciate the sticks in this physical model are needed to maintain the structural integrity of the model, and are not meant to represent part of the actual crystal structure?
- (Check) do students appreciate that in this context 'acid' means a Lewis acid?
- (Check) was the reference to the displacement reaction in terms of dance partners at the disco appreciated as an anchoring image that should not be taken too seriously?
- (Check) how students understand the symbolism of the circle used in representing benzene given that research suggests some learner interpret this as denoting something inside the hexagon.

Etcetera.

A particular issue concerns the neutral and negative parts of analogies – where the analogue has features that do not relate to anything in the target, or features that do not transfer across to the target. If the planetary model of the atom is compared to the solar system (assuming the usefully relevant features of the solar systems are already familiar – which may need to be checked), do students appreciate that there are some quite important differences between these systems, as well as appreciating the inherent limits of this model of atomic structure that means it has a limited range of useful application? They may not (Taber, 2013b).

16.6 Imagine (and Test Out) Alternative Possibilities

A sensible strategy is to offer multiple comparisons wherever possible. This can help different students in anchoring new abstract ideas on something that is well grounded in their particular thinking. Moreover, where a range of different comparisons are offered this can make it easier to see the limitations of these options as starting points for thinking – as something that we are going to develop (once initially familiar in terms of comparisons students are comfortable with) through the teaching as actually being quite distinct, and having its own nature.

So, the notion that a covalent bond is two atoms 'sharing' electrons is quite limited, but uses a familiar social notion as a metaphor than can start to make the unfamiliar familiar. Are there other anchoring points, likely resources available to these students in their thinking that could help to get them thinking in productive ways? Can we also talk about the covalent bond as acting like an arrangement of magnets? Is there any value in referring to the molecule as like a small jigsaw puzzle where the atoms with unpair electrons are like pieces that fit together?

As long as a range of comparisons that are accessible to learners is suggested, it may not be critical how strongly they reflect the target concept. If we teach 'the covalent bond is a shared pair of electrons' we teach by rote. If we offer an account of the covalent bond in terms of electrical forces (which many students will not be very familiar with at the point they are first taught about chemical bonding) but then moot various comparisons that are discussed and explored, the outcome may be that the learners see why these devices can be dismissed as strong models of the bond, as they have developed a {covalent bond} concept that has become familiar in its own terms:

> some people describe the bond as a shared pair of electrons: what might they mean by that...I've heard this compared to some magnets that both repel and attract each other at the same time, do you think that would work as a bond...someone said that we can think of the molecule as like a little jigsaw puzzle, with the atoms as pieces in the puzzle that just fit together: does that make any sense do you think? ...

The metaphors and analogies and similes used by teachers are thinking tools that help scaffold thinking towards, but are different from, the target knowledge. This process might be completed in one lesson, but, even so, would likely need regular revisiting in subsequent classes to reinforce the teaching and consolidate the learning. Progressively, the revisiting can increasingly be primarily in terms of the canonical features of the target concept – as this is no longer unfamiliar.

This example is hypothetical – but the principle behind it now seems well established, and if teaching is considered a form of scientific practice then what is needed is to have good hypotheses about how to best teach particular concepts: hypotheses that can be tested in practice. That means trying ideas out, and then modifying or even dismissing those found not to work with different classes – which of course means not just presenting the ideas, but doing so in a dialogic mode, where the teacher is constantly checking how (as well as if) students are understanding (Taber, 2014). The teacher should be always imagining, but never assuming.

16.7 Lessons for Chemistry Education Research

A fair criticism of the argument presented in the previous chapter, and indeed in some other parts of this book, might be that it is heavy on theory, but some of the detail lacks strong empirical support. A reader may reasonably ask how they can know that, for example, offering an array of comparisons of familiar objects and situations that may be understood to have something in common with the covalent bond, will actually be productive, and not just confuse students.

At one level a fair response would be 'sorry, you will just have to try it out yourself'. This is a fair answer because every class is different and every student is different, and what works in one class may not work in another. Students are not like bench reagents that are provided in standard concentrations for us to carefully divide out into identical aliquots – or strips of magnesium that we can carefully measure and cut from a roll of uniform ribbon. So, it is fair to say that even if X worked well with one class, we cannot safely generalise this to assume it will work with another class. The more similar the class (age, level, background, *etc.*) the more likely they will respond in similar way (Taber, 2012), but it is never assured. (Always imagine, but never assume!)

Although that is a fair response, readers may reasonably counter that they are not asking for evidence that specific teaching approaches WILL be effective in particular classes – but rather for empirical evidence that these approaches have been found effective in other classes that might be used as reference points to be evaluated in terms of their similarity to the class to be taught (*i.e.*, offering 'reader generalisation', Kvale, 1996).

16.8 The Research Literature

There is considerable research in chemistry education, published in *Chemistry Education Research and Practice*, in the *Journal of Chemical Education*, and in a wide range of other journals and edited volumes. However, there is certainly room for much more research to inform teaching. There is some valuable research that shows how effective teachers help learners construct scientific ideas through teaching (Lemke, 1990; Ogborn *et al.*, 1996; Jewitt *et al.*, 2001), which has informed my account in this volume, but more studies of this kind – focusing on different topics and learner groups – would be valuable.

There are many studies of students' conceptions in chemistry topics (Taber, 2018a), and a good many common alternative conceptions have been characterised (Kind, 2004). Moreover, there is material available to teachers to support classroom diagnostic assessment and, in particular, elicit some of these common conceptions (Taber, 2002). Sometimes research studies seem to be based on an assumption that if, say, 30% of your class select option C, then these students have a certain alternative conception that needs to be challenged. Diagnosing alternative conceptions is important, and such resources can be valuable, but an appreciation of the nature of chemical concepts suggests things are not so simple. Selecting a response in a diagnostic test suggests that that response has been interpreted as meaningful by a student, who *may* well entertain the alternative conception that some particular response was intended to represent (but could have interpreted the questions and response options quite differently).

This tells us little about whether a learner is highly committed to this conception; how well it is linked into the learner's conceptual network; whether this represents a formal concept or more of an intuition; or even if this is just one of several alternative ways of thinking being considered (see Chapters 4 and 14). This is important, because whilst it may sometimes be sensible for teachers to immediately commit time and effort into challenging a detected alternative conception before proceeding with a topic, at other times it is better to take the finding under advisement, and seek to find later opportunities to work with and slowly develop the learners' thinking.

There are some studies looking in detail at student's conceptions and how they might change under teaching (see Chapter 15), but such studies demand high levels of student participation, usually require multiple levels of data collection, and depend upon engagement over an extensive period, to build up a detailed picture. Thus, there has not been a great deal of this type of research. More such studies, at different academic levels, in different teaching contexts, focusing on different chemical topic areas, would be useful – both to establish the generality of the kinds of patterns reported in existing studies, and to avoid asking teachers to plan teaching based on indirect evidence from unrelated learning contexts. The general patterns and scenarios offered in this book seem to be well supported – but many of the detailed examples I have given are acknowledged as conjectural, imagining how the general patterns might apply in different specific cases. Earlier in the chapter I suggested adopting a teaching creed of 'always imagine, but never assume'. Often in preparing this volume I have imagined, and although the imagination has been based on well-supported principles, I am having to assume too much.

There are studies that explore particular teaching episodes and teaching sequences in some depth (Wightman *et al.*, 1986; Acher *et al.*, 2007; Mendonça and Justi, 2010) – and these can offer insight into what has worked in some other classrooms. Such studies offer excellent reference points for teachers in their own 'imagining' when planning their own teaching. But, again, studies reflect a limited set of combinations of chemistry topic/concept; level of study; teaching context, *etcetera*.

In the natural sciences, experimental research is useful to test hypotheses, but this relies on strict control conditions for comparison. In education, there are sometimes large-scale randomised trial of teaching approaches, such that an innovative approach is compared with some other teaching method, in a wide range of classes, allowing statistical analysis to address the variations between classrooms. Such studies are very useful – but expensive, difficult to organise, and so rare. Even when well done they suffer from several major limitations (Taber, Forthcoming):

(a) Even when the statistics suggest one approach is clearly 'better' than another in achieving some educational objective, this is nearly always only on average – so the 'better' approach will not work well in all classes, and the less effective approach may be more effective in some classes. (That said, knowing what works best 'on average' is certainly useful for teachers.)
(b) Teaching is complex, and teachers take time to hone their practice. A teacher asked to adopt a new approach or a new resource, will likely need several chances to try this out with classes, before they have optimised it and worked out effective timings with their classes, the right balance of emphasis, *etcetera*. Yet, many studies compare the first 'run-through' of an innovation against well-practised teaching. Waiting until the teachers had some level of mastery of the innovation delays the study and adds costs, so is seldom feasible.
(c) It has been shown that there are strong expectancy effects in education. When a teacher expects something to go well, it is more likely to; if a teacher expects student to learn a great deal, they tend to learn more; if a teacher seems unsure of what they are doing, students tend to learn less. In addition, people may react to novelty: students naturally, automatically pay more attention when an approach seems novel, although some may be less comfortable when there seems a good deal of novelty. These effects do not have to be considered in the natural sciences: the rate at which some zinc dissolves in some sulfuric acid is not affected by the researcher's expectations about the outcome. In educational research these effects confound experimental studies.

Most experimental studies suffer from even more severe problems as many published studies rely on very limited resources. Often, they compare two conditions in two different classes such that even when the class is assigned to the condition randomly, there is a very weak basis to make any generalisation from the individual study. (Sometimes the researchers fail to appreciate this, and seem to assume that two classes of students taught in different ways can be considered analogous to two beakers with equal amount of zinc added to acid at different temperatures.) Often, researchers use statistics to compare the patterns of performance in students' achievements in the two classes as though each student is an independent data point – where there has not been randomised assignment of students to classes (so from a statistical point of view each class in a unit of analysis, and so '$N=1$' in each condition), and where there may be weak measures of equivalence between classes (such as simply showing differences in prior attainment do not reach statistical significance).

Often, the same teacher teaches both classes, supposedly to 'control for the teacher variable' – as though a teacher can be assumed to be equally competent in applying different teaching approaches; and has no personal expectations about which approach will prove more effective; and can teach in a way that is not influenced by the differential rapport and idiosyncratic internal dynamics of different particular classes they are working with. So, whilst experimental research is indicated in chemistry (*i.e.*, it potentially has value in principle), often such studies are under-resourced and the many flawed small-scale studies offer a very limited basis for drawing general conclusions about aspects of chemistry education.

16.9 Moving the Research Programme Forward

Given the nature of teaching, then, and given the complex and indirectly accessible nature of conceptual learning, what seem most useful now are more studies that look in detail at the thinking of students in well-described learning contexts, and how that thinking shifts as teaching proceeds (and if, and how, it shifts in the months after focal teaching is complete). There will never be such studies in all topics, at all ages, in all languages, under all curricula, *etcetera*. Even if somehow such a wealth of studies were undertaken, individual classrooms, teachers, and students would always represent a somewhat different set of conditions from those reported in any particular study.

Yet a wide range of detailed studies of how student concepts change in response to teaching, studies that offer both a 'thick description' (Geertz, 1973) of the teaching context – to allow readers to judge its similarity to their own teaching context, so-called reader generalisation – and that take a nuanced approach to exploring students' concepts (seeking to test levels of tentativeness/commitment, embeddedness, range of application, manifold nature, *etc.*, as discussed in Chapter 14) and how they shift, can provide an invaluable resource base for informing teaching.

This work should not just explore the concepts themselves, but the 'meta-level' in terms of how students themselves conceptualise their concepts. For example, researchers might ask:

- Which conceptions are open to deliberate critique, rather than operating automatically without conscious awareness?
- When do learners consider their concepts as reflecting realistic accounts, rather than as models to be developed, or just imagined possibilities to be further explored?
- Do students treat manifold conceptions as alternatives they must come to choose between; or as complementary partial models reflecting different faces of multifacetted ideas; or as resources from which they are to develop a more sophisticated understanding?

It should be clear from the treatment in this book that a full understanding of someone's conceptual knowledge must include such metacognitive (Gunstone and Mitchell, 1998) aspects and epistemological commitments.

This programme of research will not reach the point of advising 'if you want to teach this concept at this level to this kind of class you present it this way' (and if teaching is to be a creative activity, most teachers would not welcome that level of prescription) but will offer insightful accounts that collectively allow us to refine our general models through providing more examples and a better idea of the range of variation we will find with different chemistry concepts and different student groups.

In chemistry, and other natural sciences, there is a tendency for research topics to become more specialised over time. A new researcher needs a strong grounding in the background to a research area; that is, to be inducted into a disciplinary tradition (Kuhn, 1970) or research programme (Lakatos, 1970); such as that exploring conceptual learning in the sciences (Taber, 2009). Then they can tackle the more challenging problems that arise when existing, well-established, theories are tested at greater resolution, or across a wider range of application. Arguably we have not seen enough of this type of progress in educational research: when basic ideas have been well established in a field, and so a good general picture is available, researchers often move on to other foci. Developing a more detailed account to support practice is difficult, and needs refined methodologies, denser data, more subtle analysis, deeper conceptualisation – we instead often see research fashions change and a shift to a new area of research (arguably, usually to one that can be addressed with more modest intellectual attention).

Concepts are at the heart of a science like chemistry. Effective teaching of chemical concepts is the core of good chemistry teaching. Learning concepts that reflect canonical chemical understandings is essential to becoming a chemist. Informing this fundamental aspect of chemistry teaching deserves a committed scientific research programme. Learners' concepts are difficult to characterise. Conceptual change is very difficult to monitor at more than a surface level. Empirically exploring how particular teaching approaches (levels of treatment, sequencing of ideas, learning activities) and 'on-line' choices in the classroom (decisions made in response to students' reactions to teaching) influence the shifts in students' conceptions is a major challenge for research. Real-time choices are at the heart of effective teaching: choices such as in the selection of moves teachers make (Taber, 2018b) in the classroom in response to on-going monitoring of learning: perhaps introducing another metaphor; maybe back-tracking to a point made earlier in the lesson; possibly reminding students of a laboratory activity recently undertaken; or changing to a different example; *etcetera*. Teachers cannot plan these actions in advance, and it may not always be obvious to the casual observer when a skilled teacher is following, modifying, or abandoning, their lesson plan. Teaching is seldom like performing a well-rehearsed classical symphony, and may be more like improvising around a well-established theme.

Such research is challenging. Yet, if research is to be of value, it needs to generate the kind of knowledge that can inform teachers by giving them real, detailed examples of what happens in the complexity of real classrooms like theirs.

References

Acher A., Arcà M. and Sanmartí N., (2007), Modeling as a teaching learning process for understanding materials: A case study in primary education, *Sci. Educ.*, **91**(3), 398–418.

Geertz C., (1973), Thick description: toward an interpretive theory of culture, in *The Interpretation of Cultures: Selected Essays*, New York: Basic Books, pp. 3–30.

Gunstone R. F. and Mitchell I. J., (1998), Metacognition and conceptual change, in Mintzes J. J., Wandersee J. H. and Novak J. D. (ed.), *Teaching Science for Understanding: A Human Constructivist View*, San Diego, California: Academic Press, pp. 133–163.

Jewitt C., Kress G., Ogborn J. and Tsatsarelis C., (2001), Exploring Learning Through Visual, Actional and Linguistic Communication: The multimodal environment of a science classroom, *Educ. Rev.*, **53**(1), 5–18.

Kind V., (2004), *Beyond Appearances: Students' Misconceptions about Basic Chemical Ideas*, 2nd edn., London: Royal Society of Chemistry.

Kuhn T. S., (1970), *The Structure of Scientific Revolutions*, 2nd edn, Chicago: University of Chicago.

Kvale S., (1996), *InterViews: An Introduction to Qualitative Research Interviewing*, Thousand Oaks, California: Sage Publications.

Lakatos I., (1970), Falsification and the methodology of scientific research programmes, in Lakatos I. and Musgrove A. (ed.), *Criticism and the Growth of Knowledge*, Cambridge: Cambridge University Press, pp. 91–196.

Lemke J. L., (1990), *Talking Science: Language, Learning, and Values*, Norwood, New Jersey: Ablex Publishing Corporation.

Mendonça P. and Justi R., (2010), Contributions of the model of modelling diagram to the learning of ionic bonding: analysis of a case study, *Res. Sci. Educ.*, 1–25.

Ogborn J., Kress G., Martins I. and McGillicuddy K., (1996), *Explaining Science in the Classroom*, Buckingham: Open University Press.

Taber K. S., (2001), The mismatch between assumed prior knowledge and the learner's conceptions: a typology of learning impediments, *Educ. Stud.*, **27**(2), 159–171.

Taber K. S., (2002), *Chemical Misconceptions – Prevention, Diagnosis and Cure: Theoretical Background* (vol. 1), London: Royal Society of Chemistry.

Taber K. S., (2004), Intuitive physics: but whose intuition are we talking about? *Physics Education*, **39**(2), 123–124.

Taber K. S., (2005), Learning quanta: barriers to stimulating transitions in student understanding of orbital ideas, *Sci. Educ.*, **89**(1), 94–116.

Taber K. S., (2009), *Progressing Science Education: Constructing the Scientific Research Programme into the Contingent Nature of Learning Science*, Dordrecht: Springer.

Taber K. S., (2012), Vive la différence? Comparing 'like with like' in studies of learners' ideas in diverse educational contexts, *Educ. Res. Int.*, **2012**(168741), 1–12.

Taber K. S., (2013a), Revisiting the chemistry triplet: drawing upon the nature of chemical knowledge and the psychology of learning to inform chemistry education, *Chem. Educ. Res. Pract.*, **14**(2), 156–168.

Taber K. S., (2013b), Upper Secondary Students' Understanding of the Basic Physical Interactions in Analogous Atomic and Solar Systems, *Res. Sci. Educ.*, **43**(4), 1377–1406.

Taber K. S., (2014), *Student Thinking and Learning in Science: Perspectives on the Nature and Development of Learners' Ideas*, New York: Routledge.

Taber K. S., (2018a), Alternative Conceptions and the Learning of Chemistry, *Isr. J. Chem.*, DOI: 10.1002/ijch.201800046.

Taber K. S., (2018b), *Masterclass in Science Education: Transforming Teaching and Learning*, London: Bloomsbury.

Taber K. S. (Forthcoming). Experimental research into teaching innovations: responding to methodological and ethical challenges.

Wightman T., Green P. and Scott P., (1986), *The Construction of Meaning and Conceptual Change in Classroom Settings: Case Studies on the Particulate Nature of Matter*, Leeds: Centre for Studies in Science and Mathematics Education, University of Leeds.

Subject Index

References to figures are given in *italic* type. References to tables are given in **bold** type.

abstractions, 20-1, 23-4, 112, 128, 134-5, 163, 194-5, 234-5, 292-4
abstract nature of concepts, 23-4, 39-41, 292
access canonical chemical concepts from non-material realm of ideas?, 185-94, *189, 190*
   does World 3 really exist?, 188-91
   do we still believe in Platonic Forms?, 187-8
   objectivity in educational questions, 191-4
accommodation, 315, **315**, 319
accredited concepts, viii, 255-7, 291
accretion – conceptual addition, 321-3
acidalogues, 98
acidals, 101
acidic oxides, 113, 236
acidity, 90, 92, 106-7, 191, 221
acidoids, 101
acids, conceptualising: reimagining a class of substances, 88-111
   acids and hydrogen, 53, 92-3
   current and historical scientific concepts, 93-108
   discriminations, 25, 151-2
   formation of a concept of acids, 60, 71, 88-9, 203, 218, 291, 306
   Lavoisier's theoretical account of acids, 90-2
   manifold concepts, 64-5
   natural kinds, as, 104-5, 115, 120, 131
   objects – entities in chemistry, 41-2, 185, 332
   proton donation, 162, 250, 257
   teaching about, 206, 209, 221, 232, 247, 255, 260
   theoretical elements of the acid concept, 89-90, 169-71
   when are chemical discoveries made?, 108-10, *109*
acids as natural kind or chemical convenience?, 104-5, 115, 120, 131
activation energy, 113-15, 268
adenine, 143-5
Ag *see* silver
air, 84, 113
Al *see* aluminium
alcohols, 44, 113, 206, 212, 244
aldehydes, 132
aldol reactions, 179
algorithmic processing, 181
aliquots, 348
alkaline earth metals, 206
alkalinity, 221
alkali(s), 88-9, 97, 103, 107, 221, 255
   carbonates, 97
   metals, 41, 163, 206, 308, 317
alkanes, 113, 206
alkenes, 113, 131, 206, 212, 247
alkynes, 113, 206

allotropes, 59
alloys, 123, 210, 246
alternative conceptions
    canonical concepts, on, 298
    common, 300
    defined, viii, 4, 19
    misconceptions, 8, 314
    prior learning, in, 341, 343–5
    students', 193, 216, 294, 304, 309–10, 316, 348–9
alternative frameworks, 4, 8, 335
alternative history – forming new concepts rather than expanding the range of existing ones, 99–102
alumina, 316
aluminium (Al), 149, 211
Amazonian tribe, 173–4
American Association for the Advancement of Science, 184
American Chemical Society, 200
amines, 113
    homopropargylic, 249
ammonia (NH_3), 23, 60, **61**, 62–3, 113, 139, 276, 290, 305
    molecules, 128, 134, 163, 277, 280
    solution, 51–2, 98
ammonium nitrate, 156
amount of substance, 40, 48
amphibians, 121
amphoteric oxides, 107, 113, 210, 236, 316
analogy with data (and evidence), 242–3, *243*
anatomy, 119
Ancient Greeks, 51
animals, 114
antiferromagnetism, 328, 330
Ar *see* argon
argon (Ar), 117, 303
aromaticity, 259
Arrhenius acid concept, 92–4, 96–8, 103–4, 109
Arrhenius model, 92
Arrhenius rate equation, 147, 151
artificial concepts, 58, 283

asking the community of chemists about canonical concepts, 176–85
    can we avoid the need to interpret representations?, 184–5
    does knowledge need to be personal or can it be distributed?, 177–8
    human knower – information resource system, 178–81, *180*
    importance of understanding, 181
    knowledge distributed across networks of people, 181–3
    sharing concepts is a process involving representation and interpretation, *183*, 183–4, 194–6
    who are the chemists who know?, 176–7
assimilation, **315**, 319
Aston, Francis, 78–9, 126
asymmetrical sign, *18*
atomic
    mass, 79
    model, Bohr, 47
    model, Dalton, 222, 224–5
    structure, 79, 208, 212–15, 217
    theory, 9
atomicity, 217
atoms
    concept of, 31, 47–8, 172–3, 224, 305
    octets or full shells, 53, 244, 300–1, 308, 333–5
    planetary model of, 333, 346
    quantum model of, 333
    rehybridisation, 43
    teaching about, 213, 221–2
    when is an atom, not an atom? (when is an ion an atom?), 222–3
Au *see* gold
Aufbau principle, 47–8, 208
Ausubel, David, 292

Subject Index

authoritative concepts, viii, 248–51, 254–6
Avogadro's constant, 139
Avogadro's law, 134, 136–40
azimuthal quantum number, 215

balvoids, 283
barium sulfate, 220
baselogues, 98
base metals, 107
bases, 92, 97–101, 105, 107, 144, 152, 162, 206–7, 250
basic oxides, 107, 113, 236
beakers, 192
behavioural
    evidence, 276
    level of description, 275–7
    outcomes, 236
bench reagents, 348
benzene (C_6H_6), 44, 53, 156, 232, 236, 280, 290, 346
Bernstein, Basil, 200
biases, 117–18, 124–6, 301
big bang theory, 53
biology, 278
birds, 121
birth defects, 70
Blondlot, Prosper-René, 93
Bohr atom model, 47
boiling temperature, 27, 116, 118–19, 244
bootstrapping, 214, 260, 287
boranes, 44
boron trichloride, 107
boundaries, 200
Boyle, Robert, 88
Boyle's law, 48, 134–5
brain, 277–8, 280, 293, 297
de Bretton-Gordon, Hamish, 130
British Science Association, 184
bromide ion, 126
Bronsted–Lowry acid concept, 90, 96, 98–9, 103–10, 151–3, 163, 250
bronze, 246
Bruner, Jerome, 212
Bunsen burners, 192, 302
burning, 42, 44, 201
butane, 156

C *see* carbon
C_2H_4 *see* ethene
C_2H_6 *see* ethane
C_6H_6 *see* benzene
Ca *see* calcium
calcium (Ca), 210
calibration, 238
caloric, 193, 332
canonical acid concept, 93, 96, 105–8, 110, 153, 233, 306
canonical associations, 31, 85, 207–8, 248
canonical chemical concepts
    can we access from a non-material realm of ideas?, 185–94, *189, 190*
    challenges of learning, 5, 67, 169
    orienting questions for the reader, 11
    so where do we find?, 194, 196
    teaching, 205, 208–10, 215, 291
canonical concepts
    alternative conceptions on, 298
    asking the community of chemists about, 176–85
    associations, 31, 85, 207–8, 248
    authoritative concepts and the impression of, 248–51
    concepts reaching canonical status, 94–6
    defined, viii
    looking for in scientific literature, 171–6
    personal concepts and, 232–5
    problem of locating, 169–71, 223
    Raoult's law (*see* Raoult's law and its deviations)
    simplification of, ix
    teaching and learning, 207, 228, 234, 341
    useful as ideal gases, as, 234
canonical conceptual content, 210
canonical knowledge, 51

canonical metal concept, 209–10, 215
canonical science, 74
canonicity: more or less alternative conceptions, 298–9
carbon (C), 77, 143, 156, 305
carbonates, 107
carbon–carbon (C–C) bonds, 36, 44, 60, 62, 131
carbon dioxide (CO_2), 139
catalysts, 221
categories, 112–13, 134
    concepts as, 22–3, 114
cathode rays, 314
caustic soda, 107. *see also* sodium hydroxide
Cavendish Laboratory, 150
C–C bonds *see* carbon–carbon bonds
CERN *see* European Organisation for Nuclear Research
CH_4 *see* methane
challenge of teaching and learning chemical concepts, vi, 3–13
    first-order approximate model of conceptual teaching, 11–12
    research is underpinned by theory, 6–8
    some things that should not be taken for granted, 9–10
    under-theorised research, 8–9
    what do we know, and what do we need to know?, 10–11
Chargaff, Erwin, 144, 182
Chargaff's rules, 136, 142–5, 149, 163
checking for shared understandings, 343–4
chelating ligands, 132
chemical bonding, 28, 31, 209, 222, 305, 307–8, 318, 332, 347
chemical change, 63–4
chemical concepts
    four types of?, 40
    kinds of things we perceive in the world, and the, 112–13
    so where do we find?, 194–6

chemical discoveries
    thinking about, 72–5
        conceptualisation is shaped by the cognitive apparatus, 73–4
        constructivism, 74–5
    when made?, 108–10, *109*
chemical education
    without a focus on concepts, 234–5
chemical enquiry, 6–7, 80
chemical inductive effect, 36
chemical knowledge, 205
chemical meta-concepts: imagining the relationships between chemical concepts, 134–65, 205
    concepts and laws (and law-like principles), 47–8, 134–48, 163, 202
    concepts and models, 47–8, 135, 148–51, 163, 203, 257
    concepts and theories?, 135, 151–63, 202–3
    conclusions on, 163–4
chemical reactions, 42, 145, 207–8, 303, 308, 326, 332
    how do chemical reactions take place in England?, 220–5
chemical reactivity, 27, 34, 77
chemistry
    deciding what counts as, 200–1
    information in, 238
    learning about the nature of, 202–3
    nature of, 205, 237, 257
    promoting intellectual development, 203–4
    social context, in its, 201–2
chemistry, choosing the, 205–8
    breadth or width?, 205–6
    organising the selection, 206–7
    teaching concepts in stages, 207–8
chemistry as a natural and empirical science, 69–72
    nature of the natural, 69–71

Subject Index

wider context of scientific discovery, 71–2
chemistry as an empirical science depends on imagination as well as benchwork, 72, *73*
chemistry education, 3, 5–10, 17, 191–2, 217, 236–7, 241, 251, 255
    lessons for (*see* lessons for chemistry education)
    research, 5–6, 348
Chemistry Education Research and Practice (journal), 259, 348
chemistry teaching *see under* teaching
Chemistry World (magazine), 184
Le Chetalier's principle, 47–8, 147
chloride ions, 222
chlorine (Cl), 78–9, 156, 223, 240, 303, 332
    atom, 128, 241, 305, 309
chlorine gas (Cl_2), 236
chromium (Cr), 156
Cl *see* chlorine
Cl_2 *see* chlorine gas
classical elements, 66
classification, 200, 209
Co *see* cobalt
CO_2 *see* carbon dioxide
cobalt (Co), 328
cognition, v, 36, 192, 232
    conscious and deliberative, 177–8, 288–9, 293
    experiences, from, 284, 301
    mental register, 279
    perception and, 74, 125, 243
    preconscious and automatic, 288–9
cognitive
    development, 315, 317
    processes, 7, 214, 280
    skills, 316
    system, 283–4
cold fusion, 93, 193
combustion, 42–3, 90, 163, 221, 248, 254, 259, 308, 332
comfort zones, 75
commitment: belief or suspicion?, 298, 304–5, 321
commonality: 'popular' and idiosyncratic alternative conceptions, 298, 299–301
communicated, how are chemical concepts?, 228–52
    authoritative concepts and the impression of canonical concepts, 248–51
    concepts subjective or objective?, 232–7
    conclusions on, 250–1
    information and understanding, 237–43
    objectivity and subjectivity, 228–31
    understanding as subjective, 243–7
communicating and teaching, 254–60
    communication of information is necessary but not sufficient for teaching, 257–8, 352
    lecturing, 258–9
    non-linearity of teaching and learning, 259–60
    teaching as moving students towards accredited conceptualisations, 255–7
communication of information is necessary but not sufficient for teaching, 257–8, 352
composition, 92
compounds, 212, 222–3, 259, 305
computer software, 286
computing, 74
concept formation, 281–8
    implicit learning in everyday life, 281–3
    inherent pattern recognition, 283–4
    phenomenological primitives, 284–8
    studying concept formation, 283

concepts, how to best understand, 20–32
    abstract nature of concepts, 23–4, 39–41, 292
    concepts act as nodes in a conceptual network, 20, 31–2, *33, 35*
    concepts are mental entities, 24–6 (*see also* mental entities, concepts as)
    concepts are only apparent when activated, 27–31
    concepts are tools used in thinking, 26–7
    concepts as categories, 22–3, 114
concepts, what kind of things are?, 17–38
    how to best understand concepts, 20–32
    making sense of Curie and and and and and Meitner, 17–20
    representing and exploring conceptual structures, 32–7
concepts, what kinds are important in chemistry?, 39–49
    concept relating to objects – entities in chemistry, 40–2, 112–13, 129, 293, 332
    concepts relating to events – processes in chemistry, 40, 42–5, 112–13, 332
    concepts relating to qualities – properties in chemistry, 40, 45–6, 112
    four types of chemical concept?, 40
    meta-concepts – concepts about chemical concepts, 39–40, 46–8, 112, 134–65, 205, 293
concepts act as nodes in a conceptual network, 20, 31–2, *33, 35*
concepts and conceptions, 19–20
concepts and laws (and law-like principles), 47–8, 134–48, 163, 202
    Avogadro's law, 134, 136–40
    Chargaff's rules, 136, 142–5, 149, 163
    conservation of mass, 136, 139–42, *141,* **142,** 144–5, 223
    other examples of law-like principles, 147–8
    Raoult's law and its deviations, 136, 145–7, 188, 218
concepts and models, 47–8, 135, 148–51, 163, 203, 257
concepts and ontology: what kind of things exist in the world of chemistry?, 112–33
    chemical concepts and the kinds of things we perceive in the world, 112–13
    concepts and typologies, 113–15, 119, 202, 217, 257, 316
    constructing typologies in chemistry, 131–2
    do chemical concepts represent natural kinds?, 115–19
    natural kinds in chemistry, 119–26
    use of the definite article in relation to chemical kinds, 126–31
concepts and theories?, 135, 151–63, 202–3
    applying the submicroscopic concepts, 161–3
    can we see salt dissolve in water?, 157–9
    has anyone seen the cat?, 159–60
    has anyone seen the Higgs boson?, 160
    micro-/macro-distinctions, 154–6
    particle theories, 153–4
    relating the two levels, 156, **162**
    shifting from observational language, 160–1
    what do we mean by an observable?, 156–7

Subject Index

concepts and typologies, 113–15, 119, 202, 217, 257, 316
    nesting of concepts, 113–15
concepts are only apparent when activated, 27–31
concepts are tools used in thinking, 26–7
concepts as categories, 22–3, 114
concepts as knowledge, 50–68
    concepts as knowledge?, 50–5
    conceptual, procedural, and episodic knowledge, 55–9
    conclusions on, 66–7
    fuzzy concepts, 59–64, 66–7, 223
    manifold concepts, 64–6, 106
conceptual change, 294, 304, 310, 313
conceptual change, studies of, 331–5
conceptual change, two types of, 319–21
    accretion – conceptual addition, 321–3
    conceptual realignment – correcting discrete conceptions, 323–8
conceptual content, ix, 210, 342
conceptual development, 276, 300, 334
conceptual ecology, metaphor of the, 316–19, 328, 334–5
    epistemological sophistication, 317–18
    rich ecologies, 318–19
conceptual frameworks, 241–2, 254, 262–4, 307–10, 316
    integrating, 330–1
    restructuring, 328–31, 334–5
conceptual gateways, 216
conceptual inductive effect, ix, 36
conceptualisation(s)
    accredited, 255–6
    alternate, 32, 106
    changes in, 283
    chemical concepts, of, 90, 96
    counterfactual, *101*
    defined, 19, 232
    levels of, 41, *140,* 155, *155*
    refining, 124
    shaped by the cognitive apparatus, 73–4
    student ideas, of, 8, 193, 344
conceptualising kinds assumes some essential properties, 119
conceptual knowledge, 55–9, 180–1, 202, 351
conceptual learning, 10, 17, 247, 275, 319, 341
    Piaget's scheme of, 315, **315**
conceptual networks, 20, 31–2, *33, 35,* 137, 309, 321–2, 349
conceptual realignment – correcting discrete conceptions, 323–8
conceptual resources, 245–6
conceptual space, *33,* 34
conceptual structures, representing and exploring, 32–7
    natural attitude – talking, and thinking, like mind readers, 36–7
    representing conceptual structures, 32–6, 253
conceptual teaching
    first-order approximate model of, 11–12
conceptual understanding, 37, 75, 238, 250–1, 276
conductivity, 124, 211, 215
congenst, ix, 30, 66, 178, 278, 291
conical flasks
    discriminations, 60, 132, 134, 151, 285
    objects – entities in chemistry, 22–4, 28, 126–7, 156, 163, 185, 188, 332
    teaching about, 264–5, 280, 317
connectivity: discrete conceptions and conceptual frameworks, 298, 307–10, 321
conservation concepts, 290
conservation of energy, 136, 139, 326
conservation of mass, 136, 139–42, *141,* **142**, 223
    law of, 140–1, **142**, 144–5
conservation of number, 290

conservation of volume, 290
constructivism, 8, 74–5, 290
coolants, 26–7, 34
copper (Cu), 62, 77, 126, 210–11, 301, 321, 323–30
  salts, 156–7
copper bromide, 141
copper carbonate, 140
copper nitride, 141
copper oxide, 140
copper sulfate ($CuSO_4$), 145, 188, 242
corrosiveness, 107
cortex, 277
Coulombic forces, 333–4
Coulomb's law, 136
covalent bonding, 60, **61**, 62, 113, 212, 305, 332, 335, 347–8
  shared pair of electrons, viii, 308–9
  types of, 64
Cr *see* chromium
creditable concepts, ix, 32, 216, 342
Cretaceous-Paleogene boundary, 121
Cretaceous-Tertiary boundary, 121
Crick, Francis, 143–4, 148–50, 163, 182
crude oil, 201
crustaceans, 114
crystalline form, 82, 210
Cu *see* copper
current and historical scientific concepts, ix, 93–108
  acids a natural kind, or a chemical convenience?, 104–5
  alternative history – forming new concepts rather than expanding the range of existing ones, 99–102
  canonical acid concept in chemistry today?, 105–8
  concepts reaching canonical status, 94–6
  extending the concept of acid, 96–7
  inventing new concepts, 93–4

  Lewis revised the acid concept, 102–4
  progressed beyond Arrhenius in discriminating acid from not acid?, 98–9
  progression from the Lavoisier concept to the Arrhenius concept, 97
curricular models, ix, 217–26, 254
  how do chemical reactions take place in England?, 220–5
  learning progression between educational stages, 224–6
  when is an atom, not an atom? (when is an ion an atom?), 222–3
curriculum, chemistry, 10, 199, 208–11, 236, 298
  bounding the, 205
curriculum, how are chemical concepts represented in the?, 199–227
  choosing the chemistry, 205–8
  curricular models, ix, 217–26, 254
  curriculum, the, 199–200
  keeping it simple, 208–17
  selection and simplification, 200–5
$CuSO_4$ *see* copper sulfate
cyclohexane, 60, 143
cytosine, 143–5

Dalton atomic model, 222, 224–5
Dalton's law of multiple proportions, 147
Darwin, Charles, 56, 103, 122, 182, 233
Darwinian natural selection, 103, 122
data, 279
  -collection methods, 333
  evidence, with, 242–3, *243*
dative bonds, 212
Davy, Humphry, 75–8, 80–3, 108, 179, 291
deduction, 117, 151

Subject Index

definite article in relation to
   chemical kinds, 126–32
      no methane molecule is
         tetrahedral, 130–1
      talk about ideal prototypes,
         128–9
      which methane molecule is
         tetrahedral?, 129–30
deliberative
   cognition, 177–8, 288–9, 293
   concepts, 292–3
Democritus, 103, 224
density, 83–4, 124, 137
Descartes, René, 188
Dewar structures, 44
diamagnetism, 328–30
Diels–Alder reaction, 206
diethyl ether, 116, 118
diffraction, 82
diffusion, 145
digital technologies, 181
dihydrogen (H_2), 60, **61**
dinitrogen (N_2), 60, **61**
dinosaurs, 121
dipole moment, 244, 259
disaccharides, 113
disciplinary matrix, 7
disequilibration, 315, **315**
dispersion forces, 309
displacement reactions, 221, 346
dissolve, 64, 91–2, 97–8, 112, 156–9,
   248–9, 281, 350
distillation, 23–4, 31, 56, 113–14
diversity, 204
d-level splitting, 209
DNA, 107, 143–5, 156
      structural model, 48, 148–51,
         163, 182
double bonds, 134, 344
ductility, 45, 210, 322
Dulong–Petit law, 92, 147–8

ears, 279
EBID *see* electron beam imaging
   device
echinoderms, 114
efflorescence, 103

Einstein, Albert, 139, 306
electrical conductivity, 45, 116,
   156–7, 210, 278, 322
electricity, 105, 332
electrolysis experiments, 179
electrolytes, 248
electromagnetic radiation, 105
electromagnetism, 105
electron beam imaging device
   (EBID), 154
electronegativity, 45–7, 211, 332
electronic configurations, 328–9
electron(s)
   cloud, 211, 213
   -cloud model, 333
   concept, 31, 41
   density, 224
      patterns, 213
   distribution, 244
   donation, 45–6, 222
   microscope, 159–60, 229
   particle
      behaviour of, 25, 29
      theories, 156, 162
   rehybridisation, 43
   sharing, viii, 220, 222, 225,
      308, 335, 347
   shells, 53, 213, 215, 244, 300,
      303, 308–9
   spin, 329
   transfer, 220, 225, 256, 335
   unpaired, 185
elements
   concept, 23, 31–2, 65, 172,
      284
   fuzzy concepts, 59, 222–3
   heavy, 77
   main block, 131
   metallic, 321–2
   modern chemical, 66
   natural kinds, as, 119, 123–5
   potassium, 77–8
   teaching about, 247, 249, 259
   transition, 29, 31, 34, 131, 206,
      261
   transuranic, 70, 132
   typologies, 113, 163

empirical
  evidence, 348
  sciences, 69–72, 257, 316
ENC *see* English National Curriculum
endosomolytic polymer nanoparticles, 249
energy, 31–2, 139, 240–1, 265–6, 270–1, 303, 310
  activation, 113–15, 268
  levels, 48, 213
engineering, 182
English National Curriculum (ENC), 64, 205, 220–4, 229
enthalpy, 139, 259
  ionisation, 240–1
  lattice, 232, 235
  neutralisation, of, 259
  reaction, of, 268
entropy, 24, 32, 176–7
environmental issues, 201
enzymes, 29, 303
episodic knowledge, 55–9
epistemological
  assumptions, 9, 191, 317–18, 352
  sophistication, 317–18
  terms, 217
epistemology, 17, 192
equilibration, **315**
equilibrium, 29, 62, 305
ethane (C_2H_6), 36, 60, 236
ethanoic acid, 36
ethanol, 62
ethene (C_2H_4), 156, 236
  molecule, 126
ethylene *see* ethene
European Organisation for Nuclear Research (CERN), 160
evaporation, 222
events – processes in chemistry, concepts relating to, 40, 42–5, 112–13, 332
evidence, 242–3, *243*
evolution, 70, 125
Examinations Institute, 200

experimental research, 351
expert knowledge, 265
explicit concepts, 284
explicitness, 298, 301–4, 321, 343

F *see* fluorine
false
  negatives, 74, 96, 126
  positives, 74, 96, 126
Fe *see* iron
Fedyakin, Nikolai, 93
fermionic spin, 330
ferromagnetic materials, 329–31
ferromagnetism, 329–30
ferrous metals, 328
fires, 42
first-order approximate model of conceptual teaching, 11–12
fish, 114, 121
flame test, 307–8
flasks, 192
Fleischmann, Martin, 93
fluorine (F), 47, 132, 303, 320, 332
formal concepts, 284–5, 316
formative assessment, 264
fossil fuels, 201
fractional distillation, 56
framing information to allow meaningful understanding, 260–72
  making unfamiliar concepts familiar, 264–72
  modelling the other person's sense-making, 261–4, *263*
Franklin, Rosalind, 182
Friedel–Crafts reactions, 132
frogs, 119–20
fugacity, 147
fuzzy concepts, 59–64, 66–7, 223

gas, 48, 123, 134–8, 156, 249, 313
  -phase reactions, 290
Gedanken experiment, 189
Geiger–Müller counter, 79
generalisations, 116, 118–19
gestalt, 302

Subject Index

gold (Au), 77
Gosling, Raymond, 182
Graham's law of diffusion, 145
granted, some things that should not be taken for, 9–10
graphite, 156
gravitation, 135, 186, 303, 319
    universal, 135
Grignard reagents, 131
guanine, 143–5
guinea pigs, 121

H *see* hydrogen
H_2 *see* dihydrogen
H_2O *see* water
H_2SO_3 *see* sulfurous acid
H_2SO_4 *see* sulfuric acid
Haber process, 201
Hahn, Otto, 182
half-life, 27
halogens, 29, 131–2, 200, 206, 317
hardness, 45–6
HCl *see* hydrochloric acid
He *see* helium
heat, 116, 332
    capacity, 124
heavy elements, 77
Heisenberg's matrices, 106
helium (He), 65, 77, 123, 223
heterocyclic compounds, 143–4
5,6,12,13,19,20-hexaimidazolium-ethoxy-cyclotriveratrylene, 249
Hg *see* mercury
hidden curriculum, 199
historical concepts, ix, 93, 95–6, **96**
history, 5
Homo erectus, 84
Homo habilis, 84
Homo heidelbergensis, 123
homolytic bond fission, 42
homopropargylic amines, 249
Homo sapiens, 84, 123
Hubble space telescope, 229
human perceptive-conceptual system, 73
Hund's rules, 48, 213

hybridisation, 31
    sp^3 hybridisation, 213
hybrid orbitals, 21
hydrides, 244
hydrocarbons, 113, 131
hydrochloric acid (HCl), 91, 94, 103, 221, 236
hydrogen (H)
    acids, and, 53, 92–3
    bonding, 62, 149, 214, 244
    concept, 23, 65, 163
    element, 77, 123
    gas, 139, 156
    ions, 52–3, 221–2, 255, 306
    orbitals, 43
    oxidation of, 132
    teaching about, 305, 317, 320
hydronium ion, 221
hydroxides, 107
hydroxyl ions (OH^-), 221–2
hyperconjugation, 208–9
hypotheses, 72, 80, 118, 122, 149, 170, 194, 264, 325, 347, 349

I *see* iodine
ideal gases, 234
ideal gas law, 29, 40, 47, 146, 163, 234
ideal prototypes, 128–9
ideas, 9, 185–94, *189, 190,* 253
    big, 216–17
imagine, but never assume, always, 344–6
imagine (and test out) alternative possibilities, 346–7
implicit knowledge, 56–7, 285, 301–2
    engaging in building deliberative concepts, 292–3
implicit learning, 230
    everyday life, in, 281–3
indefinite article, 127
indicators, 88, 221
induction, 115, 117, 119, 172
    bias, prejudice, and, 117–18
    community of practice, into a, 235–6

inductive
    effect, 259, 310, 344
    sciences, 115
inert gases, 135, 313
information, 174, 253, 257–8, 272, 298
    access, 245
information and understanding, 228, 237–43
    analogy with data (and evidence), 242–3, *243*
    information in chemistry, 238
    understanding and meaning, 239–42
information processing, 181, 285
    level of description, 275, 279
information resource system – the human knower, 178–81, *180*
information technology, 253
innovation, 350
inorganic chemistry, 206
input transducer, 279
intellectual development, chemistry promoting, 203–4
intermediate structure, 266–7
interpretation, 183–5
introspection, 293
intuition, 233, 349
intuitive theories, 8
inventing new concept, 93–4
iodide, 223
    ion, 132
iodine (I), 132
iodine pentoxide, 91
ionic bonding, **61**, 64, 113, 215, 255, 305, 308–9, 332, 335
ionisation, 42, 47
    energy, 211–12, 303, 309, 335
    enthalpy, 240–1
iron (Fe), 62, 132, 210–11, 321, 323, 328
isomerism, 45
isotopes, 27, 78, 81–2, 274, 284, 313
    radioactive, 185, 274

Journal of Chemical Education, 348

K *see* potassium
Kekulé, August, 280, 306
Kekulé structures, 44, 53, 280, 306
Kendrew, John, 182
keto–enol tautomerism, 44, 212
ketones, 44, 206, 212
KF *see* potassium fluoride
kinetics, 96, 265
Knoevenagel condensation, 249
knower, human – information resource system, 178–81, *180*
knowledge, v–vi, ix, 17, 171–6, 318
    bases, 177
    breadth or width?, 205–6
    distributed across networks of people, 181–3
    personal or distributed?, 177–8
    texts and the nature of, 173
knowledge, concepts as?, 50–5
    conceptual, procedural, and episodic knowledge, 55–9
    fuzzy concepts, 59–64, 66–7
    knower and the known, 54–5
    knowledge, a more relevant notion of, 54
    knowledge, belief, and truth, 51–2
    manifold concepts?, 64–6
    we should not believe in scientific knowledge, 52–4
    why does it matter?, 66–7
knowledge, implicit, 56–7
knowledge, two systems of, 288–94
    engaging implicit knowledge in building deliberative concepts, 292–3
    learning scientific concepts, 290–2
    melded concepts, 292
    spontaneous concepts, 289–90
    teaching informed by the knowledge-in-pieces model, 293–4, 318
knowledge – conceptual, procedural, and episodic, 55–9
    can conceptual knowledge be tacit?, 58–9

Subject Index

implicit knowledge, 56–7
mighty oaks from ignorant acorns grow, 57–8
knowledge elements, 284, 289
Kr *see* krypton
krypton (Kr), 313

labels, concept, 32, 34, 66
laboratories, 72, 89, 170, 190, 222, 283
   chemistry, 6, 23–4
   teaching, 72
Lamarkianism, 103
language, 175, 247, 279, 286, 292
   natural, 257, 283
Large Hadron Collider, 229
larynx, 279
lattice enthalpy, 232, 235
Lavoisier, Antoine, 75, 90–3, 181, 291, 306
Lavoisier, Marie-Anne Pierrette Paulze, 90, 181
Lavoisier acid concept, 92–4, 96–7, 104, 109
   progression to the Arrhenius concept, 97
Lavoisier's theoretical account of acids, 90–2
laws *see* concepts and laws (and law-like principles)
lead (Pb), 323
learning, v–vi, 9, 214, 280, 318
   black box, as a, 236–7, 243, 276, 280
   challenges of, 5, 67, 169
   journey, person on a, 297
   nature of chemistry, about the, 202–3
   progression between educational stages, 224–6
   progressions, big ideas, and threshold concepts, 216–17
   scientific concepts, 290–2
learning demand, 216, 256
learning pathway, 333

learning progressions (LP), 216–17
   empirically observed LP, 216
   hypothetical LP, 216
learning-to-the-test, 237
lecturing, 258–9, 271
Leibig condenser, 92
lessons for chemistry education, 341–54
   always imagine, but never assume, 344–6
   checking for shared understandings, 343–4
   imagine (and test out) alternative possibilities, 346–7
   lessons for chemistry education research, 348
   lessons for chemistry teachers, 341–2
   moving the research programme forward, 351–3
   research literature, 348–51
   responsibilities of the chemistry teacher, 342
   understanding differently, 342–3
lessons for chemistry education research, 348
lessons for chemistry teachers, 341–2
Lewis acid, 60, 346
   concept, 90, 96, 98–9, 103–10, 163
   model, 47
Lewis revised the acid concept, 102–4
Li *see* lithium
von Liebig, Justus, 92
ligand, 185
   field theory, 47–8
liquids, 31–2, 145–6
   range, 26
lithium (Li), 65, 123
litmus test, 106–7
logic, 115–17, 123, 125
lower anchor, 216
LP *see* learning progressions
lustre, 18, 45–6, 210–11

magnesium (Mg), 210, 348
magnetic
    materials, 323–31
    properties, 328
magnetism, 37, 105, 146, 319, 330
main block elements, 131
making sense of Curie and and and and and Meitner, 17–20
    concepts and conceptions, 19–20
making the familiar unfamiliar, 264
making the unfamiliar familiar, 264, 318, 344
malleability, 210, 322
mammals, 121
manganese (Mn), 59
manifold acid concept, 107
manifold concepts, 64–6, 106, 305–7, 327, 351
mass, 48, 124, 139, 224, 284
    conservation of, 136, 139–42, *141,* **142,** 144–5
mass defect, 139
mass-energy, 139
mass spectroscopy, 78, 83, 123, 214, 229, 238, 314–15
mass spectrum, 235, 238, 245–6, 314–15
mathematics, 5
matter, 31
mauveine, 72
meaningful learning, 245–7, 292, 307
    revisited, 315–16
Meitner, Lise, 182, 306
melded concepts, 292
melting, 31–2, 42, 44, 46, 64, 112, 116, 158
    point, 84
    temperature, 45–6, 62–3, 83, 112, 124, 215, 235
memory, 279–80
    working, 75
mental bias, 124–6
mental entities, concepts as, 21, 24–6, 32, 95, 137, 194–5
    communicating, 228, 232, 235, 247, 253, 293

mental models, 217, 261, 264
mental register, 9
    revisited, 279–81
mercury (Hg), 323
metacognitive knowledge, 272, 352
meta-concepts – concepts about chemical concepts, 39–40, 46–8, 112, 134–65, 205, 293
metal carbonates, 255
metal concept, 84–5, 260, 327
    what details shall we teach about?, 211–13
    what shall we teach about?, 209–10
metallic bonding, 113, 309
metallic element, 321–2
metalloids, 21, 100
metal oxides, 255
metals, 18–19, 26, 28, 31, 37, 41, 123, 221, 278, 332
metaphysical assumptions, 191
methane (CH_4), 46–7, 132, 139, 236
    molecules, 126, 129–31, 156, 332
methodology, 7–8, 352
Mg *see* magnesium
Michelangelo's David, 188, 231
microgenetic studies, 36, 283
micro-/macro-distinctions, 154–6
micronutrients, 278
microscope, 153–4, 159
Millikan's oil-drop experiment, 57
misconceptions, 4, 8, 314
misinformation, 238
mixture, 31
Mn *see* manganese
modelling process, 261–4
modelling the other person's sense-making, 261–4, *263*
models *see* concepts and models
molar mass, 139
molecular
    docking, 249
    geometries, 235–7
    mass, 139, 145, 244

molecules
  ammonia (NH_3), 128, 134, 163, 277, 280
  concept, 24–5, 31, 41, 44, 137–8, 223
  electronic structures of, 276
  interactions, 147
  molecular geometries, 235–7
  particle models, 162
  polar, 244
moles, 136, 138, 241
molluscs, 114
monosaccharides, 113
mooted concepts, ix, 95–6, **96**, 265, 280
multifacetted concepts, 30–1, 64–6
multiplicity: unitary and manifold conceptions, 298, 305–7, 321
muriatic acid, 91, 97, 103

N *see* nitrogen
N_2 *see* dinitrogen
Na *see* sodium
Na^+ *see* sodium ion
Na_2O *see* sodium oxide
NaCl *see* sodium chloride
nanochemistry, 155
nanoflowers, 249
nanoparticles
  endosomolytic polymer, 249
nanoscience, 155
NaOH *see* sodium hydroxide
natural attitude – talking, and thinking, like mind readers, 36–7
natural kinds, do chemical concepts represent?, 115–19
  conceptualising kinds assumes some essential properties, 119
  importance of natural kinds in science, 115–17, 128–9
  induction, bias, and prejudice, 117–18
natural kinds in chemistry, 119–26
  elements as natural kinds, 119, 123–5
  natural kinds just the operation of a mental bias?, 124–6
  species as natural kinds? a warning from biology, 121–3
natural philosophy, 72
natural products, 70
natural sciences, 17, 69–72, 119, 190–2, 349–50, 352
natural selection, 103, 122
natural theology, 192
nature of the chemical concept
  accessing chemical concepts for teaching and learning, 169–98
  challenge of teaching and learning chemical concepts, 3–13
  chemical meta-concepts: imagining the relationships between chemical concepts, 39–40, 46–8, 112, 134–65, 205, 293
  concepts and ontology: what kind of things exist in the world of chemistry?, 112–33
  concepts as knowledge, 50–68
  conceptualising acids: reimagining a class of substances, 88–111
  how are chemical concepts communicated?, 228–52
  how are chemical concepts represented in teaching?, 253–73
  how are chemical concepts represented in the curriculum?, 199–227
  how do students acquire concepts?, 274–96
  how do students' concepts develop?, 313–37
  lessons for chemistry education, 341–54
  origin of the chemical concept: ongoing discovery of potassium, 69–87

nature of the chemical concept (*continued*)
    what is the nature of students' conceptions?, 297–312
    what kind of things are concepts?, 17–38
    what kinds of concepts are important in chemistry?, 39–49
nature of the natural, 69–71
Nb *see* niobium
Ne *see* neon
neon (Ne), 274, 303, 313–14
Nernst equation, 147
nerve agent, 130
nesting of concepts, 113–15
neural
    circuits, 280
    network, 283
neurones, 277, 279
neuroscience, 277
neurosurgery, 328
neutralisation, 42–3, 101, 103, 107, 151–2, 221–2, 225, 248, 259, 309
neutral oxides, 113, 236
neutrons, 26, 163, 213
NH_3 *see* ammonia
Ni *see* nickel
nickel (Ni), 328
niobium (Nb), 246–7, 322–3, 330
nitrogen (N), 59, 62, 76–7, 143, 326
    group, 206
nitrogen–oxygen atmosphere, 190
noble gases, 113, 117, 135, 179, 223, 306
N-rays, 93, 95, 193
nuclear
    atom, 224
    power station, 34
    reactions, 53
    reactor, 26
    science, 139
nucleophiles, 344
nucleophilic
    attack, 303
    substitution, 176, 303
nucleus, 31, 213, 284–5, 299

O *see* oxygen
O_2 *see* oxygen molecule
objective
    knowledge, 190–1
    reality, 191
objectivity and subjectivity, 228–31, 276
    objectivity in science, 229–30, 243
    subjective reports, 230–1
objectivity in educational questions, 191–4
objects – entities in chemistry, concepts relating to, 40–2, 112–13, 129, 293, 332
observational language, shifting from, 160–1
observations, 72–3, 112, 132, 156–9, 170, 230, 243
    conceptual interpretations, 161, **161**
    empirical, 237
octane, 132, 156
OH^- *see* hydroxyl ions
Ohm's law, 138
ontological
    claim, 9, 191
    terms, 217, 332
ontology, 17, 41
optimum level of simplification, ix, 214–16
orbitals, 43, 48, 136, 162, 213, 224, 233, 259
    3p orbitals, 156
    atomic, 31, 211
    hybrid, 21
    model, 217, 317
    p orbitals, 41, 163, 303
    s orbitals, 303
    sp^3 hybrid orbitals, 59
order, 32
organic
    chemistry, 206
    compounds, 113, 232
    foods, 70

Subject Index

origin of the chemical concept:
  ongoing discovery of potassium,
  69–87
    chemistry as a natural and
      empirical science, 69–72
    chemistry as an empirical
      science depends on imagin-
      ation as well as
      benchwork, 72
    discovery of potassium: im-
      agining a new substance,
      75–85
    thinking about chemical
      'discoveries', 72–5
output measures, 10
oversimplification, 215
oxidation, 23, 25, 31, 71, 132
  communicating about, 221,
    233, 248
  states, 156, 213, 256
    variable, 32, 213
  teaching about, 176, 209, 211,
    260, 317, 332
oxides, 107, 113
oxidising agents, 60, 131, 151, 185,
  303, 332
oxyacids, 91
oxygen (O), 62, 90–2, 113, 139, 256,
  305, 308, 326, 332
  theory, 331
    combustion, of, 103
oxygen molecule (O_2), 60, **61**
ozone, 59

paramagnetism, 328, 330
particle
  models, 161–2, 213
  theories, 153–4, 156
patina, 211
pattern recognition, inherent, 283–4
Pauling, Linus, v, 182
Pauling, Peter, 182
Pavlovian conditioning, 43
Pb *see* lead
PCK *see* pedagogical content
  knowledge

pedagogical content knowledge
  (PCK), 11, 260
pedagogic principles, general, 260
perceptions, 39, 73–4, 135, 161,
  186, 279
periodicity, 208
periodic table, 31–2, 113, 131,
  211, 322
personal
  concepts, 232–5, 249, 255
  knowledge, ix, 55, 232
Perutz, Max, 182
pH, 221, 255
phenomena, 186, 320
  chemical, 45, 54, 148, 163, 331
  natural, 69, 170, 302, 316
  social, 69, 71
phenomenological primitives
  (P-prims), 284–8, 292–4, 301–2
philosophy, v, 72
  science, of, 81
phlogiston, 90, 193, 332
  theory, 103, 306, 308, 331
phonemes, 286
phosphorus, 76, 123
photons, 211
physical
  change, 64–5
  chemistry, 206
  properties, 104, 132
  science, 211
physics, v, 70, 182, 278, 303
  learning, 333
physiological–anatomical level of
  description, 275, 277–9
physiology, 119
Piaget, Jean, 289, 315
Picasso, Pablo, 281
pie chart, 245
planetary model of the atom, 333, 346
plasticity, 293
Plato, 186–7
Platonic Forms, 186
  do we still believe in?, 187–8
plum-pudding model, 224
Polanyi, Michael, 56

polar bonding, **61,** 62, 309, 332
polarity, 244
polywater, 93, 95
Pons, Stanley, 93
Popper, Karl, 53, 80, 115, 173–4, 186, 188, 193
positive feedback cycle, 263, *263*
positive-ray spectrum, personal conception of
    implicit concept formation, 274–5, 291
    reflective concept development, 313–15
positrons, 315
potash (potassium hydroxide), 75–6, 82, 84, 248
potassium (K), 41, 124, 134, 248, 308
potassium, discovery of, 75–85
    construction and development of the potassium concept, 78–80, 92, 96, 154, 169–71, 203, 291, 307
    creation of potassium, 76–7, 108–9, 179
    developing the metal concept, 84–5, 102, 323
    natural kinds, as, 120
    potassium *sans* isotopes, 81–4, *83*
    shifting potassium concept, 80–1
potassium fluoride (KF), 60, **61**
potassium hydroxide *see* potash
potassium iodide, 162
potassium salts, 308
P-prims *see* phenomenological primitives
practical kinds, 128–9
precipitation reactions, 162, 220, 222
prejudice, 117–18
prerequisite
    knowledge, 343
    learning, 264, 298, 341, 345
pressure, 40, 48, 134, 136–8, 145, 163
pre- to post-test design, 10

principles *see* concepts and laws (and law-like principles)
prior
    knowledge, 271
    learning, 292, 298, 341, 343
probability, 224
    -orbit model, 333
problem of locating canonical concepts, 169–71
procedural knowledge, 55–9
processes in chemistry *see* events – processes in chemistry, concepts relating to
processing, 74
    system, 275, 279, 285
products, 268–9, 305
propene, 247
properties in chemistry *see* qualities – properties in chemistry, concepts relating to
protoacids, 107
proton(s), 213
    acceptance, 152, 162
    donation, 98, 102, 152, 157, 162, 222, 250
    transfer, 220, 225
Proust's law of definite composition, 147
psychology, 5, 7, 27, 118
    behaviourist school in, 276
public knowledge, ix, 51, 55, 232, 234
purines, 143–5, 149
purity, 253
pyrimidines, 143–5, 149

qualities – properties in chemistry, concepts relating to, 40, 45–6, 112
quanticles, 25, 217, 291
quantum
    mechanics, 106
    model of the atom, 333
    numbers, 213, 329–30
        azimuthal, 215
questions for the reader, orienting, 11

Subject Index

radiation, 123
    beta, 79
radioactive emissions, 274, 313
radioactivity, 27, 78–9, 108, 246, 315
Raoult's law and its deviations, 136, 145–7, 188, 218
rate of reactions, 24, 206, 209, 275
reactants, 29, 268–9, 305
reaction, 259–60
    profile, 245, 266–71
reactivity, 84, 92
reagents
    bench, 348
    Grignard, 131
realism, 281
recall, 179
recognition, 179
redox reactions, 132, 206, 344, *345*
reducing agents, 132, 163, 185
refutation, 80
rehybridisation, 42–4
relativism, 204, 317–18
reptiles, 121
research
    literature, 348–51
    methods, v
    programme forward, moving the, 351–3
    scientific, 6–8, 80
    support, 36
    underpinned by theory, 6–8
    under-theorised, 8–9
        mental register, 9
resistivity, 211
resonance, 23, 71
responsibilities of the chemistry teacher, 342
reversible reaction, 62
Rh *see* rhodium
rhodium (Rh), 59
ribose, 149
RNA, 143
rote learning, 245–7, 315
Royal Society of Chemistry, 200, 259
Rutherford, Ernest, 306

S *see* sulfur
saltals, 100–1
salting out, 248
saltoids, 100–1
salts
    coloured, 31
    concept, 131
    dissolve, 281
    DNA, of, 148
    reaction to form, 34, 97, 99–101, 103, 210–11, 221–2
    salting out, 248–9
    solubility, 156
    solution, 158
samarium (Sm), 211
scaffolding, 265–6, 347
scanning-tunnelling microscope (STM), 28–9, 153–4, 160, 187
schematics, 34
Schrödinger's wave equation, 71, 106
science, technology, engineering, and mathematics (STEM) subjects, 212
Science Museum in London, 150
scientific community, 194–6, 235–6, 243, 248–51, 255
scientific concepts, x, 95–6, **96**, 292
scientific discoveries, 291, 306
    wider context of, 71–2
scientific enquiry, 80
scientific experiments, 202, 230
scientific knowledge, 194, 196, 254
    we should not believe in, 52–4
scientific law, 135–6, 146, 163
scientific literature, looking for canonical concepts in the, 171–6
    criticisms of this position, 175–6
    lack of coherence in the literature, 171–3
    texts and the nature of knowledge, 173–4
    texts contain only representations of concepts, 174–5
scientific models, 217–18
scientific progress, 103

scientific research, 6–8, 80
selection, organizing the, 206–7
selection and simplification, 200–5
    bounding the chemistry curriculum, 205
    chemistry in its social context, 201–2
    chemistry promoting intellectual development, 203–4
    deciding what counts as chemistry, 200–1
    learning about the nature of chemistry, 202–3
sensory
    data, 74
    input, 279, 288
separation, 31
sesterterpenoids, 249
sharing concepts
    process involving representation and interpretation, *183*, 183–4, 194–6
    scientific community, in the, 248–51
shell-fish, 121
shell model of the atom, 40, 53, 217, 224
silver (Ag), 77
silver chloride, 220
simple, keeping it, 208–17
    spiral curriculum, 212–14, 216
    what details shall we teach about the metal concept?, 211–12
    what shall we teach about the metal concept?, 209–10
simplification
    optimum level of, ix, 214–16
    selection and, 200–5
Sm *see* samarium
social sciences, 17, 69–70, 119
sociology, 71, 81
Socrates, 117, 186, 254
sodium (Na)
    atom, 63, 309
    concept, 26–7, 34, 47
    metallic element, as, 76–7, 84–5, 210
    teaching about, 240–1, 299, 305, 323
sodium chloride (NaCl), **61**, 63, 156–7, 215, 221–3, 236, 309
sodium hydroxide (NaOH), 221. *see also* caustic soda
sodium ion (Na^+), 63, 156, 222–3
sodium oxide (Na_2O), 236
software design, 182
solid, 31–2
solubility, 112, 156, 215
    gas, 248
solution, 248–9
solvent, 98, 158, 215, 222, 259, 309
sonority, 210
sounds, 286
sourness, 88–90, 93, 97, 103, 106–7
space-time, 186, 235
species as natural kinds? A warning from biology, 121–3
specifications, 219
spectrum *see* mass spectrum
speech, 279
spiny-metal phenolic coordination crystals, 249
spiral curriculum, 212–14, 216
spontaneous concepts, 289–90, 292–3
square planar complex ion, 233
stability, chemical, 63, 232, 303, 309
standards, 256
state-electron model, 333
statistical analysis, 349
steel, 247, 328
STEM subjects *see* science, technology, engineering, and mathematics subjects
steric hinderance, 262
STM *see* scanning-tunnelling microscope
Strassmann, Fritz, 182
student conceptions in chemistry, dimensions of, 298

Subject Index

students, how to acquire concepts?, 274–96
    concept formation, 281–8
    personal conception of positive-ray spectrum: implicit concept formation, 274–5
    three levels of description, 275–81
    two systems of knowledge, 288–94

students' conceptions, what is nature of?, 297–312
    canonicity: more or less alternative conceptions, 298–9
    commitment: belief or suspicion?, 298, 304–5, 321
    commonality: 'popular' and idiosyncratic alternative conceptions, 298, 299–301
    conclusions on, 310
    connectivity: discrete conceptions and conceptual frameworks, 298, 307–10, 321
    dimensions of student conceptions in chemistry, 298
    explicitness: learners are aware of some, but not all, of their thinking about chemical topics, 298, 301–4, 321, 343
    multiplicity: unitary and manifold conceptions, 298, 305–7, 321
    person on a learning journey, 297

students' concepts, how to develop?, 313–37
    meaningful learning revisited, 315–16
    metaphor of the conceptual ecology, 316–19
    personal conception of positive-ray spectrum: reflective concept development, 313–15
    restructuring conceptual frameworks, 328–31, 334–5
    studies of conceptual change, 331–5
    two types of conceptual change?, 319–21

subjective experiences, 278
subjective language, 231
subjective or objective, are concepts?, 232–7, 247
    canonical concepts are as useful as ideal gases, 234
    chemical education without a focus on concepts, 234–5
    induction into a community of practice, 235–6
    learning as a black box, 236–7, 243
    personal concepts and canonical concepts, 232–4
subjective reports, 230–1
subjective understanding, 188
subject knowledge, 341, 343
submicroscopic concepts, applying the, 161–3
substance
    amount of, 40, 48
    chemical, 220
    concept, 31, 42, 99, 113–15, 142
    elemental, 77, 172
    teaching about, 207–8, 212, 217, 253, 259–60
substitution, 259
sugar(s), 113, 156
    phosphates, 143
sulfate ion, 45, 47
sulfur (S), 40–1, 45, 76–7, 123, 156
sulfuric acid (H_2SO_4), 90, 106, 350
sulfurous acid (H_2SO_3), 90
superacids, 106
supernovae, 77
survival, 125–6
syllabus, 219
syllogism, 117, 316
symbolic language, 245
symbols, strings of, 239–42, 247, 253

synapses, 277, 279–80
synaptic
    changes, 277–8, 328
    connections, 30, 277, 280, 330–1
    level, 277
synergistic proteome modulations, 249

Ta *see* tantalum
tacit
    concepts, 58–9
    knowledge, x, 58–9, 230
tantalum (Ta), 238–41
target
    concepts, 283, 347
    knowledge, x, 219–20, 254–6, 298, 345, 347
tautomerism, 254
teaching, how are chemical concepts represented in?, 253–73
    communicating and teaching, 254–60
    framing information to allow meaningful understanding, 260–72
teaching and communicating *see* communicating and teaching
teaching and learning, accessing chemical concepts for, 168–98, 228, 234
    asking the community of chemists about canonical concepts, 176–85
    can we access canonical chemical concepts from a non-material realm of ideas?, 185–94
    looking for canonical concepts in the scientific literature, 171–6
    problem of locating canonical concepts, 169–71
    so where do we find chemical concepts?, 194–6
teaching and learning, non-linearity of, 259–60

teaching as moving students towards accredited conceptualisations, 255–7
teaching concepts in stages, 207–8
teaching informed by the knowledge-in-pieces model, 293–4
teaching-to-the-test, 219, 237, 256
technical knowledge, 202
temperature, 26, 40, 48, 124, 134–8, 145, 156, 163, 254, 350
tensile strength, 116
terminology, 8, 113
tetrahedral geometry, 45, 47, 129–31
tetraoxosulfate(VI) ion, 45
texts
    contain only representations of concepts, 174–6
    nature of knowledge, and the, 173–4
thalidomide, 70
theoretical
    frameworks, 5
    knowledge, 46
theories *see* concepts and theories?
theory of mind, 36–7
thermal
    conductivity, 26, 210, 322
    decomposition, 221
    energy, 211
thermodynamics, 200, 265
thermometric properties, 123–4
thermometry, 123–4
thinking, 9, 26–7, 278, 280, 289, 301, 316
Thomson, J.J., 78
thought experiment, 189
threshold concepts, 216–17
thymine, 143–5
toxicity, 27
transition
    elements, 29, 31, 34, 131, 206, 261
    metals, 246–7
    salts, 56
    state, 71, 185, 262, 264–70, 272, 344

Subject Index

transuranic elements, 70, 132
tungsten (W), 321-3
Turing test, 189
typologies
    concepts and, 113-15, 119, 202, 217, 316
    constructing in chemistry, 131-2

U *see* uranium
understanding, vi, 280, 297, 319, 341
    checking for shared, 343-4
    differently, 342-3
    gap in, 256
    importance of, 181
    information, and, 228, 237-43, 298
    meaning, and, 239-42
understanding as subjective, 243-7
    meaningful and rote learning, 245-7
    understanding and explanation, 244
unfamiliar concepts familiar, making, 264-72
    Scenario 1, 267
    Scenario 2, 267-70
unitary concepts, 305-7
universality, 136
University Chemistry Education (journal), 259
unsaturated, 259
upper anchor, 216
uracil, 143
uranium (U), 172, 211
US National Academy of Sciences, 212

vacuum tubes, 314
valence shell electron pair repulsion theory (VSEPRT), 47, 277
van der Waals forces, 28, 146, 244, 309
vapour pressure, 145, 147
vat, 190
vegetable dyes, 88, 103
verdigris, 211
vertebrates, 121
vinegar, 51-2, 88-9, 93-4
virtual reality, 188-9
visible-UV spectrum, 232
volume, 40, 48, 124, 134, 136-8, 146, 156, 163
VSEPRT *see* valence shell electron pair repulsion theory
Vygotsky, Lev, 290-2

W *see* tungsten
Wallace, Alfred Russel, 182
warrants, 54
water (H_2O), 28, 60, **61**, 62-3, 84, 157-9, 221-2, 244, 305
Watson, James, 143-4, 148-50, 163, 182
Wedgwood, Josiah, 56
whales, 114, 121, 128-9
what do we know, and what do we need to know?, 10-11
    some orienting questions for the reader, 11
Wilkins, Maurice, 182
wood, 113
World 1 of objects, 185-8
World 2 of human subjective experience, 186, 188
World 3 of ideas, does it really exist?, 186-91

Xe *see* xenon
xenon (Xe), 313
X-ray
    data, 83
    diffraction, 82, 149

zinc (Zn), 301, 323, 350
Zn *see* zinc

www.ingramcontent.com/pod-product-compliance
Lightning Source LLC
Chambersburg PA
CBHW071554080526
44588CB00010B/906